HEMATOPOIETIC STEM CELLS VI

ANNALS OF THE NEW YORK ACADEMY OF SCIENCES
Volume 1106

HEMATOPOIETIC STEM CELLS VI

*Edited by Lothar Kanz, Katja C. Weisel, John E. Dick,
and Willem E. Fibbe*

Published by Blackwell Publishing on behalf of the New York Academy of Sciences
Boston, Massachusetts
2007

Library of Congress Cataloging-in-Publication Data

International Conference and Workshop on Hematopoietic Stem Cells (6th : 2006 : Tubingen, Germany) Hematopoietic stem cells VI / edited by Lothar Kanz . . . [et al.].
   p. ; cm. – (The annals of the New York Academy of Sciences, ISSN 0077-8923)
  Includes bibliographical references.
  ISBN-13: 978-1-57331-676-7 (paper : alk. paper)
  ISBN-10: 1-57331-676-8 (paper : alk. paper)
  1. Hematopoietic stem cells–Congresses. I. Kanz, Lothar. II. New York Academy of Sciences. III. Title. IV. Title: Hematopoietic stem cells 6. V. Series.
  [DNLM: 1. Hematopoietic Stem Cells–Congresses. 2. Hematopoiesis–Congresses. 3. Stem Cells–Congresses. W1 AN626YL 2007 / WH 380 I6052h 2007]
  QP92.I556 2006
  612.4'1–dc22
                                                                                    2007005116

The *Annals of the New York Academy of Sciences* (ISSN: 0077-8923 [print]; ISSN: 1749-6632 [online]) is published 28 times a year on behalf of the New York Academy of Sciences by Blackwell Publishing with offices at 350 Main St., Malden, MA 02148 USA; 9600 Garsington Road, Oxford, OX4 2ZG UK; and 600 North Bridge Rd, #05-01 Parkview Square, 18878 Singapore.

Information for subscribers: For new orders, renewals, sample copy requests, claims, changes of address and all other subscription correspondence please contact the Journals Department at your nearest Blackwell office (address details listed above). UK office phone: +44 (0)1865 778315, fax +44 (0)1865 471775; US office phone: 1-800-835-6770 (toll free US) or 1-781-388-8599; fax: 1-781-388-8232; Asia office phone: +65 6511 8000, fax; +44 (0)1865 471775, Email: customerservices@blackwellpublishing.com

Subscription rates:
Institutional Premium   The Americas: $4043   Rest of World: £2246
The Premium institutional price also includes online access to full-text articles from 1997 to present, where available. For other pricing options or more information about online access to Blackwell Publishing journals, including access information and terms and conditions, please visit www.blackwellpublishing. com/nyas
*Customers in Canada should add 6% GST or provide evidence of entitlement to exemption.
**Customer in the UK or EU: add the appropriate rate for VAT EC for non-registered customers in countries where this is applicable. If you are registered for VAT please supply your registration number.

Mailing: The *Annals of the New York Academy of Sciences* is mailed Standard Rate. Mailing to rest of world by DHL Smart & Global Mail. Canadian mail is sent by Canadian publications mail agreement number 40573520. **Postmaster:** Send all address changes to *Annals of the New York Academy of Sciences*, Blackwell Publishing Inc., Journals Subscription Department, 350 Main St., Malden, MA 02148-5020.

Membership information: Members may order copies of *Annals* volumes directly from the Academy by visiting www.nyas.org/annals, emailing membership@nyas.org, faxing 212-298-3650, or calling 800-843-6927 (US only), or 212-298-8640 (International). For more information on becoming a member of the New York Academy of Sciences, please visit www.nyas.org/membership. Claims and inquiries on member orders should be directed to the Academy at email: membership@nyas.org or Tel: 212-298-8640 (International) or 800-843-6927 (US only).

Copyright and Photocopying:
© 2007 The New York Academy of Sciences. All rights reserved. No part of this publication may be reproduced, stored, or transmitted in any form or by any means without the prior permission in writing from the copyright holder. Authorization to photocopy items for internal and personal use is granted by the copyright holder for libraries and other users registered with their local Reproduction Rights Organization (RRO), e.g. Copyright Clearance Center (CCC), 222 Rosewood Drive, Danvers, MA 01923, USA (www.copyright.com), provided the appropriate fee is paid directly to the RRO. This consent does not extend to other kinds of copying such as copying for general distribution, for advertising or promotional purposes, for creating new collective works, or for resale. Special requests should be addressed to Blackwell Publishing at: journalsrights@oxon.blackwellpublishing.com.

Printed in the USA. Printed on acid-free paper.

Disclaimer: The Publisher, the New York Academy of Sciences and the Editors cannot be held responsible for errors or any consequences arising from the use of information contained in this publication; the views and opinions expressed do not necessarily reflect those of the Publisher, the New York Academy of Sciences, or the Editors.

Annals are available to subscribers online at the New York Academy of Sciences and also at Blackwell Synergy. Visit www.blackwell-synergy.com or www.annalsnyas.org to search the articles and register for table of contents e-mail alerts. Access to full text and PDF downloads of *Annals* articles are available to nonmembers and subscribers on a pay-per-view basis at www.blackwell-synergy.com and www.annalsnyas.org.

The paper used in this publication meets the minimum requirements of the National Standard for Information Sciences Permanence of Paper for Printed Library Materials, ANSI Z39.48-1984.

ISSN: 0077-8923 (print); 1749-6632 (online)
ISBN-10: 1-57331-676-8 (paper); ISBN-13: 978-1-57331-676-7 (paper)

A catalogue record for this title is available from the British Library.

ANNALS OF THE NEW YORK ACADEMY OF SCIENCES

Volume 1106
June 2007

HEMATOPOIETIC STEM CELLS VI

Editors
LOTHAR KANZ, KATJA C. WEISEL, JOHN E. DICK,
AND WILLEM E. FIBBE

This volume is the result of a conference entitled **Sixth International Symposium and Workshop on Hematopoietic Stem Cells**, held on September 14–16, 2006, in Tübingen, Germany.

CONTENTS

Introduction. *By* KATJA C. WEISEL, JOHN E. DICK, WILLEM E. FIBBE, AND LOTHAR KANZ xi

Part I. Stem Cell Function

AMD3100 and CD26 Modulate Mobilization, Engraftment, and Survival of Hematopoietic Stem and Progenitor Cells Mediated by the SDF-1/CXCL12-CXCR4 Axis. *By* HAL E. BROXMEYER, GIAO HANGOC, SCOTT COOPER, TIMOTHY CAMPBELL, SHIGEKI ITO, AND CHARLIE MANTEL 1

The Stem Cell Continuum: Cell Cycle, Injury, and Phenotype Lability. *By* PETER J. QUESENBERRY, GERALD COLVIN, GERRI DOONER, MARK DOONER, JASON M. ALIOTTA, AND KEVIN JOHNSON 20

Inference, Validation, and Dynamic Modeling of Transcription Networks in Multipotent Hematopoietic Cells. *By* SHAMIT SONEJI, SUI HUANG, MATTHEW LOOSE, IAN JOHN DONALDSON, ROGER PATIENT, BERTHOLD GÖTTGENS, TARIQ ENVER, AND GILLIAN MAY 30

Part II. Stem Cell Niche

Maintenance of Quiescent Hematopoietic Stem Cells in the Osteoblastic Niche. *By* FUMIO ARAI AND TOSHIO SUDA 41

Cytokine Signaling, Lipid Raft Clustering, and HSC Hibernation. *By*
SATOSHI YAMAZAKI, ATSUSHI IWAMA, YOHEI MORITA, KOJI ETO,
HIDEO EMA, AND HIROMITSU NAKAUCHI 54

Dormant and Self-Renewing Hematopoietic Stem Cells and Their Niches. *By*
ANNE WILSON, GABRIELA M. OSER, MAIKE JAWORSKI,
WILLIAM E. BLANCO-BOSE, ELISA LAURENTI, CHRISTELLE ADOLPHE,
MARIEKE A. ESSERS, H. ROBSON MACDONALD, AND
ANDREAS TRUMPP 64

Part III. Stem Cell Fate

Cartography of Hematopoietic Stem Cell Commitment Dependent upon a
Reporter for Transcription Factor Activation. *By* KOICHI AKASHI 76

Gradients of Antigen Expression and Developmental Potential in
Hematopoiesis. *By* JAE-YONG KWAK, SCOTT CHO, AND
GERALD J. SPANGRUDE 82

Biological and Molecular Evidence for Existence of Lymphoid-Primed
Multipotent Progenitors. *By* SIDINH LUC, NATALIJA BUZA-VIDAS, AND
STEN EIRIK W. JACOBSEN 89

Part IV. Malignant Hematopoiesis

Insertional Mutagenesis by Replication-Deficient Retroviral Vectors Encoding
the Large T Oncogene. *By* ZHIXIONG LI, OLGA S. KUSTIKOVA,
KENJI KAMINO, THOMAS NEUMANN, MATHIAS RHEIN, ELKE GRASSMAN,
BORIS FEHSE, AND CHRISTOPHER BAUM 95

NUP98 Dysregulation in Myeloid Leukemogenesis. *By* M. A. S. MOORE,
K. Y. CHUNG, M. PLASILOVA, J. J. SCHURINGA, J.-H. SHIEH, P. ZHOU, AND
G. MORRONE 114

Part V. Regulators

LEF-1 Is a Decisive Transcription Factor in Neutrophil Granulopoiesis. *By*
JULIA SKOKOWA AND KARL WELTE 143

Role of Thrombopoietin in Mast Cell Differentiation. *By*
ANNA RITA MIGLIACCIO, ROSA ALBA RANA, ALESSANDRO M. VANNUCCHI,
AND FRANCESCO A. MANZOLI 152

Thrombopoietic Cells and the Bone Marrow Vascular Niche. *By*
H. G. KOPP AND S. RAFII 175

Differential Effects of G Protein–Coupled Receptors on Hematopoietic
Progenitor Cell Growth Depend on their Signaling Capacities. *By*
XINGKUI XUE, ZHEN CAI, GABRIELE SEITZ, LOTHAR KANZ,
KATJA C. WEISEL, AND ROBERT MÖHLE 180

Effect of FLT3 Inhibition on Normal Hematopoietic Progenitor Cells. *By*
KATJA C. WEISEL, SEDAT YILDIRIM, ERIC SCHWEIKLE, LOTHAR KANZ, AND
ROBERT MÖHLE 190

Part VI. Embryonic Stem Cells and Plasticity

The Cdx-Hox Pathway in Hematopoietic Stem Cell Formation from
  Embryonic Stem Cells. *By* CLAUDIA LENGERKE,
  SHANNON MCKINNEY-FREEMAN, OLAIA NAVEIRAS, FRANK YATES,
  YUAN WANG, DIMPLE BANSAL, AND GEORGE Q. DALEY 197

Differentiation Potential of Histocompatible Parthenogenetic Embryonic
  Stem Cells. *By* CLAUDIA LENGERKE, KITAI KIM, PAUL LEROU, AND
  GEORGE Q. DALEY 209

Hematopoiesis from Human Embryonic Stem Cells. *By* MICKIE BHATIA 219

Emergence of Human Angiohematopoietic Cells in Normal Development and
  from Cultured Embryonic Stem Cells. *By* ELIAS T. ZAMBIDIS, LIDIA SINKA,
  MANUELA TAVIAN, VENTA JOKUBAITIS, TEA SOON PARK, PAUL SIMMONS,
  AND BRUNO PÉAULT 223

Part VII. Stem Cell Exhaustion

Epigenetic Control of Hematopoietic Stem Cell Aging: The Case of Ezh2. *By*
  GERALD DE HAAN AND ALICE GERRITS 233

Telomere Length in Human Natural Killer Cell Subsets. *By* QIN OUYANG,
  GABRIELA BAERLOCHER, IRMA VULTO, AND PETER M. LANSDORP 240

Part VIII. Mesenchymal Stem Cells and Transplantation

Flt3 in Regulation of Type I Interferon-Producing Cell and Dendritic Cell
  Development. *By* NOBUYUKI ONAI, AYA OBATA-ONAI,
  MICHAEL A. SCHMID, AND MARKUS G. MANZ 253

Novel Markers for the Prospective Isolation of Human MSC. *By*
  HANS-JÖRG BÜHRING, VENKATA LOKESH BATTULA, SABRINA TREML,
  BERNHARD SCHEWE, LOTHAR KANZ, AND WICHARD VOGEL 262

Modulation of Immune Responses by Mesenchymal Stem Cells. *By*
  WILLEM E. FIBBE, ALMA J. NAUTA, AND HELENE ROELOFS 272

Feasibility and Outcome of Reduced-Intensity Conditioning in Haploidentical
  Transplantation. *By* RUPERT HANDGRETINGER, XIAOHUA CHEN,
  MATTHIAS PFEIFFER, INGO MUELLER, TOBIAS FEUCHTINGER,
  GREGORY A. HALE, AND PETER LANG 279

Index of Contributors 291

Financial assistance was received from:

- Amgen Europe/Germany

> The New York Academy of Sciences believes it has a responsibility to provide an open forum for discussion of scientific questions. The positions taken by the participants in the reported conferences are their own and not necessarily those of the Academy. The Academy has no intent to influence legislation by providing such forums.

Introduction

The Sixth Biennial International Symposium and Workshop on Hematopoietic Stem Cells was held at the University of Tübingen, Germany, from September 14–16, 2006. The conference was again hosted by the University of Tübingen and took place in the historical rooms of the Protestant Collegiate (*Evangelisches Stift*), which has belonged to the theological faculty of the University of Tübingen since the sixteenth century, and where famous scientist-philosophers, like Johannes Kepler, Friedrich Hegel, and Friedrich Hölderlin have lived and studied. These surroundings created a unique and inspiring atmosphere for the meeting. Thirty speakers, leading scientists in the field of hematopoiesis, and a few young investigators, came together to discuss the most recent developments in hematopoietic stem cell (HSC) science and their clinical implications. This year's major topics were stem cell function, stem cell niche and stem cell fate, regulators, malignant hematopoiesis, embryonic stem cells and plasticity, and stem cell exhaustion, as well as mesenchymal stem cells and transplantation (meeting program: www.onkologie-tuebingen.de).

Hematopoietic stem cell biology is a fascinating field and continuously in progress. New technologies, intriguing concepts, and crosslinks perpetuate our interest and excitement. Lively discussions of these ideas with other scientists represent an invaluable stimulus for all, in particular the young investigators in the field. Following the tradition of previous HSC meetings, only a small group of scientists was invited, facilitating an open exchange and intensive discussion of current findings, mostly including unpublished data. We are thankful to all of the speakers, who fully appreciated this approach and, by sharing their data, made possible an open exchange of thoughts and hypotheses in the field of HSC biology.

Data presented at this conference demonstrated the progress being made toward understanding mechanisms of stem cell function involving genetic events and cytokine–receptor interaction. It was shown that signal transducers, activators of transcription-3 as well as Mad-2, are required for optimal HSC function in mice. Also mathematical models of transcription factor interaction can elucidate their role in various stages of hematopoietic commitment. The role of the cell cycle in hematopoietic cell commitment was shown in a model of conversion of cytokine-stimulated hematopoietic cells to cells with a lung phenotype.

The influence and role of the stem cell niche was discussed in several presentations. Results obtained with an upregulation of WNT-inhibitory factor-1 in osteoblasts suggested a role of WNT-signaling in hematopoietic stem cell decisions. Furthermore, it was shown that the role of cell adhesion in the maintenance and quiescence of hematopoietic stem cells in the niche is crucial. One key regulator of this adhesion process is N-cadherin. In addition, oxidative stress plays an important role for maintenance of stem cell–niche interaction. Most primitive hematopoietic stem cells reside as single cells in the endosteal niche. Lipid raft clustering induced by cytokines is essential for the reentry of quiescent hematopoietic stem cells into the cell cycle.

In discussions about stem cell fate, it was proposed that the timing of key transcription factor expression primarily drives the lineage-commitment sequences from hematopoietic stem cells, resulting in the formation of more complex differentiation steps within the hematopoietic developmental tree. During lineage differentiation, T-cell development is dependent upon exposure of developing progenitor cells to specific levels of IL-7 at distinct stages of development. Stem cell factor mediates both expression of progenitors and a developmental switch between T and NK lineages. In neutrophil granulopoiesis, LEF-1 is the key regulator of neutrophil granulopoiesis and plays a critical role in maturation arrest of myeloid progenitors from congenital neutropenia patients. In mast cell differentiation, however, it was shown that thrombopoietin is one of the essential regulating cytokines. In dendritic cell differentiation, Flt3 is one of the key regulators. Development of dendritic cells can be rescued by artificial Flt3 overexpression in Flt3-negative progenitor cells by inducing a dendritic cell development transcription factor profile, while suppression of Flt3 signaling reduces dendritic cell numbers in lymphatic tissues.

In elucidating mechanisms of malignant hematopoiesis, it was demonstrated that retrovirus vector insertion sites show a significant overrepresentation of insertions in genes involved in cytokine signaling, apoptosis, and transcriptional regulation. Furthermore, it was reported that neoplastic B cells show plasticity, and this dedifferentiation process offers an explanation for the unique phenotype of classical Hodgkin's lymphoma. In development of acute myeloid leukemia, NUP98 plays an important role in pathogenesis and prognosis. Disruption of the nuclear pore by NUP98 depletion either by LOH or nuclear body sequestration with NUP98-HOX fusion proteins is a leukemogenic event leading to enhanced stem cell expansion and impaired differentiation.

New aspects of stem cell exhaustion and senescence were described. Overexpression of the polycomb group gene Ezh2 results in a delayed myeloproliferative syndrome in serial transplantations. In order to study genes that regulate telomere length, novel live cell–imaging approaches were presented using transgenic murine embryonic stem cells expressing fusion genes encoding proteins such as TRF1, histone 2B, and PCNA fused to blue, green, and red fluorescent proteins.

In discussions about hematopoietic differentiation of embryonic stem cells, it was shown that embryonic stem cells generated by parthenogenesis in mice display *in vitro* hematopoietic activity similar to regular ES cell lines, and blood derivatives can repopulate hematopoiesis in irradiated adult mouse recipients. For human embryonic stem cells, repopulation activity of differentiated hematopoietic cells remains critical. Here, it was shown that Jagged-1 and Wnts are able to induce more appropriate mesodermal patterning and gene expression. Detecting hematopoietic stem cell populations in the embryo, BB9 is a novel marker of human angiohematopoietic stem cells expressed in the earliest stages of embryonic blood and blood vessel development to the adult hematovascular system.

Mesenchymal stem cells (MSCs) provide a unique potential as immunomodulating agents, especially in patients undergoing stem cell transplantation. Recent studies have indicated that MSCs inhibit the differentiation of monocytes into immature dendritic cells. However, cotransplantation assays have also shown that MSCs are not intrinsically immunosuppressive and may induce graft rejection under appropriate nonmyeloablative conditions.

We hope that the informal, open, and intense spirit of this very fruitful and successful meeting can be carried forward to the reader of this book, so that it will help to stimulate new and outstanding investigations in experimental haematology.

We thank the New York Academy of Sciences for publishing this proceedings. Our special thanks are extended to Linda Mehta, acquisitions editor, who also participated in the meeting, and Steven Bohall, project manager of the *Annals*. We are grateful for the rapid and easy communication that made this publication possible in a very timely manner. Finally, we want to extend our very special gratitude to Amgen International Europe/Germany for their generous contribution in support of the conference.

—KATJA C. WEISEL
University of Tübingen, Tübingen, Germany

—JOHN E. DICK
University of Toronto, Toronto, Ontario, Canada

—WILLEM E. FIBBE
Leiden University Medical Center, Leiden, the Netherlands

—LOTHAR KANZ
University of Tübingen, Tübingen, Germany

AMD3100 and CD26 Modulate Mobilization, Engraftment, and Survival of Hematopoietic Stem and Progenitor Cells Mediated by the SDF-1/CXCL12-CXCR4 Axis

HAL E. BROXMEYER, GIAO HANGOC, SCOTT COOPER, TIMOTHY CAMPBELL, SHIGEKI ITO, AND CHARLIE MANTEL

Department of Microbiology and Immunology, and the Walther Oncology Center, Indiana University School of Medicine, Indianapolis, Indiana 46202, USA

Walther Cancer Institute, Indianapolis, Indiana 46208, USA

ABSTRACT: The chemokine stromal cell-derived factor-1 (SDF-1/CXCL12) and its receptor, CXCR4, are involved in a number of facets of the regulation of hematopoiesis at the level of hematopoietic stem (HSCs) and progenitor (HPCs) cells. Modulation of this ligand–receptor interaction may be of clinical utility. We now report that: (1) the CC chemokine, macrophage inflammatory protein-1α (MIP-1α/CCL3) synergizes with AMD3100 (an antagonist of the binding of SDF-1/CXCL12 to CXCR4) to rapidly mobilize HPCs to the blood of mice; moreover, the combination of granulocyte colony-stimulating factor (G-CSF) with AMD3100 and MIP-1α/CCL3, given in a specific sequence, mobilizes the greatest number of HPCs compared to any combination of two of these mobilizing agents; (2) pretreatment of recipient mice with Diprotin A, an inhibitor of CD26/Dipeptidylpeptidase IV (DPPIV), enhances the competitive HSCs repopulating capacity of untreated donor cells; (3) the survival-enhancing effects of SDF-1/CXCL12 on HPCs subjected *in vitro* to delayed addition of growth factors (GFs) are mediated in part through the cell cycle-related proteins $p21^{cip1/waf1}$ (as assessed using $p21^{cip1/waf1}$ $-/-$ and $+/+$ mice) and Mad2 (using Mad2 $+/-$ and $+/+$ mice); and (4) deletion of CD26/DPPIV on mouse bone marrow cells increases the survival-enhancing effects of SDF-1/CXCL12 on HPCs. These results demonstrate the means to increase the mobilization of HPCs, the engrafting capability of HSCs, and responsiveness of HPCs to the survival-enhancing activity of SDF-1/CXCL12, effects that may be of practical value.

Address for correspondence: Hal E. Broxmeyer, Ph.D., Walther Oncology Center, Indiana University School of Medicine, 950 West Walnut Street, R2–302, Indianapolis, IN 46202–5181. Voice: 317-274-7510; fax: 317-274-7592.

hbroxmey@iupui.edu

KEYWORDS: AMD3100; CD26/DPPIV; SDF-1/CXCL12; CXCR4; hematopoietic stem cells; progenitor cells

INTRODUCTION

Hematopoiesis is a tightly regulated process, in large part controlled by interactions between specific cytokines/chemokines produced/released from accessory cells, and the receptors for these biomolecules that are found on hematopoietic stem (HSCs) and progenitor (HPCs), and accessory cells.[1,2] These ligand–receptor interactions trigger intracellular signaling effects that mediate proliferation, self-renewal, survival, and migration of HSCs and HPCs. While single cytokines or chemokines can influence functions of HSCs and HPCs, combinations of cytokines are likely highly influential in enhancing or suppressing these functions. Understanding the regulation of HSCs and HPCs *in vitro* and *in vivo* is dependent on assays that detect not only the phenotype of these cells, but most importantly their functions.[3,4] While phenotyping HSCs and HPCs have been informative and helpful in identifying and tracking HSCs and HPCs,[5,6] phenotype does not always recapitulate function, especially after manipulation of these cells *ex vivo*. Hence functional verification of a specific phenotype is crucial for a complete understanding of HSCs and HPCs, and their responsiveness to the actions of cytokines and chemokines.

The capacity of HSCs and HPCs to migrate and survive is of great relevance to HSCs–HPCs transplantation. Mobilizing these cells from bone marrow, where they are produced, into circulating blood allows for their collection from the blood in quantities necessary for their successful use in transplantation to treat malignant and nonmalignant disorders.[7,8] It is generally believed, with a few exceptions, that after infusion into recipients, HSCs do not home with absolute efficiency to the marrow microenvironment where they will be retained, and nurtured for optimal engraftment.[9] Thus, means to enhance mobilization, homing, and engraftment of HSCs and HPCs are important for the treatment of a large number of blood and other disorders.

The chemokine, stromal cell-derived factor-1 (SDF-1/CXCL12), which binds and activates its receptor, CXCR4, has been implicated as an important ligand–receptor interaction for movement into, retention within, and survival of HSCs/HPCs in the bone marrow microenvironment.[10–13] Modulating the SDF-1/CXCL12-CXCR4 interaction has allowed for enhanced mobilization of HSCs/HPCs to the blood, and in this context, AMD3100, a specific antagonist of SDF-1/CXCL12 binding to CXCR4, has been of practical use for inducing HSC–HPC mobilization in man,[14–18] mice,[14] dogs,[19] and monkeys,[20] and for synergizing with granulocyte colony-stimulating factor (G-CSF) to greatly enhance G-CSF-induced mobilization of HSCs/HPCs.[14,21,22]

CD26 is a cell-surface protein, which is a dipeptidylpeptidase IV (DPPIV). DPPIV has the capacity to truncate SDF-1/CXCL12.[9,12] The truncated

SDF-1/CXCL12 is both inactive as a chemotactic molecule, and also blocks the chemotaxis of full-length SDF-1/CXCL12.[23] Inhibition of CD26 on HSCs/HPCs by specific small peptide inhibitors of DPPIV, such as Diprotin A or Val-Pyr, enhances chemotaxis of these cells to SDF-1/CXCL12, presumably by blocking CD26 truncation of SDF-1/CXCL12.[23] This treatment also enhances homing of HSCs *in vivo* in lethally irradiated mice in competitive and noncompetitive HSC assays.[9,24,25] Consistent with this, treatment of mice with Diprotin A or Val-Pyr, or deletion of CD26 (in CD26 −/− mice) results in decreased mobilization of HPCs in response to exogenous administration of G-CSF.[26,27]

In this article, we demonstrate that: (1) another chemokine, macrophage inflammatory protein (MIP)-1α/CCL3, synergizes with AMD3100 to enhance *in vivo* mobilization of HPCs in mice; (2) pretreatment of recipient mice with Diprotin A enhances the engraftment of unmanipulated and untreated mouse bone marrow cells into lethally irradiated mice; and (3) SDF-1/CXCL12 enhancement of the survival of HPCs *in vitro*, is mediated in part through cell cycle regulators $p21^{cip1/waf1}$ and mitotic arrest deficiency (Mad2), and is enhanced by the treatment of target cells with Diprotin A.

MATERIALS AND METHODS

Mice

C3H/HeJ, C57Bl/6, and B6.BoyJ mice were purchased from Jackson Laboratories (Bar Harbor, ME) USA. The $p21^{cip1/waf1}$ −/−,[28] Mad2 +/−,[29] and CD26 −/−[9,27] mice were previously described.

Cells

Unseparated or separated mouse bone marrow cells were used as described in the legends to the Table and Figures. $CD34^+$ cord blood cells were purified as described elsewhere.[23]

HPCs

The assays for mouse bone marrow granulocyte macrophage (CFU-GM), erythroid (BFU-E), and multipotential (CFU-GEMM) progenitor cells have been described elsewhere.[3,4]

Competitive Repopulating Mouse Bone HSC Assays

Primary competitive repopulation and secondary noncompetitive transplants have been described elsewhere.[3,9]

Details of Experimental Procedures

This can be found in the Table and Figure legends, and within the text.

Statistical Analysis

This was done using the Student's *t*-test.

RESULTS AND DISCUSSION

Effect of MIP-1α/CCL3 on the Mobilization of HPCs, Alone and in Combination with AMD3100 and/or G-CSF

Several members of the chemokine family, including MIP-1α/CCL3,[30–32] IL-8/CXCL8,[33–35] MIP-2,[36] and GRO-β/CXCL2[37,38] mobilize HPCs into the blood of mice. MIP-1α/CCL3 was one of the first members of the chemokine family to be used to mobilize HPCs and HSCs to the blood of mice.[30] This mobilization is mediated through chemokine receptor CCR1,[32] which is one of the receptors that bind MIP-1α. Mobilization of HPCs in mice with MIP-1α/CCL3 is potent and rapid, with maximal effects occurring within 1 h. We reasoned that mobilization with MIP-1α/CCL3 might be different in mechanism than that of AMD3100, which works through antagonizing the binding of SDF-1/CXCL12 to its receptor, CXCR4,[14] and also that of G-CSF, which occurs as a much more prolonged response after days of multiple administrations of G-CSF.[39,40] The protocol, shown in FIGURE 1, was used to assess the effects of MIP-1α/CCL3 alone, AMD3100 alone, G-CSF alone, the combination of AMD3100 plus MIP-1α, the combination of AMD3100 plus G-CSF, and the combination of AMD3100, MIP-1α/CCL3, and G-CSF.

As shown in FIGURE 2, for the average results of a total of three independent experiments, MIP-1α/CCL3 or AMD3100, each alone, enhanced by 2.3- to 5.1-fold the mobilization of CFU-GM, BFU-E, and CFU-GEMM within 1 h of their administration to the circulating blood. Simultaneous administration of MIP-1α/CCL3 and AMD3100 together resulted in an 8.8- to 13.2-fold enhancement in blood HPCs levels. This was greater than the additive induction of mobilization of these HPCs than what was seen with MIP-1α/CCL3 alone plus that of AMD3100 alone.

In two separate experiments, shown in FIGURE 3, we assessed the effects of G-CSF, AMD3100, and MIP-1α/CCL3, alone and in combination, on the mobilization of CFU-GM, BFU-E, and CFU-GEMM. The combination of G-CSF with either AMD3100 or MIP1α resulted in the enhancement of circulating HPCs/mL blood compared to that of G-CSF, effects previously reported

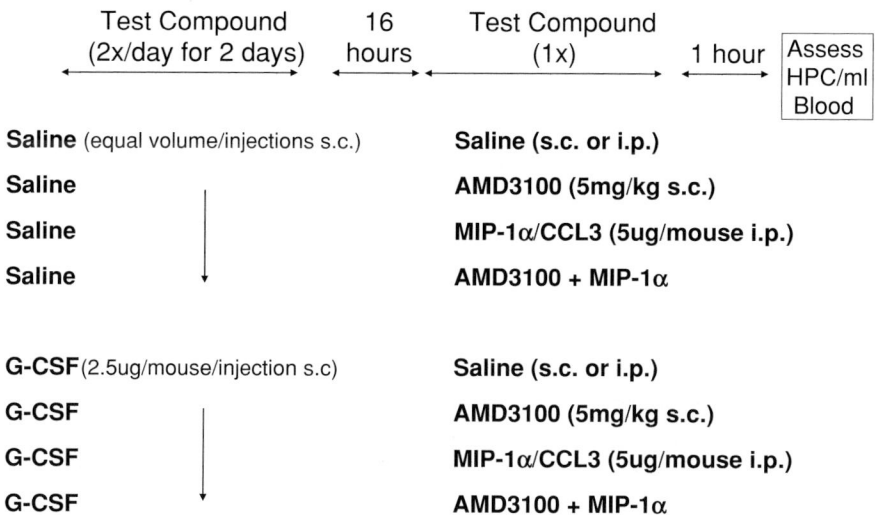

FIGURE 1. Protocol for the mobilization of HPCs to the blood by G-CSF, AMD3100, and MIP-1α/CCL3, each alone and in various combinations.

by us.[14,32] The largest fold increase in circulating levels of HPCs was detected when the combination of G-CSF, AMD3100, and MIP-1α/CCL3 was administered to the mice. However, the triple combination, while inducing the greatest number of HPCs to the circulation, was only borderline significant, or not statistically different from the numbers of HPCs mobilized by the combination of G-CSF plus AMD3100.

When the MIP-1α/CCL3 analog BB10010 was tested for mobilization in patients with breast cancer, this compound was not a very active mobilizer of HPCs.[31] Thus, MIP-1α/CCL3 or BB10010 may not be efficacious for the mobilization of HSCs/HPCs in man, alone or in combination with AMD3100 or G-CSF. Perhaps, other chemokines may be more useful in such a context. One such candidate may be GRO-β/CXCL2 or the truncated form of GRO-β.[37,38] Both bind the receptor CXCR2, and mobilize HSCs and HPCs. It has been reported that AMD3100 plus GRO-β/CXCL2 mobilize two- to fourfold more competitive repopulating HSCs in mice than that of G-CSF,[41] and that combined use of G-CSF, AMD3100, and GRO-β/CXCL2 induced suprasynergistic mobilization of CFU-GM.[42]

Not all patients respond well to the mobilizing effects of G-CSF alone.[39,40,43] Efforts to enhance the mobilization of HSCs/HPCs are warranted, and may be especially of value for use in patients who do not respond well to the mobilizing effects of G-CSF alone.

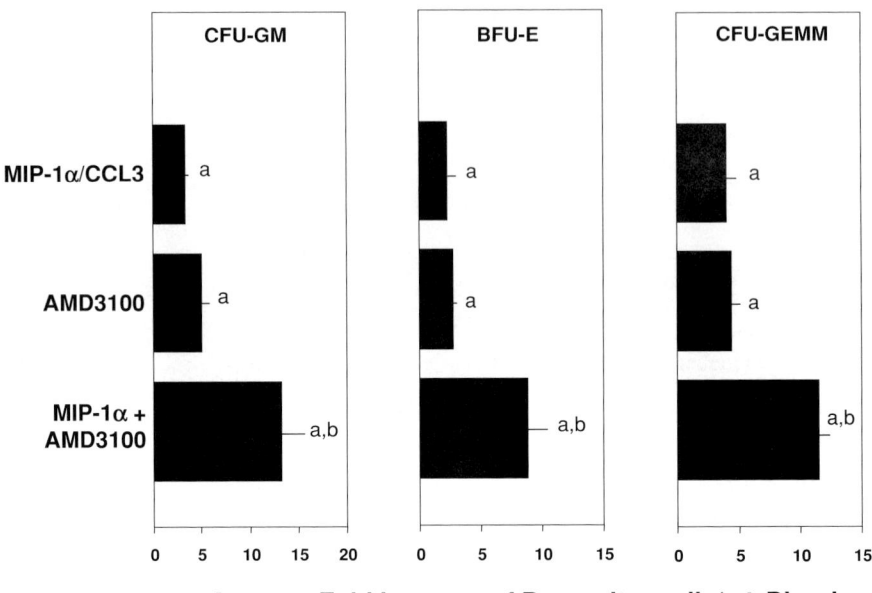

Average Fold Increase of Progenitor cells/mL Blood

FIGURE 2. Influence of MIP-1α/CCL3, and AMD3100, alone and in combination, on the mobilization of CFU-GM, BFU-E, and CFU-GEMM to the blood of C3H/HeJ mice. Results are shown as the mean fold increase ±SD for a total of three experiments. Mice were injected with either AMD3100, MIP-1α/CCL3, or their combination, at the dosages and routes of administration shown in FIGURE 1 and blood collected for analysis 1 h later. For two of the three experiments, the mice received saline s.c. 2×/day for 2 days prior to the administration of AMD3100 and/or MIP-1α/CCL3. Colonies were stimulated *in vitro* with 1 U/mL recombinant human erythropoietin (Epo), 50 ng/mL recombinant mouse stem cell factor (SCF), 1% $^{vol}/_{vol}$ pokeweed mitogen mouse spleen cell conditioned medium, and 0.1 mM hemin as reported elsewhere.[14] The fold increase in circulating progenitor cells was based respectively for the three experiments on control numbers ± 1 SD (saline injection) per milliliter of 92 ± 10, 304 ± 30, and 560 ± 151 for CFU-GM; 53 ± 16, 116 ± 25, and 186 ± 35 for BFU-E; and 12 ± 3, 35 ± 14, and 58 ± 7 for CFU-GEMM. $^{a}P < 0.05$ compared to saline control; $^{b}P < 0.05$ compared to either MIP-1α/CCL3 or AMD3100.

Effects of Inhibiting CD26/DPPIV Activity in Recipients on the Engraftment of Donor Cells

Inhibition of deletion of CD26/DPPIV in mice decreases the responsiveness of these mice to the induction of HPCs mobilization by G-CSF,[26,27] and inhibition or deletion of CD26/DPPIV on donor mouse bone marrow HSCs enhances the homing and engrafting capability of these cells in both a noncompetitive and competitive situation in lethally irradiated mouse recipients.[9,24,25] Since CD26 is expressed on a number of different cell types, in addition to that of HSCs and HPCs, we reasoned that pretreatment of recipient mice with Diprotin

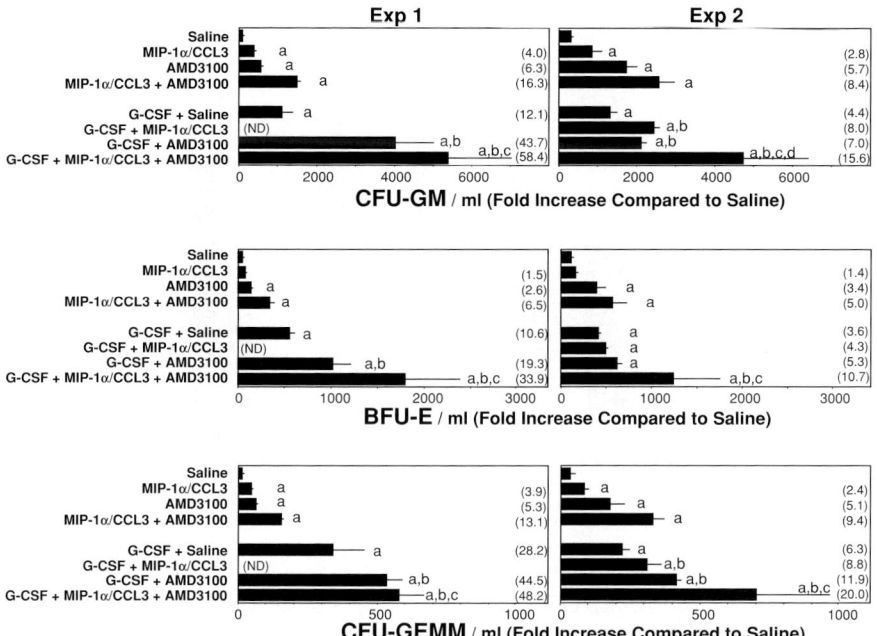

FIGURE 3. Influence of either G-CSF, MIP-1α/CCL3, or AMD3100 each alone, or the combination of MIP-1α/CCL3 plus AMD3100, G-CSF plus MIP-1α/CCL3, G-CSF plus AMD3100, or G-CSF, MIP-1α/CCL3, and G-CSF on the mobilization of CFU-GM, BFU-E, and CFU-GEMM in C3H/HeJ mice. Colonies were stimulated as denoted in the legend to FIGURE 2. Results are shown each for two separate experiments, done as described in FIGURE 1, with results presented as progenitors per milliliter ± 1 SD, with fold changes for each point shown in parenthesis to the right of the bars. $^{a}P < 0.05$ compared to saline control; $^{b}P < 0.05$ compared to G-CSF alone; $^{c}P < 0.05$ compared to MIP-1α/CCL3 and AMD3100; $^{d}P < 0.05$ compared to G-CSF and AMD3100.

A, a specific inhibitor of DPPIV activity, may also result in enhanced engraftment of donor HSCs into lethally irradiated mice. To test this hypothesis, we set up a preliminary experiment, the results of which are seen in FIGURE 4. Recipient B6BoyJ (CD45.1^{+}) mice were pretreated on day -2 with either Diprotin A (5 μM/recipient, 2×/day) or an equal volume of phosphate-buffered saline (PBS, sterile and pyrogen-free) 2×/day. On day -1, the mice received Diprotin A (5 μM) or PBS, followed 5 h later by an irradiation dose of 950 cGy. On day 0, mice were give Diprotin A (5 μM) or PBS once, followed 5 h later by intravenous (i.v.) injection of 5×10^{5} unseparated and untreated C57Bl/6 (CD45.2^{+}) and 5×10^{5} competitor (nonirradiated B6.BoyJ; CD45.1^{+}) bone marrow cells. The donor cells were injected into recipient mice 20 h after irradiation of the recipients. As seen in FIGURE 4, pretreatment of recipient B6.BoyJ mice resulted in enhanced competitive repopulation of donor HSCs at 2, 4,

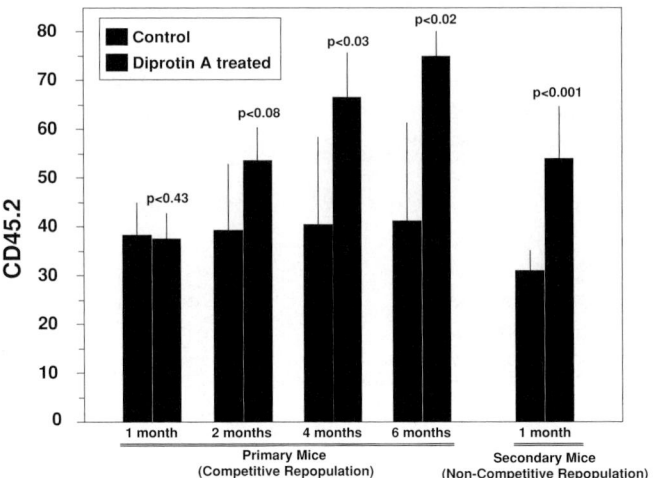

FIGURE 4. Competitive repopulation of C57Bl/6 (CD45.2$^+$) mouse bone marrow cells in Diprotin A pretreated B6.BoyJ (CD45.1$^+$) recipients. Results shown are the mean ±1 SD engraftment from pooled C57Bl mice injected at 5×10^5 cells in a 1:1 ratio with B6.Boy J cells into lethally irradiated (950 cGy) B6.BoyJ mice that were untreated ($n = 5$) or pretreated with Diprotin A ($n = 3$). Bone marrow cells from three primary mice each were transplanted into three lethally irradiated secondary recipients in a noncompetitive assay for a total of nine secondary recipients for each group. Transplants into primary and secondary mice was as reported by us.[9]

and 6 months post transplantation. That enhanced engraftment in Diprotin A treated mice was for self-renewing stem cells is suggested by enhanced repopulation of secondary mice with bone marrow cells from the Diprotin A treated primary mice. Repopulation of primary mice is slower than the enhanced engraftment previously reported by us in which pretreatment of donor cells with Diprotin A, or Val-Pyr, or use of donor CD26−/− cells demonstrated enhanced engraftment at 1-month post transplant of cells into irradiated recipients.[9] It is emphasized that this is a preliminary attempt at evaluating the means to enhance HSCs transplantation by pretreating the recipient mice. While these results are encouraging, we may not have used optimal pretreatment dosing and timing. Moreover, we consider it is possible that pretreatment of both donor cells and recipients may yield enhanced engraftment beyond that of treating only donor cells or only the recipients.

Our original studies demonstrating that inhibition of CD26 on donor mouse bone marrow cell populations enhances engraftment of these cells into mice[9] has been repeated by other groups,[24,25] and preliminary studies have shown that inhibition of CD26 or human CD34$^+$ cord blood cells enhances their engraftment of sublethally irradiated mice with a non-obese diabetic severe combined immunodeficient genotype.[44] Efforts are under way to evaluate the effects of CD26 inhibition of donor cells for enhancement of engraftment in dogs and monkeys, and it may be useful to also evaluate the effectiveness in these larger animal models of inhibiting CD26 in the large animal recipients.

It is hoped that the means to enhance the homing and engrafting capabilities of HSCs will be of clinical use in the future for transplantation in which only limiting numbers of donor cells are available for transplantation. To compensate for these limiting numbers of cells, cord blood transplantation in adults is currently using infusion of multiple cord blood collections from different donors for engraftment of adults,[45,46] to compensate for inadequate cell numbers usually found in single cord blood collections when adults are the transplant recipients.

A Positive Role for $p21^{cip1/waf1}$ and Mad2 in SDF-1/CXCL12 Enhancement of the Survival of HPCs in Vitro

SDF-1/CXCL12 enhances survival and decreases apoptosis of relatively unseparated, and also, highly purified populations of human bone marrow, cord blood, and mouse bone marrow CFU-GM, BFU-E, and CFU-GEMM during withdrawal and then delayed the addition of growth factors (GFs).[47–49] The survival-enhancing effects are directly acting as assessed on colony formation by single isolated human bone marrow and cord blood $CD34^{+++}$ cells.[48] These latter results reflected a colony readout in semisolid culture medium 14 days after the initiation of the single cell cultures. We followed up on this by a more in-depth analysis of the effects of SDF-1/CXCL12 on single cells in suspension culture in Terraski–Well plates early after the initiation of the single cell cultures. As shown in FIGURE 5, the majority of $CD34^+$ cells that died in the absence of GFs did so within the first 24 h (FIG. 5 left panel compared to center). SDF-1/CXCL12 had no protective/survival effect in this context at any time in the absence of GFs where the cells remained undivided. However, if the single cells divided at least once, the survival-enhancing effects of SDF-1/CXCL12 were seen (FIG. 5, right). In contrast, when GFs were present, the survival-enhancing effects of SDF-1 were observable at all times. At first glance, the GFs seemed to have a death-enhancing effect on single cells in the first 24 h, but due to the large variation between the experiments, this was not statistically significant ($P > 0.05$). These data lead us to conclude that survival-enhancing effects of SDF-1/CXCL12 on human cord blood (CB)-derived $CD34^+$ cells occur if the cells are in a state of rapid division, due to GF-induced cell cycle progression. SDF-1 also protected single $CD34^+$ cells if they were in an extremely slow state of cycling (FIG. 5, right, in which the cells only divided once in 6 days).

In our search for cell cycle-related intracellular molecules that may play a role in survival-enhancing events triggered by SDF-1/CXCL12, we focused on $p21^{cip1/waf1}$[28] and Mad2.[29] $p21^{cip1/waf1}$ plays a role in proliferation of HPCs

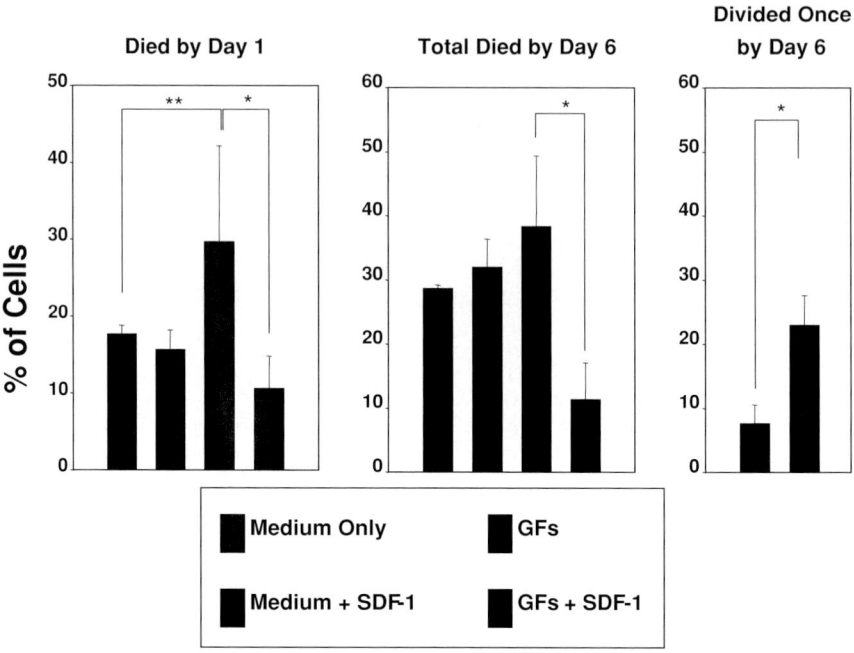

FIGURE 5. SDF-1/CXCL12 survival effects on single human CB-derived CD34$^+$ cells. CD34$^+$ cells were sorted directly into Terrasaki–Well plates containing complete medium without GFs, or with GFs (50 ng/mL SCF, and 10 ng/mL GM-CSF) and with or without 100 ng/mL rhuSDF-1. Cells were then incubated at 37°C in humidified 5% CO$_2$ with the lids on tight and plates removed once/day and examined microscopically. Percentage of cells represents the percentage of cells (verified as a single cell/well on day 0) that either died (as determined by the presence of cellular fragments and no intact cells) or remained quiescent, or that proliferated. Wells/cells that were marked as "died on day 1" were repeatedly examined on subsequent days to ensure that the cell was not obscured or unseen and had in fact died. Results are from three independent experiments using cells from three different CB collections. Each experiment was designed so that three separate Terrasaki–Well plates were set up per condition and on average contained a single verifiable cell in 55 of the 60 wells in each plate, or on average 165 cells analyzed per condition per experiment. Numbers are the mean ± 1 SD from the three experiments. *$P < 0.05$ and **$P < 0.06$.

and HSCs.[28,50] As shown in TABLE 1, SDF-1/CXCL12 did not prevent loss of recovery of CFU-GM colonies from p21$^{cip1/waf1}$ −/− mouse bone marrow cells after 1 day delayed GF addition. However, SDF-1/CXCL12 did significantly enhance recovery of the +/+ bone marrow cells. That SDF-1/CXCL12 did not have survival-enhancing effects on HPCs from p21$^{cip1/waf1}$ −/− mice implicates p21$^{cip1/waf1}$ in the positive regulation of cell survival mediated by SDF-1/CXCL12. Exactly what role p21$^{cip1/waf1}$ plays in these SDF-1/CXCL12 effects and how they are mechanistically mediated remains to be determined.

TABLE 1. Influence of SDF-1/CXCL12 on colony formation of CFU-GM from p21$^{cip1/waf1}$ (−/−) and control mouse (+/+) bone marrow in the context of delayed addition of GFs

	GFs Added			
	Day 0		Day 1	
Mice[GFs]	Control	+SDF-1/ CXCL12	Control	+SDF-1/ CXCL12
Part#1				
+/+[GM-CSF]	37 ± 2	35 ± 4 (95)	12 ± 2 (32)^a	23 ± 2 (62)^{a,b}
p21^{cip1} −/−[GM-CSF]	18 ± 4	18 ± (100)	5 ± 1 (28)^a	6 ± 1 (33)^a
Part#2				
+/+[GM-CSF+SCF]	59 ± 4	59 ± 4 (100)	10 ± 2 (17)^a	21 ± 2 (36)^{a,b}
p21^{cip1} −/−[GM-CSF+SCF]	36 ± 2	37 ± 2 (103)	5 ± 2 (14)^a	5 ± 1 (14)^a

Results are shown as mean colonies ± 1 SD [in parenthesis are shown: percentage control (survival) compared to day 0, control (= no SDF-1/CXCL12) for one experiment in which GF added were either GM-CSF (part 1) or the combination of GM-CSF plus SCF (part 2).] Bone marrow cells were plated at 5×10^4/mL in 0.3% agar culture medium with 10% fetal bovine serum (FBS) in the absence or presence of SDF-1/CXCL12 (200 ng/ml). GFs [GF: murine GM-CSF (10 ng/mL) or the combination of 10 ng/mL GM-CSF plus murine SCF (50 ng/mL)] were added at day 0, or day 1 and colonies scored 7 days after the addition of GFs. ^{a}significant difference from day 0 control, $P < 0.001$; $^b P < 0.01$ compared to minus SDF-1/CXCL12 on that day.

It is possible that these effects may in part be linked to our report that the p21$^{cip1/waf1}$ Chk1 pathway monitors G$_1$ phase microtubule integrity and is crucial for restriction point transition.[51]

We recently identified a role for Mad2 in the regulation of hematopoiesis.[29] Mad2+/− bone marrow and spleen manifested decreased absolute numbers and cycling status of immature, but not mature, HPCs. This may relate to the fact that bone marrow CFU-GM from Mad2+/− mice did not respond to the synergistic proliferation of combined stimulation with SCF plus GM-CSF.[29] Moreover, the percentage of Annexin V$^+$ c-kit$^+$ Lin$^-$ bone marrow cells was higher in Mad2+/−, than Mad2+/+ cells after culture with SCF and GM-CSF. Our results suggested that Mad2 was involved in the synergistic growth of immature HPCs (those HPCs responsive to synergistic stimulation by multiple GFs), effects that may be mediated via physical association of Mad2 with c-kit.[29] Immunoprecipitation assays done with the human GF-dependent cell line MO7e showed that c-kit physically associated with Mad2. Confocal microscopy demonstrated Mad2 co-localization with c-kit in the cytoplasm of MO7e cells. Stimulation of MO7e cells with the combination of SCF and GM-CSF resulted in disassociation of Mad2 from c-kit.

As seen in FIGURE 6, CFU-GM, BFU-E, and CFU-GEMM from Mad2+/− mouse bone marrow survived less well than +/+ bone marrow HPCs after 1 day of delayed GF addition. While SDF-1/CXCL12 either completely or greatly prevented this loss of HPCs from +/+ cells, it did not influence the loss of

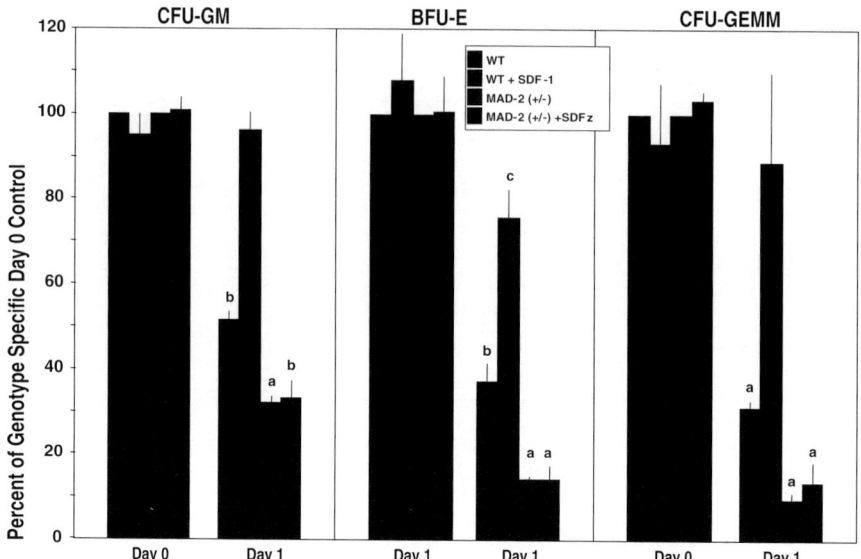

FIGURE 6. Survival-enhancing effects of SDF-1/CXCL12 on CFU-GM, BFU-E and CFU-GEMM cells from wild-type or Mad2+/− bone marrow. Cells were plated in methylcellulose culture medium with 30% FBS and 0.1 mM hemin in the absence or presence of SDF-1/CXCL12 (200 ng/mL). A GF cocktail consisting of Epo (1 μ/mL), PWMSCM (5% v/v), SCF (50 ng/mL) was added to plates at day 0 or day 1 and colonies scored 7 days after the addition of GFs. Results are given as mean ± 1 SD for three experiments in which control colony numbers at day 0 without SDF-1 averaged 65 ± 3 CFU-GM, 5 ± 1 BFU-E, and 4 ± 1 CFU-GEMM per 5×10^4 cells plated. Significance values refer to day 0 control colony numbers for WT or Mad2+/− cells. $^a P < 0.001$; $^b P < 0.002$; $^c P < 0.04$.

HPCs from the marrow of Mad2+/− mice. These studies implicate Mad2 as a positive regulator in SDF-1/CXCL12 enhanced recovery of clonogenic cells in the context of delayed addition of GFs and adds to our previous studies in which Mad2+/− cells demonstrated enhanced apoptosis under selected conditions.[29] Mad2+/− and +/+ Sca-1+Lin− cells both expressed CXCR4 as determined by multivariable flow cytometry (FIG. 7), and HPCs from Mad2+/− and Mad2+/+ bone marrow cells migrated toward SDF-1/CXCL12 in a chemotaxis assay (data not shown), suggesting that the defective survival-enhancing effects of SDF-1/CXCL12 on HPCs did not relate to expression of CXCR4, or to the ability of CXCR4 to respond functionally to SDF-1/CXCL12. Confirmation that Mad2 expression is lower in Mad2+/− compared to Mad2+/+ Sca-1+Lin− cells is shown in FIGURE 8.

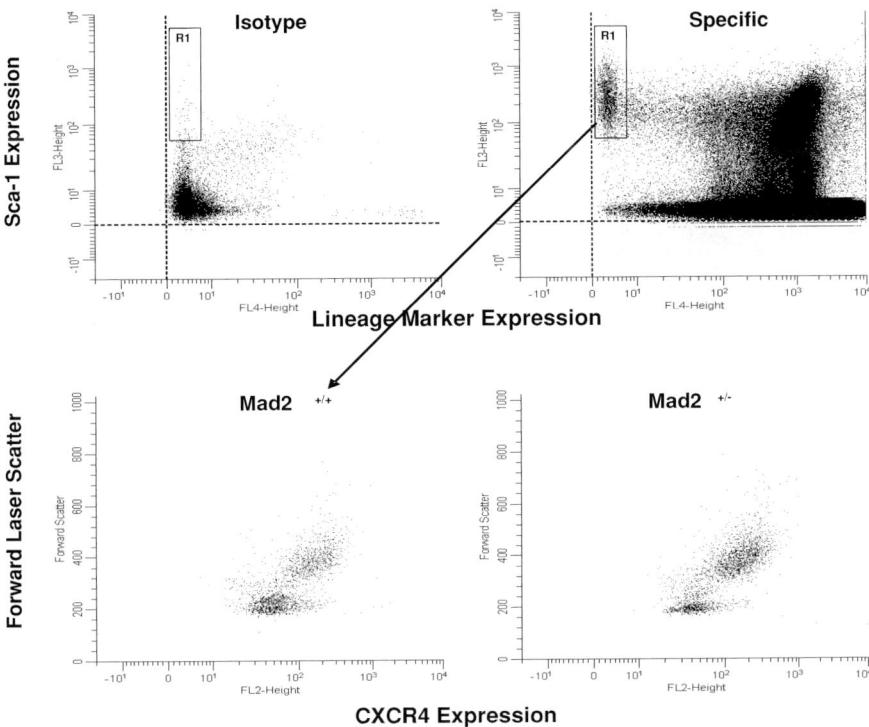

FIGURE 7. CXCR4 expression is similar in Mad2+/+ and Mad2+/− Sca-1+Lin− bone marrow cells. The upper panels are dot plots of Sca-1 versus lineage marker expression in Mad2+/+ bone marrow cells. Gate R1 contains the Sca-1+Lin− cells. Expression of CXCR4 is shown in these gated cells only in the lower dot plots, which compare the two genotypes (CXCR4 isotype control Ab fluorescence was below 10 FL).

Influence of CD26/DPPIV on the HPCs Survival-Enhancing Activity of SDF-1/CXCL12

Since CD26/DPPIV truncates SDF-1/CXCL12 and inactivates the chemotactic effects of SDF-1/CXCL12,[23] and deletion or inhibition of CD26 results in enhanced chemotaxis of HPCs to SDF-1/CXCL12,[9,23] we felt that by inhibiting/deleting CD26 on target cells we would also enhance the sensitivity of target cells to the survival-enhancing effects of SDF-1/CXCL12 in the context of delayed addition of GFs. As seen in FIGURE 9, deletion of CD26 on mouse bone marrow cells resulted in significantly increased survival-enhancing effects of SDF-1/CXCL12 on CFU-GM, BFU-E, and CFU-GEMM in the setting of delayed addition of GFs.

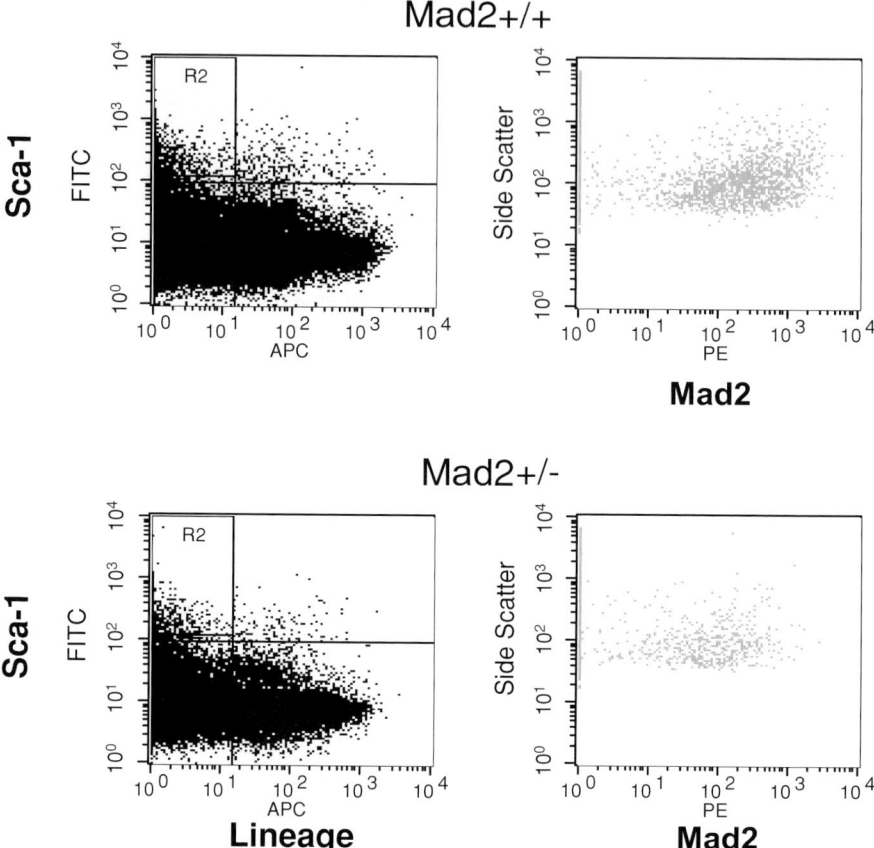

FIGURE 8. Sca-1+Lin− bone marrow cells from Mad2+/− mice have lower Mad2 protein levels compared to wild-type mice. The left dot plots compare the proportions of Sca-1+/Lin- cells in total bone marrow from the two genotypes. The right dot plots show only the cells in gate R2 from the left panels.

CONCLUDING REMARKS

It is clear that the SDF-1/CXCL12-CXCR4 axis plays a number of important roles in the maintenance and proper functioning of HSCs/HPCs and hematopoiesis. While there is still much to be learned regarding this ligand-receptor interaction, and how it is mechanistically mediated for multiple functions, modulation of this pathway at various levels has the potential to enhance treatment modalities for the mobilization and engraftment of HSCs/HPCs.

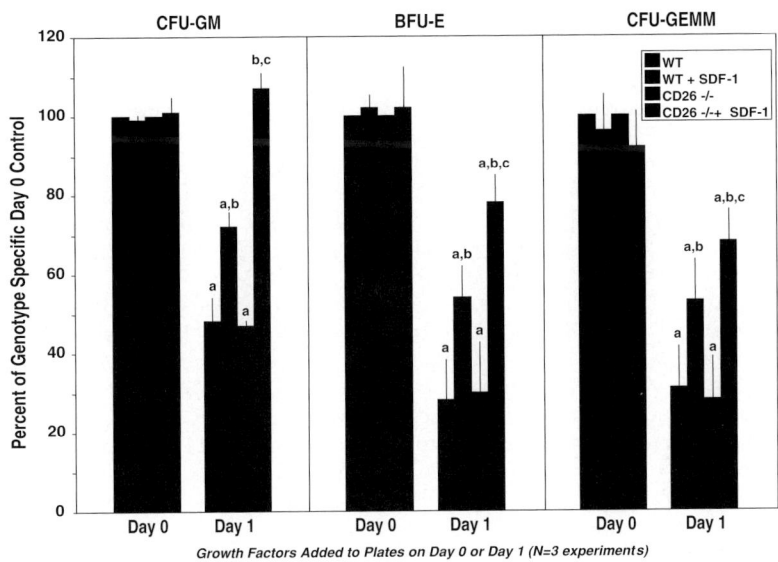

FIGURE 9. Deletion of CD26 on murine bone marrow cells increases the survival-enhancing effects of SDF-1/CXCL12 on CFU-GM, BFU-E, and CFU-GEMM subjected to delayed additions of GFs. Unseparated C57Bl/6 mouse bone marrow cells were plated at 5 × 10^4 cells/mL in 1% methylcellulose culture medium with 30% FBS and 0.1 mM hemin at time 0 in the absence or presence of 100 ng/mL SDF-1/CXCL12. GFs (1 U/mL recombinant human Epo, 5% v/v pokeweed mitogen mouse spleen cell conditioned medium, 50 ng/mL recombinant murine SCF) were added to plates either at time 0, or 24 h later. Cells were incubated at 5% CO_2 in lowered (5%) O_2 and colonies were scored 7 days after the addition of GFs. Results are expressed as the mean percentage of colonies ± 1 SD at each point compared to the time 0 control (+/+) colonies grown in the absence of SDF-1/CXCL12 but in the presence of the GFs. The mean ±1 SD refers to results of three separate experiments in which each point per experiment was the average of triplicate plates. [a]significant decrease ($P < 0.01$) compared to day 0, +/+ cells in the absence of SDF-1/CXCL12; [b]significant increase ($P < 0.05$) in survival/growth of MPC in the presence versus absence of SDF-1/CXCL12 for each strain of mouse; [c]significant increase in survival/growth of MPC of CD26−/− cells compared with survival/growth of MPC of CD26−/− cells with SDF-1/CXCL12 ($P < 0.01$ for CFU-GM and BFU-E; $P < 0.05$ for CFU-GEMM). Control colony numbers (mean ± 1 SD for time 0, +/+ cells without SDF-1/CXCL12 for three experiments were as follows: CFU-GM: 57 ± 5, 64 ± 7, 63 ± 3; BFU-E: 4 ± 0.5, 3 ± 0.6, 6 ± 1; CFU-GEMM: 4 ± 0.6, 4 ± 1, 5 ± 0.3.

ACKNOWLEDGMENTS

These studies were supported by U.S. Health Service Grants RO1 HL56416, RO1 HL67384, and a Project on PO1 HL53586 to HEB from the National Institutes of Health.

REFERENCES

1. SHAHEEN, M. & H.E. BROXMEYER. 2005. The humoral regulation of hematopoiesis. *In* Hematology: Basic Principles and Practice, 4th ed. R. Hoffman, E. Benz, S. Shattil, B. Furie, H. Cohen, L. Silberstein P. McGlave, Eds.: Chapter 19, 233–265. Elsevier Churchill Livingston. Philadelphia, PA.
2. BROXMEYER, H.E. 2004. Proliferation, self-renewal, and survival characteristics of cord blood hematopoietic stem and progenitor cells. *In* Cord Blood: Biology, Immunology, Banking, and Clinical Transplantation. H.E. Broxmeyer, Ed.: Chapter 1, 1–21. Amer. Assoc. Blood Banking. Bethesda, MD.
3. BROXMEYER, H.E., E. SROUR, C. ORSCHELL, *et al.* 2006. Cord blood stem and progenitor cells. *In* Method in Enzymology, Vol 419. I. Klimanskaya and R. Lanza, Eds.: 439–473. Academic Press, Elsevier Science. San Diego, CA.
4. COOPER, S. & H.E. BROXMEYER. 1996. Measurement of interleukin-3 and other hematopoietic growth factors, such as GM-CSF, G-CSF, M-CSF, erythropoietin and the potent co-stimulating cytokines steel factor and Flt-3 ligand. *In* Current Protocols in Immunology. J.E. Coligan, A.M. Kruisbeek, D.H. Margulies, E.M. Shevach, W. Strober, R. Coico, Eds.: Suppl 18, 6.4.1–6.4.12. John Wiley & Sons. New York.
5. KONDO, M., A.J. WAGERS, M.G. MANZ, *et al.* 2003. Biology of hematopoietic stem cells and progenitors: implications for clinical application. Ann. Rev. Immunol. **21:** 759–806.
6. MANZ, M.G., K. AKASHI & I.L. WEISSMAN. 2004. Biology of hematopoietic stem and progenitor cells. *In* Thomas' Hematopoietic Cell Transplantation, 3rd ed. K.G. Blume, S.J. Forman F.R. Appelbaum, Eds.: 69–95. Blackwell Publishing. Malden, MA.
7. PAPAYANNOPOULOU, T. 2004. Current mechanistic scenarios in hematopoietic stem/progenitor cell mobilization. Blood **103:** 1580–1588.
8. WINKLER, I.G. & J.P. LÉVESQUE. 2006. Mechanisms of hematopoietic stem cell mobilization: when innate immunity assails the cells that make blood and bone. Exp. Hematol. **34:** 996–1009.
9. CHRISTOPHERSON, K.W., II, G. HANGOC, C. MANTEL, *et al.* 2004. Modulation of hematopoietic stem cell homing and engraftment by CD26. Science **305:** 1000–1003.
10. AIUTI, A., L.J. WEBB, C. BLEUL, *et al.* 1997. The chemokine SDF-1 is a chemoattractant for human $CD34^+$ hematopoietic progenitor cells and provides a new mechanism to explain the mobilization of $CD34^+$ progenitors to peripheral blood. J. Exp. Med. **185:** 111–120.
11. KIM, C.H. & H.E. BROXMEYER. 1998. *In vitro* behavior of hematopoietic progenitor cells under the influence of chemoattractants: stromal cell-derived factor-1, steel factor and the bone marrow environment. Blood **91:** 100–110.
12. BROXMEYER, H.E. & K.W. CHRISTOPHERSON II. 2004. Stromal cell-derived factor-1/CXCL12, CXCR4 and CD26 in the mobilization and homing of hematopoietic stem and progenitor cells. Curr. Med. Chem. Anti-Inflammatory Anti Allergy Agents (Special Issue: Lymphocyte Migration and Chemokines. C.H. Kim, Guest Ed.). **3:** 303–311.
13. DAR, A., O. KOLLET & T. LAPIDOT. 2006. Mutual, reciprocal SDF-1/CXCR4 interactions between hematopoietic and bone marrow stromal cells regulate human stem cell migration and development in NOD/SCID chimeric mice. Exp. Hematol. **34:** 967–975.

14. BROXMEYER, H.E., C.M. ORSCHELL, D.W. CLAPP, et al. 2005. Rapid mobilization of murine and human hematopoietic stem and progenitor cells with AMD3100, a CXCR4 antagonist. J. Exp. Med. **201:** 1307–1318.
15. LILES, W.C., H.E. BROXMEYER, E. RODGER, et al. 2003. Mobilization of hematopoietic progenitor cells in healthy volunteers by AMD3100, a CXCR4 antagonist. Blood **102:** 2728–2730.
16. HUBEL, K., W.C. LILES, H.E. BROXMEYER, et al. 2004. Leukocytosis and mobilization of $CD34^+$ hematopoietic progenitor cells by AMD3100, a CXCR4 antagonist. Support. Cancer Ther. **1:** 165–172.
17. DEVINE, S.M., N. FLOMENBERG, D.H. VESOLE, et al. 2004. Rapid mobilization of $CD34^+$ cells following administration of the CXCR4 antagonist AMD3100 to patients with multiple myeloma and non-Hodgkin's lymphoma. J. Clin. Oncol. **22:** 1095–1102.
18. GRIGNANI, G., E. PERISSINOTTO, G. CAVALLONI, et al. 2005. Clinical use of AMD3100 to mobilize $CD34^+$ cells in patients affected by non-Hodgkin's lymphoma or multiple myeloma. Clin. Oncol. **23:** 3871–3872.
19. BURROUGHS, L., M. MIELCAREK, M.T. LITTLE, et al. 2005. Durable engraftment of AMD3100-mobilized autologous and allogeneic peripheral-blood mononuclear cells in a canine transplantation model. Blood **106:** 4002–4008.
20. LAROCHELLE, A., A. KROUSE, M. METZGER, et al. 2006. AMD3100 mobilizes hematopoietic stem cells with long-term repopulating capacity in nonhuman primates. Blood **107:** 3772–3778.
21. LILES, W.C., E. RODGER, H.E. BROXMEYER, et al. 2005. Augmented mobilization and collection of $CD34^+$ hematopoietic cells from normal human volunteers stimulated with G-CSF by single-dose administration of AMD3100, a CXCR4 antagonist. Transfusion **45:** 295–300.
22. FLOMENBERG, N., S.M. DEVINE, J.F. DIPERSIO, et al. 2005. The use of AMD3100 plus G-CSF for autologous hematopoietic progenitor cell mobilization is superior to G-CSF alone. Blood **106:** 1867–1874.
23. CHRISTOPHERON, K.W., G. HANGOC & H.E. BROXMEYER. 2002. Cell surface peptidase CD26/DPPIV regulates CXCL12/SDF-1a mediated chemotaxis of human $CD34^+$ progenitor cells. J. Immunol. **169:** 7000–7008.
24. TIAN, C., J. BAGLEY, D. FORMAN, et al. 2006. Inhibition of CD26 peptidase activity significantly improves engraftment of retrovirally transduced hematopoietic progenitors. Gene Ther. **13:** 652–658.
25. PERANTEAU, W.H., M. ENDO, O.O. ADIBE, et al. 2006. CD26 inhibition enhances allogeneic donor-cell homing and engraftment after in utero hematopoietic-cell transplantation. Blood **108:** 4268–4274.
26. CHRISTOPHERSON, K.W., II, S. COOPER & H.E. BROXMEYER. 2003. Cell surface peptidase CD26/DPPIV mediates G-CSF mobilization of mouse progenitor cells. Blood **101:** 4680–4686.
27. CHRISTOPHERSON, K.W., S. COOPER & H.E. BROXMEYER. 2003. CD26 is essential for normal G-CSF-induced progenitor cell mobilization as determined by CD26 -/- mice. Exp. Hematol. **31:** 1126–1134.
28. MANTEL, C., Z. LUO, J. CANFIELD, et al. 1996. Involvement of $p21^{cip}$-1 and p27kip-1 in the molecular mechanisms of steel factor induced proliferative synergy *in vitro* and of p21cip-1 in the maintenance of stem/progenitor cells *in vivo*. Blood **88:** 3710–3719.
29. ITO, S., C. MANTEL, M.-K. HAN, et al. 2007. Mad2 is required for optimal hematopoiesis: Mad2 associates with c-Kit MO7e cells. Blood **109:** 1923–1930.

30. LORD, B.I., L.B. WOOLFORD, L.M. WOOD, *et al.* 1995. Mobilization of early hematopoietic progenitor cells with BB-10010: a genetically engineered variant of human macrophage inflammatory protein-1 alpha. Blood **85:** 3412–3415.
31. BROXMEYER, H.E., A. ORAZI, N.L. HAGUE, *et al.* 1998. Myeloid progenitor cell proliferation and mobilization effects of BB10010, a genetically engineered variant of human macrophage inflammatory protein-1a, in a phase I clinical trial in patients with relapsed/refractory breast cancer. Blood Cells Mol. Dis. **24:** 14–30.
32. BROXMEYER, H.E., S. COOPER, G. HANGOC, *et al.* 1999. Dominant myelopoietic effector functions mediated by chemokine receptor CCR1. J. Exp. Med. **189:** 1987–1992.
33. LATERVEER, L., I.J. LINDLEY, M.S. HAMILTON, *et al.* 1995. Interleukin-8 induces rapid mobilization of hematopoietic stem cells with radioprotective capacity and long-term myelolymphoid repopulating ability. Blood **85:** 2269–2275.
34. FIBBE, W.E., J.F. PRUIJT, G.A. VELDERS, *et al.* 1999. Biology of IL-8-induced stem cell mobilization. Annls. N. Y. Acad. Sci. **872:** 71–82.
35. LATERVEER, L., J.M. ZIJLMANS, I.J. LINDLEY, *et al.* 1996. Improved survival of lethally irradiated recipient mice transplanted with circulating progenitor cells mobilized by IL-8 after pretreatment with stem cell factor. Exp. Hematol. **24:** 1387–1393.
36. WANG, J., N. MUKAIDA, Y. ZHANG, *et al.* 1997. Enhanced mobilization of hematopoietic progenitor cells by mouse MIP-2 and granulocyte colony-stimulating factor in mice. J. Leuk. Biol. **62:** 503–509.
37. KING, A.G., D. HOROWITZ, S.B. DILLON, *et al.* 2001. Rapid mobilization of murine hematopoietic stem cells with enhanced engraftment properties and evaluation of hematopoietic progenitor cell mobilization in rhesus monkeys by a single injection of SB-251353, a specific truncated form of the human CXC chemokine GRObeta. Blood **97:** 1534–1542.
38. PELUS, L.M., H. BIAN, A.G. KING, *et al.* 2004. Neutrophil-derived MMP-9 mediates synergistic mobilization of hematopoietic stem and progenitor cells by the combination of G-CSF and the chemokines GRObeta/CXCL2 and GRObetaT/CXCL2delta4. Blood **103:** 110–119.
39. NG-CASHIN, J. & T. SHEN. 2004. Mobilization of autologous peripheral blood hematopoietic cells for support of high-dose cancer therapy. *In* Thomas' Hematopoietic Cell Transplantation, 3rd ed. K.G. Blume, S.J. Forman F.R. Appelbaum, Eds.: 576–587. Blackwell Publishing. Malden, MA.
40. SCHMITZ, N. 2004. Peripheral blood hematopoietic cells for allogeneic transplantation. *In* Thomas' Hematopoietic Cell Transplantation, 3rd ed. K.G. Blume, S.J. Forman F.R. Appelbaum, Eds.: 588–598. Blackwell Publishing. Malden, MA.
41. PELUS, L.M., S. FUKUDA & G. BRIDGER. 2006. The CXCR4 antagonist AMD3100 and the CXCR2 agonist GRObsynergistically mobilize hematopoietic stem cells (HSCs) with short and long term repopulating activity. Blood **108** (Suppl Part 1):105a (Abstract 339).
42. PELUS, L.M., D.W. CLAPP & G. BRIDGER. 2006. Suprasynergistic peripheral blood stem cell mobilization in normal and Fanconi anemia knockout mice by the combination of G-CSF plus the CXCR4 antagonist AMD3100 and the CXCR2 agonist GROb. Blood **108**(Suppl. Part 1): 909a (Abstract 3185).
43. CROOP, J.M., R. COOPER, C. FERNANDEZ, *et al.* 2001. Mobilization and collection of peripheral blood CD34$^+$ cells from patients with Fanconi anemia. Blood **98:** 2917–2921.

44. CAMPBELL, T., G. HANGOC & H.E. Broxmeyer. 2005. The role of CD26 in human umbilical cord blood CD34$^+$ cell engraftment of NOD/SCID mice. Blood **106**(Part 1): 487a (Abstract 1708).
45. BARKER, J.N., D. J. WEISDORF, T.E. DEFOR, *et al.* 2005. Transplantation of 2 partially HLA-matched umbilical cord blood units to enhance engraftment in adults with hematologic malignancy. Blood **105:** 1343–1347.
46. BALLEN, K.K., T.R. SPITZER, B.Y. YEAP, *et al.* 2007. Double unrelated reduced intensity umbilical cord blood transplantation in adults. Biol. Blood Marrow Transpl. **13:** 82–89.
47. LEE, Y., A. GOTOH, H.-J. KWON, *et al.* 2002. Enhancement of intracellular signaling associated with hematopoietic progenitor cell survival in response to SDF-1/CXCL12 in synergy with other cytokines. Blood **99:** 4307–4317.
48. BROXMEYER, H.E., L. KOHLI, C.H. KIM, *et al.* 2003. Stromal cell derived factor-1/CXCL12 enhances survival/anti-apoptosis of hematopoietic stem and myeloid progenitor cells: direct effects mediated through CXCR4 and Gai proteins. J. Leuk. Biol. **73:** 630–638.
49. BROXMEYER, H.E., S. COOPER, L. KOHLI, *et al.* 2003. Transgenic expression of stromal cell derived factor-1/CXCL12 enhances myeloid progenitor cell survival/anti-apoptosis *in vitro* in response to growth factor withdrawal and enhances myelopoiesis *in vivo*. J. Immunol. **170:** 421–429.
50. CHENG, T., N. RODRIGUES, H. SHEN, *et al.* 2000. Hematopoietic stem cell quiescence maintained by p21cip1/waf1. Science **287:** 1804–1808.
51. MANTEL, C.R., V.M. GELFANOV, Y.-J. KIM, *et al.* 2002. P21waf-1-Chk1 pathway monitors G1-phase microtubule integrity and is crucial for restriction point transition. Cell Cycle **1:** 327–336.

The Stem Cell Continuum
Cell Cycle, Injury, and Phenotype Lability

PETER J. QUESENBERRY, GERALD COLVIN, GERRI DOONER,
MARK DOONER, JASON M. ALIOTTA, AND KEVIN JOHNSON

*Division of Hematology and Oncology, Rhode Island Hospital, Providence,
Rhode Island 02903, USA*

ABSTRACT: The phenotype of the hematopoietic stem cell is intrinsically labile and impacted by cell cycle and the effects of tissue injury. In published studies we have shown that there are changes in short- and long-term engraftment, progenitor numbers, gene expression, and differentiation potential with cytokine-induced cell cycle transit. Critical points here are that these changes are reversible and not unidirectional weighing, heavily against a hierarchical model of stem cell regulation. Furthermore, a number of studies have now established that stem cells separated by lineage depletion and selection for Sca-1 or c-kit or low rhodamine and Hoechst staining are in fact a cycling population. Last, studies on Hoechst separated "cycling" stem cells indicates that the observed phenotype shifts relate to phase of cell cycle and are not due to in vitro exposure to cytokines. These data suggest a continuum model of stem cell regulation and further indicate that this model holds for in vivo situations. Observations that marrow cells can convert to various tissue cells under different injury conditions continue to be published despite a small, but influential, number of negative studies. Our studies and those of others indicate that conversions of marrow-derived cells to different tissue cells, such as skeletal muscle and lung, is critically dependent upon multiple variables, the most important of which is the presence of tissue injury. Variables which affect conversion of marrow cells to nonhematopoietic cells after in vivo transplantation include the nature and timing of the injury; marrow mobilization; the marrow cell type infused; the timing of cell infusion and the number of cells infused; the cell cycle state of the marrow cells, and other functional alterations in the marrow cells the treatment of the host mouse separate from specific injury; the mode of cell delivery; and possibly the presence of microvesicles from injured tissue. At least some of the highlighted negative reports on stem cell plasticity appear to be due to a failure to address these variables. Recently, we have observed that irradiated lung releases microvesicles which can enter marrow cells and lead to the marrow cells

Address for correspondence: Peter J. Quesenberry, M.D., Division of Hematology and Oncology, Rhode Island Hospital, 593 Eddy Street, George 3, Providence, RI 02903. Voice: 401-444-4830; fax: 401-444-4184.
pquesenberry@lifespan.org

expressing lung-specific mRNA and protein. This could provide an underlying mechanism for many of the plasticity phenomena. Altogether, marrow appears to represent a highly flexible ever-changing cell system with the capacity to respond to products of injured cells and top repair a broad range of tissues.

KEYWORDS: stem cell; continuum; plasticity

INTRODUCTION

The stem cell continues to be characterized by a variety of approaches including functional assays, cell-surface markers, and genetic characteristics. The wide variety of distinguishing features have not clarified the picture, rather they have suggested the existence of a variety of stem cells, which vary in different ways. The gold standard, of course, has been long-term repopulation into the lethally irradiated mouse, but even here there are difficulties. The timing of long-term repopulation may identify different classes of stem cells, as may the capacity for secondary and tertiary repopulation in serial transplants. In addition, there are clear circadian rhythms for stem cell engraftment[1] raising the question of whether the basic phenotype changes with different times of the day or with different seasons of the year.

Hopes have continued that studies on a single cell basis could definitively characterize the marrow stem cell, but this has not happened. Clonal studies have established that the most purified stem cell populations, while pure, based on the separative features employed, are totally heterogeneous when other functional characteristics, such as differentiation in response to cytokines, are evaluated.[2]

The pioneering studies of Till, McCulloch, and Siminovitch[3] identifying colony-forming unit-spleen (CFU-s) as the first clonal stem cell, speak to the above problem. They noted that the self-renewal capacity of the CFU-s was totally heterogeneous and commented that CFU-s might be very much like radioactive isotopes. These latter had decay rates, which, when considered on a population basis, could be precisely determined and were reproducible, while when individual nuclei were considered, the decay rates were not predictable and were heterogeneous. This suggests, from this early work, that attempts to characterize the stem cell on a single cell basis are probably doomed to failure. Rather, they probably have to be characterized on a population basis, in order to be understood.

In a similar related vein, a different aspect of stem cell heterogeneity or definition has more recently been discovered; that of stem cell plasticity. There has been much unneeded controversy over the term plasticity. It simply implies the capacity of marrow cells to give rise to cells in nonhematopoietic lineages, regardless of the mechanisms involved. It is clear now that marrow cells, after transplantation *in vivo*, can give rise to hepatocytes, epithelial lung cells,

TABLE 1. Hematopoietic changes with cell cycle progression

Short- and long-term *in vivo* marrow engraftment.
Progenitor numbers and progenitor/stem cell inversions.
Expression of adhesion proteins and cytokine receptors.
Global gene expression.
Potential for megakaryocyte, granulocyte, and lymphocyte differentiation.

---

skeletal and cardiac myocytes, skin cells, pancreatic cells, endothelial cells, and neural cells to cite a noninclusive list.[4–38] The mechanisms underlying such plasticity could relate to differentiation from marrow cells with different potentials, dedifferentiation, transdifferentiation, epigenetic changes in marrow stem cells, or cell fusion. A critical aspect of the observations of plasticity is that in order for plasticity to occur cell injury is necessary.[8,9,13]

We have also shown that the phenotype of marrow stem cells changes with position in cell cycle (or time after *in vitro* cytokine exposure) and that these phenotypic changes are reversible.[2,39–43] This cycle-related reversibility holds for both hematopoietic and nonhematopoietic epithelial cells[44] and has formed the basis for a continuum model of stem cell biology in which the potential and characteristics of the stem cell are continually changing with cell cycle passage and in which changes are reversible until a terminal-differentiating stimulus is encountered at a cycle-susceptible time. These observations indicate that cell cycle status and microenvironmental exposure to the products of injured cells may play key roles in determining the differentiated expression of a marrow stem cell. TABLE 1 presents critical points on the definition of a marrow stem cell.

CELL CYCLE-RELATED STEM CELL LABILITY IN THE HEMATOPOIETIC SYSTEM

In previous work, we have shown that many stem cell characteristics show reversible alterations with cell cycle passage. Initially, using a model of induction of stem cell cycle transit by exposure to interleukin-3 (IL-3), IL-6, IL-11, and steel factor, we showed reversible changes in short- and long-term engraftment, expression of various adhesion proteins and cytokine receptors, differentiation, and global gene expression.[2,39–45] We also showed that when lineage-negative rhodamine low Hoechst low (LRH) marrow stem cells were exposed to these cytokines, they reproducibly progressed through G1 and entered S phase at about 18–20 h, the first population doubling was then seen at 36–38 h.[46] These stem cells were truly quiescent at isolation, since the separation was based on quiescence, and they progress through cycle in a highly synchronized fashion. After the first long cell cycle, subsequent synchronous cycles are approximately 12 h in length, and this synchrony is maintained out to six total cycles. These cells *in vivo*, however, are slowly cycling as demonstrated by *in vivo* BrdU labeling showing that after oral administration of BrdU, over 70% of isolated LRH stem cells label.[47-49] The cycling nature of stem cells is also illustrated by studies showing that 22% of lineage-negative Sca-1$^+$, c-kit$^+$, thy-1^{10} stem cells are in S phase at isolation,[50] similar to the number of Lin$^-$Sca-1$^+$ cells showing approximately 20% in S phase.

Studies on hematopoietic differentiation of stem cells with cycle passage have shown an inverse relationship of progenitors with engraftable stem cells during the first cycle passage of LRH from quiescence.[2] In these studies, the inducing cytokines were thrombopoietin, Flt3L, and steel factor. Subsequent work has now shown reversible peaks of cytokine-induced differentiation at specific phases of cytokine-stimulated cycle passage.[45] When LRH cells were exposed to G-CSF, GM-CSF, and steel factor at different points in a cytokine-induced cell cycle transit, there was a marked increase in megakaryocyte differentiation at the G1/S interface and in granulocyte differentiation at several points in cycle. More recently, preliminary experiments have also indicated the presence of B and T cell hotspots at specific points in cycle. These observations indicate cell cycle-related alterations in differentiation potential with cell cycle passage. Recent data assessing Lin$^-$Sca-1$^+$ cells separated into different phases of cycle (G1 and three components of S/G2) have shown a megakaryocyte hotspot at G1/S, indicating that the changes noted with cytokine-stimulated cells were due to position in cell cycle, and not to an independent *in vitro* action of cytokines. It is important to note that since cells require cytokine exposure to progress through cycle, the exposure to cytokines and cell cycle progression are tightly linked, and the observation that certain stem cell classes at isolation show S phase cells indicates previous *in vivo* exposure. These data, in highly synchronized cytokine-stimulated stem cells or stem cells separated into cell cycle phases by Hoechst staining, indicate a system in which the phenotype of the stem cell reversibly varies with the stage of cell cycle. It suggests a model in which chromatin access for transcription factors varies with cycle state, and this in turn determines the capacity of a primed stem cell to respond to differentiating inducing stimuli. In a synchronous population as represented by LRH cells stimulated *in vitro* with thrombopoietin, Flt3L, and steel factor there would be a massive differentiation at a specific cell cycle point. However, in an asynchronous population of stem cells there would always be a small percentage of cells receptive or primed for a specific differentiation induction, such as megakaryocyte differentiation at any particular time. They can only respond, however, when that cell is in a chromatin-responsive state, which occurs at specific points in cell cycle. With cycle progression, different cohorts of stem cells would always be moving into a receptive state and, with a continued inductive stimulus, megakaryocyte production would continue. We have termed this a flow-response model, in that cells would continually flow into a responsive state (FIG. 1).

The number of hematopoietic stem cell phenotype changes with cell cycle progression is summarized in TABLE 1.

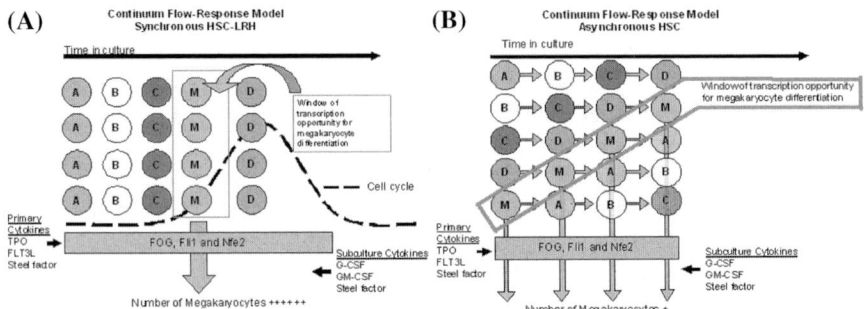

FIGURE 1. Continuum-flow response. Models of marrow stem cell regulation. (**A**) Synchronous cell model of megakaryocyte differentiation of stem cells synchronously progressing through cell cycle. (**B**) Asynchronous cell model of megakaryocyte differentiation of asynchronous stem cells progressing through cycle. M = stem cell with chromatin/histone configuration giving access to promoters responsive to FOG, Fli1 and Nfe2 and capable of inducing megakaryocyte differentiation. *Circles* with letters represent stem cells in different phases of cell cycle. (Reproduced with permission of Exp. Hematol. **35**: 96–107, 2007.[45])

STEM CELL PLASTICITY—A REASONABLE BUT ABUSED CONCEPT

Stem cell plasticity implies a capacity of the marrow stem cell to produce a wide variety of cell types, including nonhematopoietic cells. It does not imply an underlying mechanism. One mechanism put forward has been transdifferentiation. This is highly unlikely, since it has never been conclusively demonstrated in any experimental setting. In every case, differentiation from a rare population of stem cells has never been excluded.

There has been discussion of dedifferentiation, although the distinction between this and fluctuating differentiation potential as described above, is not clear. Epigenetic cell changes are probably involved and may be mediated by signals received from injured cells. Our own bias is toward differentiation in response to specific signals, especially from injured cells, which may be delivered in unique fashions (see below—microvesicles). There has been much discussion about fusion, and this appears to be a mechanism in some reports of marrow plasticity, but not others.

Regardless of the mechanism, it has now been established that marrow cells, probably marrow hematopoietic stem cells, have the capacity to produce nonmarrow cells in many tissues after engraftment into irradiated mice; the need for additional injury appears to depend upon the specific tissue under study. Lagasse and colleagues[4] provided proof-of-principle, in studies on the fumaryl acetate hydroxylase-deficient mouse, a tryosinemic mouse that dies without administration of a bypass drug. When these animals were transplanted with highly purified stem cells, provided by the Weissman group, and then subjected to repeated withdrawal of the drug, a number were eventually cured with significant replacement of diseased hepatocytes by donor marrow-derived hepatocytes. The final mechanism here was clearly cell fusion, although whether initial events represent cell fusion is not totally clear. In addition, in other liver plasticity models the mechanism was not fusion.[51,52]

We have focused on the capacity of marrow cells from green-fluorescent protein-positive (GFP+) transgenic mice to engraft and produce non-hematopoietic epithelial cells in the lungs of irradiated wild-type mice. In these studies, we have found "robust" levels of donor marrow-derived pulmonary epithelial cells routinely in the 2–3% range, which with added cell mobilization with repeated G-CSF injections, reach the 5–7% range.[13] These studies showed that the level of irradiation, which correlated with chimerism, determined how many marrow to lung conversion events were seen. The phenotype of the converting cell was that of a hematopoietic stem cell; lineage-negative, Sca-1$^+$; and c-kit$^+$. Further studies showed that cytokine-treated cells (IL-3, IL-6, IL-11, and steel factor) at the G1/S interface had a threefold increase in conversion events, and this was reversible. In preliminary studies with Lin$^-$Sca-1$^+$ stem cells separated by Hoechst staining and FACS on the basis of DNA content, S phase cells had an increased conversion capacity suggesting that the increases in conversion capacity seen previously with cytokine-treated cells were, in fact, due to cell cycle position and not a separate *in vitro* effect of cytokines. These observations indicated that cell cycle-related stem cell phenotype shifts were seen with both hematopoietic and nonhematopoietic (pulmonary) differentiation. This suggested that the differentiation potential of marrow hematopoietic stem cells might be continually changing based on cell cycle state, and raised the possibility that differentiation hotspots might exist for many different non-hematopoietic cell types. Furthermore, these studies, plus studies on marrow to skeletal muscle conversions, indicated, that there were many variables that could determine the outcome of a stem cell plasticity study. These are summarized in TABLE 2. A failure to address many of these variables probably accounts for at least some of the negative reports on stem cell plasticity.

The mechanism of cell change when marrow cells interact with injured tissue was an open question. Cell fusion had been addressed with cross-sex transplants and did not appear to explain the lung conversion results. Jang *et al.*[52] had previously carried out co-cultures of injured liver separated from marrow cells by a cell impermeable membrane, and found that the marrow cells began to express genes for albumin and make the protein. We used a similar system, culturing irradiated lung opposite unseparated murine marrow cells. We found that lungs from mice exposed to 500 cGy 5 days previously induced marrow cells to express genes (as determined by real time RT-PCR) for lung-specific proteins. These were clara cell-specific protein, surfactant B, and surfactant C. Other stem cell-specific genes were also induced. Marrow cells co-cultured in this fashion produced prosurfactant C protein 3 weeks after lung exposure and showed an increased capacity to convert to type II pneumocytes after

TABLE 2. Variables effecting conversion of marrow cells to nonhematopoietic cells after *in vivo* transplantation

1. Nature of tissue injury.
2. Repetitive nature of injury.
3. Marrow mobilization.
4. Marrow cell type infused.
5. Timing of engraftment and of injury.
6. Marrow cell number infused.
7. Cell cycle state of marrow cells.
8. Other functional alterations in marrow cells.
9. Presence of microvesicles from injured tissue in marrow cells.
10. Treatment of host mouse separate from specific tissue injury, especially irradiation.
11. Mode of cell delivery.

in vivo transplantation into lethally irradiated mice. When 500 cGy exposed lung was used to condition media, the conditioned media also induced these lung-specific genes and, with ultracentrifugation, it was found that the converting activity resided in microvesicles in the pellet from the centrifugation. These microvesicles contained high levels of lung-specific mRNA and, on incubation with marrow cells, showed entry into subpopulations of these cells. These evolving data suggest that microvesicle derived from injured tissue might mediate marrow cell phenotype change, and further suggest that part of a cell cycle effect might relate to the ease of microvesicle cell entry at different points in cell cycle.

Altogether, these studies indicate a highly flexible ever-changing marrow cell system with capacity to respond to products of injured cells and to repair a broad range of tissues.

REFERENCES

1. D'HONDT, L., C. MCAULIFFE, J. DAMON, *et al.* 2004. Circadian variations of bone marrow engraftability. J. Cell. Physiol. **200:** 63–70.
2. COLVIN, G.A., J.F. LAMBERT, B.E. MOORE, *et al.* 2004. Intrinsic hematopoietic stem cell/progenitor plasticity: inversions. J. Cell. Physiol. **199:** 20–31.
3. TILL, J.E., E.A. MCCULLOCH & L. SIMINOVITCH 1964. A stochastic model of stem cell proliferation, based on the growth of spleen colony-forming cells. Proc. Natl. Acad. Sci. USA **51:** 29–36.
4. LAGASSE, E., H. CONNORS, M. AL-DHALIMY, *et al.* 2000. Purified hematopoietic stem cells can differentiate into hepatocytes *in vivo*. Nat. Med. **6:** 1229.
5. LABARGE, M.A. & H.M. BLAU 2002. Biological progression from adult bone marrow to mononucleate muscle stem cell to multinucleate muscle fiber in response to injury. Cell **111:** 589–601.
6. BITTNER, R.E., C. SCHOFER, K. WEIPOLTSHAMMER, *et al.* 1999. Recruitment of bone-marrow-derived cells by skeletal and cardiac muscle in adult dystrophic mdx mice. Anat. Embryol. (Berl.) **199:** 391–396.

7. GUSSONI, E., Y. SONEOKA, C.D. STRICKLAND, et al. 1999. Dystrophin expression in the mdx mouse restored by stem cell transplantation. Nature **401:** 390–394.
8. ABEDI, M., D.A. GREER, G.A. COLVIN, et al. 2004. Tissue injury in marrow transdifferentiation. Blood Cells Mol. Dis. **32:** 42–46.
9. ABEDI, M., D.A. GREER, G.A. COLVIN, et al. 2004. Robust conversion of marrow cells to skeletal muscle with formation of marrow-derived muscle cell colonies: a multifactorial process. Exp. Hematol. **32:** 426–434.
10. ORTIZ, L.A., F. GAMBELLI, C. MCBRIDE, et al. 2003. Mesenchymal stem cell engraftment in lung is enhanced in response to bleomycin exposure and ameliorates its fibrotic effects. Proc. Natl. Acad. Sci. USA **100:** 8407–8411.
11. ABE, S., G. LAUBY, C. BOYER, et al. 2003. Transplanted BM and BM side population cells contribute progeny to the lung and liver in irradiated mice. Cytotherapy **5:** 523–533.
12. GROVE, J.E., C. LUTZKO, J. PRILLER, et al. 2002. Marrow-derived cells as vehicles for delivery of gene therapy to pulmonary epithelium. Am. J. Respir. Cell. Mol. Biol. **27:** 645–651.
13. ALIOTTA, J., P. KEANEY, M. PASSERO, et al. 2006. Bone marrow production of lung cells: the impact of G-CSF, cardiotoxin, graded doses of irradiation, and subpopulation phenotype. Exp. Hematol. **34:** 230–241.
14. LAGASSE, E., H. CONNORS, M. AL-DHALIMY, et al. 2000. Purified hematopoietic stem cells can differentiate into hepatocytes *in vivo*. Nat. Med. **6:** 1229–1234.
15. KRAUSE, D.S., N.D. THEISE, M.I. COLLECTOR, et al. 2001. Multi-organ, multi-lineage engraftment by a single bone marrow-derived stem cell. Cell **105:** 369–377.
16. PEREIRA, R., K. HALFORD, M. O'HARA, et al. 1995. Cultured adherent cells from marrow can serve as long-lasting precursor cells for bone, cartilage, and lung in irradiated mice. Proc. Natl. Acad. Sci. USA **92:** 4857–4861.
17. NILSSON, S.K., M.S. DOONER, H.U. WEIER, et al. 1999. Cells capable of bone production engraft from whole bone marrow transplants in nonablated mice. J. Exp. Med. **189:** 729–734.
18. BADIAVAS, E.V. & V. FALANGA. 2003. Treatment of chronic wounds with bone marrow-derived cells. Arch. Dermatol. **139:** 510–516.
19. BADIAVAS, E.V., M. ABEDI, J. BUTMARC, et al. 2003. Participation of bone marrow derived cells in cutaneous wound healing. J. Cell. Physiol. **196:** 245–250.
20. KATAOKA, K., R.J. MEDINA, T. KAGEYAMA, et al. 2003. Participation of adult mouse bone marrow cells in reconstitution of skin. Am. J. Pathol. **163:** 1227–1231.
21. DIREKZE, N.C., S.J. FORBES, M. BRITTAN, et al. 2003. Multiple organ engraftment by bone-marrow-derived myofibroblasts and fibroblasts in bone-marrow-transplanted mice. Stem Cells **21:** 514–520.
22. NYGREN, J.M., S. JOVINGE, M. BREITBACH, et al. 2004. Bone marrow-derived hematopoietic cells generate cardiomyocytes at a low frequency through cell fusion, but not transdifferentiation. Nat. Med. **10:** 494–501.
23. ORLIC, D., J. KAJSTURA, S. CHIMENTI, et al. 2001. Bone marrow cells regenerate infracted myocardium. Nature **410:** 701–705.
24. MURRY, C.E., M.H. SOONPAA, H. REINECKE, et al. 2004. Haematopoietic stem cells do not transdifferentiate into cardiac myocytes in myocardial infarcts. Nature **428:** 664–668.
25. JACKSON, K.A., S.M. MAJKA, H. WANG, et al. 2001. Regeneration of ischemic cardiac muscle and vascular endothelium by adult stem cells. J. Clin. Invest. **107:** 1395–1402.

26. KOCHER, A.A., M.D. SCHUSTER, M.J. SZABOLCS, *et al.* 2001. Neovascularization of ischemic myocardium by human bone-marrow-derived angioblasts prevents cardiomyocyte apoptosis, reduces remodeling, and improves cardiac function. Nat. Med. **7:** 430–436.
27. IANUS, A., G.G. HOLZ, N.D. THEISE & M.A. HUSSAIN. 2003. *In vivo* derivation of glucose-competent pancreatic endocrine cells from bone marrow without evidence of cell fusion. J. Clin. Invest. **111:** 843–850.
28. HESS, D., L. LI, M. MARTIN, *et al.* 2003. Bone marrow-derived stem cells initiate pancreatic regeneration. Nat. Biotechnol. **21:** 763–770.
29. HOI, J.B., H. UCHINO, K. AZUMA, *et al.* 2003. Little evidence of transdifferentiation of bone marrow-derived cells into pancreatic beta cells. Diabetologia **46:** 1366–1374.
30. MATHEWS, V., P.T. HANSON, E. FORD, *et al.* 2004. Recruitment of bone marrow-derived endothelial cells to sites of pancreatic beta-cell injury. Diabetes **53:** 91–98.
31. JIANG, Y., B.N. JAHAGIRDAR, R.L. REINHARDT, *et al.* 2002. Pluripotency of mesenchymal stem cells derived from adult marrow. Nature **418:** 41–49.
32. KOSHIZUKA, S., S. OKADA, A. OKAWA. *et al.* 2004. Transplanted hematopoietic stem cells from bone marrow differentiate into neural lineage cells and promote functional recovery after spinal cord injury in mice. J. Neuropathol. Exp. Neurol. **63:** 64–72.
33. GRINNEMO, K.H., A. MANSSON, G. DELLGREN, *et al.* 2004. Xenoreactivity and engraftment of human mesenchymal stem cells transplanted into infracted rat myocardium. J. Thorac. Cardiovasc. Surg. **127:** 1293–1300.
34. AIREY, J.A., G. ALMEIDA-PORADA, E.J. COLLETTI, *et al.* 2004. Human mesenchymal stem cells form Purkinje fibers in fetal sheep heart. Circulation **109:** 1401–1407.
35. MCBRIDE, C., D. GAUPP & D.G. PHINNEY. 2003. Quantifying levels of transplanted murine and human mesenchymal stem cells *in vivo* by real-time PCR. Cytotherapy **5:** 7–18.
36. WANG, J.F., Y.F. WU, J. HARRINGTON & I.K. MCNIECE. 2004. *Ex vivo* expansions and transplantations of mouse bone marrow-derived hematopoietic stem/progenitor cells. J. Zhejiang Univ. Sci. **5:** 157–163.
37. ORLIC, D., J. KAJSTURA, S. CHIMENTI, *et al.* 2001. Bone marrow cells regenerate infracted myocardium. Nature **410:** 701–705.
38. MURRY, C.E., M.H. SOONPAA, H. REINECKE, *et al.* 2004. Haematopoietic stem cells do not transdifferentiate into cardiac myocytes in myocardial infarcts. Nature **428:** 664–668.
39. HABIBIAN, H.K., S.O. PETERS, C.C. HSIEH, *et al.* 1998. The fluctuating phenotype of the lymphohematopoietic stem cell with cell cycle transit. J. Exp. Med. **188:** 393–398.
40. LAMBERT, J.F., M. LIU, G.A. COLVIN, *et al.* 2003. Marrow stem cells shift gene expression and engraftment phenotype with cell cycle transit. J. Exp. Med. **197:** 1563–1572.
41. CERNY, J., M.S. DOONER, C.I. MCAULIFFE, *et al.* 2002. Homing of purified murine lymphohematopoietic stem cells: a cytokine induced defect. J. Hematother. Stem Cell Res. **11:** 913–922.
42. BERRIOS, V.M., G.J. DOONER, G. NOWAKOWSKI, *et al.* 2001. The molecular basis for the cytokine-induced defect in homing and engraftment of hematopoietic stem cells. Exp. Hematol. **29:** 1326–1335.

43. BECKER, P.S., S.K. NILSSON, Z. LI, *et al.* 1999. Adhesion receptor expression by hematopoietic cell lines and murine progenitors. Exp. Hematol. **27:** 533–541.
44. DOONER, M., J. CERNY, G. COLVIN, *et al.* 2004. Homing and conversion of murine hematopoietic stem cells to lung. Blood Cells Mol. Dis. **32:** 47–51.
45. COLVIN, G.A., M.S. DOONER, G.J. DOONER, *et al.* 2007. Stem Cell continuum: directed differentiation hotspots. Exp. Hematol. **35:** 96–107.
46. REDDY, G.P., C.Y. TIARKS, L. PANG, *et al.* 1997. Cell cycle analysis and synchronization of pluripotent hematopoietic progenitor stem cells. Blood **90:** 2293–2299.
47. CHESHIER, S.H., S.J. MORRISON, X. LIAO & I.L. WEISSMAN. 1999. *In vivo* proliferation and cell cycle kinetics of long-term self-renewing hematopoietic stem cells. Proc. Natl. Acad. Sci. USA **96:** 3120–3125.
48. BRADFORD, G.B., B. WILLIAMS, R. ROSSI & I. BERTONCELLO. 1997. Quiescence, cycling, and turnover in the primitive hematopoietic stem cell compartment. Exp. Hematol. **25:** 445–453.
49. PANG, L., P.V. REDDY, C.I. MCAULIFFE, *et al.* 2003. Studies on BrdU labeling of hematopoietic cells: stem cells and cell lines. J. Cell. Physiol. **197:** 251–260.
50. FLEMING, H.F., E.J. ALPERN, N. UCHIDA, *et al.* 1993. Functional heterogeneity is associated with the cell cycle status of murine hematopoietic stem cells. J. Cell Biol. **122:** 897–902.
51. KOGLER, G., S. SENSKEN, J.A. AIREY, *et al.* 2004. A new human somatic stem cell from placental cord blood with intrinsic pluripotent differentiation potential. J. Exp. Med. **200:** 123–135.
52. JANG, Y.-Y., M.I. COLLECTOPR, S.B. BAYLIN, *et al.* 2004. Hematopoietic stem cells convert into liver cells within days without fusion. Nat. Cell Biol. **6:** 532–539.

Inference, Validation, and Dynamic Modeling of Transcription Networks in Multipotent Hematopoietic Cells

SHAMIT SONEJI,[a] SUI HUANG,[b] MATTHEW LOOSE,[c]
IAN JOHN DONALDSON,[d] ROGER PATIENT,[a] BERTHOLD GÖTTGENS,[e]
TARIQ ENVER,[a] AND GILLIAN MAY[a]

[a]*Weatherall Institute of Molecular Medicine, Molecular Haematology Unit, University of Oxford, John Radcliffe Hospital, Oxford OX3 9DS, United Kingdom*

[b]*Department of Surgery and Vascular Biology Program, Children's Hospital, Harvard Medical School, and Harvard Stem Cell Institute, Children's Hospital, Boston, MA 02115, USA*

[c]*Institute of Genetics, The University of Nottingham, Queen's Medical Centre, Nottingham NG7 2UH, United Kingdom*

[d]*Faculty of Life Sciences, University of Manchester, Michael Smith Building, Oxford Road, Manchester M13 9PT, United Kingdom*

[e]*Department of Haematology, Cambridge Institute for Medical Research, University of Cambridge Hills Road, Cambridge CB2 2XY, United Kingdom*

ABSTRACT: Identifying the transcription factor interactions that are responsible for cell-specific gene expression programs is key to understanding the regulation of cell behaviors, such as self-renewal, proliferation, differentiation, and death. The rapidly increasing availability of microarray-derived global gene expression data sets, coupled with genome sequence information from multiple species, has driven the development of computational methods to reverse engineer and dynamically model genetic regulatory networks. An understanding of the architecture and behavior of transcriptional networks should lend insight into how the huge number of potential gene expression programs is constrained and facilitates efforts to direct or redirect cell fate.

KEYWORDS: network inference; transcriptional networks; dynamic modeling; hematopoietic progenitors; Gata; Pu1

Because of their unique biological properties and potential medical importance, hematopoietic stem cells have attracted considerable scientific, commercial,

Address for correspondence: Tariq Enver, Molecular Haematology Unit, The Weatherall Institute of Molecular Medicine, University of Oxford, John Radcliffe Hospital, Headington, Oxford OX3 9DS, UK. Voice: 0044 (0)1865 222 425; fax: 0044 (0)1865 222 449.
tariq.enver@imm.ox.ac.uk

and public interest.[1] The cell fate options that confront stem cells include self-renewal, differentiation and lineage-specification, programmed cell death, and quiescence. To varying degrees, these fates also extend to the progenitor cells that reside at different levels of the hematopoietic differentiation hierarchy. Understanding how the different cell fates that confront stem and progenitor cells are selected and coordinated is a key challenge for the future; positive adoption of a given fate must be coupled to appropriate negative regulation of the alternative pathways. Although much still remains to be understood about the nature of the molecular pathways involved in the regulation of stem and progenitor cell fate, it is generally accepted that transcription factors are key intrinsic regulators of these fate decisions[2,3] and that fate choice involves modulation of linked networks of transcription factors.[4] The potential number of different gene expression patterns is combinatorially vast, and yet the number of distinct cell types, each defined by a distinct gene expression pattern, is relatively few, with cells making very precise transitions between apparently stable "network states" or gene expression patterns. Moreover, it seems that alteration of the level of a single transcription factor within a network may be sufficient to alter cell fate; this is highlighted by the experimental phenomenon of lineage reprogramming (FIG. 1). It is not clear how any of this is achieved, and thus a description of the underlying transcription factor networks and understanding of their dynamic regulation is an important, albeit daunting, task. Herein, we outline some of the experimental and computational challenges involved, with particular reference to fate decisions within the hematopoietic hierarchy.

Looking to model organisms, such as yeast, provides a rubric for how one might begin to go about delineating and testing transcription factor networks. An elementary aim is to determine the "map" or "wiring diagram" of the network that is collectively established by the regulatory interactions between the genes. Such maps represent what are now called Genetic Regulatory Networks (GRNs). Information about the GRN architecture will pertain to both the topology (which genes interact with which) but also the directionality, modality, and logic of the interactions (regulator versus target, activation versus inhibition, and logical functions that define the joint effect of multiple regulators, etc.).

One approach to determine GRN topology is to build out from key nodal regulators through mapping their interactions and target genes. Information of this type may be assembled and used to build GRNs of the sort elegantly described by Davidson and colleagues[5,6] and also to identify network motifs, such as feedback or feedforward switches, based on transcription factor cross-regulation.[7] An example of this approach in the context of hematopoiesis has recently been reported for B cells.[8] Another strategy is to obtain global gene expression profiles from cells with different cell fate status and under different conditions. Using various computational approaches to analyze such data allows the prediction of regulatory relationships between individual transcription factors, also referred to as network inference.[9] We, and others, have obtained

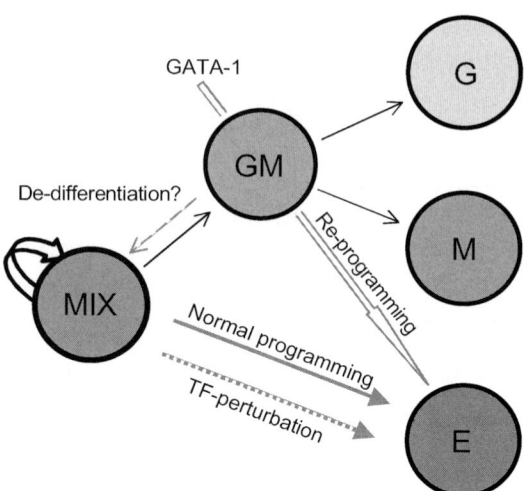

FIGURE 1. Schematic of a hematopoietic lineage hierarchy, exemplifying the discrete cell types or states and the ability of a single transcription factor (TF), in this case the erythroid-associated factor GATA-1, to influence the cell fate of a multi-potential progenitor cell (MIX) during normal programming or as a consequence of experimental manipulation (TF-perturbation). Forced expression of GATA-1 in the granulocyte-monocyte (GM) compartment is able to over-ride the GM program and promote the development of erythroid (E) cells.

global gene expression profiles to obtain baseline expression information that may be used for this purpose.[10–12] These data provide a framework for then using a combined approach in which specific transcription factors are manipulated, and the consequences for both cell behavior and the expression of all other transcription factors recorded. More recently, advances in combining chromatin immunoprecipitation with microarray technology have opened up the possibility of comprehensively identifying all directly bound target genes of a given transcription factor, providing a "core" network of direct interactions that can be integrated with gene expression data (FIG. 2).

GENETIC REGULATORY NETWORKS

GRNs depict the regulatory interactions between individual genes, representing the available information in an accessible form and providing insight into the molecular mechanisms underlying developmental decisions and processes. Detailed analysis of the topology of GRNs of *Escherichia coli* and yeast has resulted in the identification of a number of different regulatory modules or "network motifs." These can be considered as small subnetworks of a particular structure that together define the transcriptional program of a

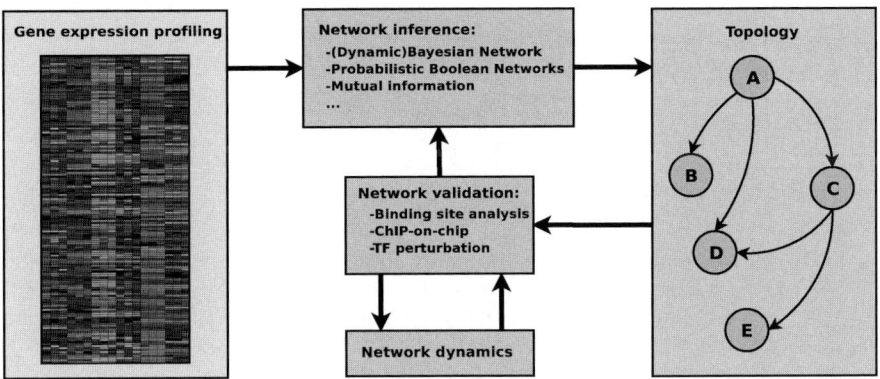

FIGURE 2. Example of a possible work-flow leading from initial global gene expression profiling by microarray through network inference to derive a topological network identifying, in this instance, genes B-E which are regulated directly or indirectly by gene A. The proposed topology can then be validated by a number of approaches including the identification of binding sites for transcription factor A in sequences flanking target genes B, C and D but not E, chromatin immunoprecipitation, and analysis of the transcriptional consequences of perturbing the expression of each regulatory factor in turn.

cell. Some such motifs are relatively simple, such as an autoregulation motif where a transcription factor directly regulates its own expression. Others are more complex, involving several transcription factors cross-regulating multiple targets. Temporal control within an overall network is afforded by transcription factors acting in series either in a simple regulatory chain (FIG. 3A) or a more complex feedforward loop (FIG. 3B). Cross-antagonism of factors is also frequently encountered (FIG. 3C).

Currently available literature on the activities of transcription factors within different hematopoietic compartments and processes has been collated and reviewed by Matthew Loose and Roger Patient and integrated into a series of GRNs underlying the initial specification of the hematopoietic stem cell and its subsequent differentiation to the erythroid lineage.[13] Information was derived from published expression profiles, transcription factor perturbation experiments, chromatin immunoprecipitation, and knowledge of cis-regulatory elements (FIG. 4). Key regulatory molecules, such as GATA1, GATA2, SCL, FOG-1, NF-E2, and PU.1, can be seen to be interconnected either positively (▶) or negatively (pathways are represented by and/or logic gates, and the requirement for protein–protein interactions to achieve activation or repression is also indicated (⊣). An interactive version of this regulatory network can be seen at www.nottingham.ac.uk/genetics/mouse.

Of course, the quality of the GRN obtained is limited by both the quality and quantity of the available experimental data. Moreover, the relative importance of a regulator may be underestimated if there is a lack of information about its function. Some transcription factors and links may be missing entirely from the

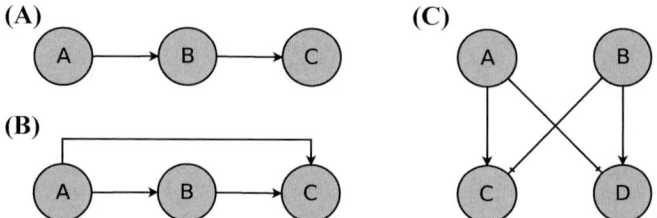

FIGURE 3. Regulatory modules or motifs operating within transcription factor networks. (**A**) Regulatory chain: three or more factors (A, B and C) operate in series. (**B**) Feedforward motif: transcription factor A regulates factor B, and both A and B feedforward to regulate the target gene C. The regulation exerted by each of A and B can be positive or negative. (**C**) Cross-antagonistic switch: transcription factor A positively regulates a target gene while negatively regulating the target of a second factor B, and vice versa.

compiled network due to insufficient observations. Nevertheless, these GRNs present a first step toward predicting the mechanisms underlying hematopoietic cell fate decisions. In many cases, these predictions support the current experimental observations, but novel predictions can also be made that are open to experimental verification. In addition, such GRNs provide a valuable point of reference for transcriptional networks derived from mathematical analysis of gene expression data.

COMPUTATIONAL APPROACHES FOR DEVELOPING TRANSCRIPTIONAL NETWORKS

The quantity of publicly available microarray data sets (e.g., at the Web-based gene expression omnibus [GEO] depository) has driven the development of computational methods to infer the structure of transcriptional networks from their dynamics. The type of data determines the type of analysis. Early attempts focused on cell cycle time-course data from yeast, since cultures could be synchronized yielding very high-quality data.[14] Relatively simple techniques, such as correlation analysis, were extended to allow subsections of profiles to be aligned instead,[15,16] acknowledging the fact that a time delay may be present between the differential expression of a transcription factor and its target. Model-based methods, such as Bayesian nets, have been implemented,[17] but these are static and produce acyclic graphs, so to analyze time course data Dynamic Bayesian Networks (DBNs)[18] have been developed. A drawback of both is that computation is complicated, which means that analyzing all genes from an array experiment is not feasible. This is further complicated by the fact that the number of genes is considerably greater than the number of conditions/time points. An attractive alternative is given by Probabilistic Boolean Networks (PBNs), which allow the preservation of cyclic behavior when

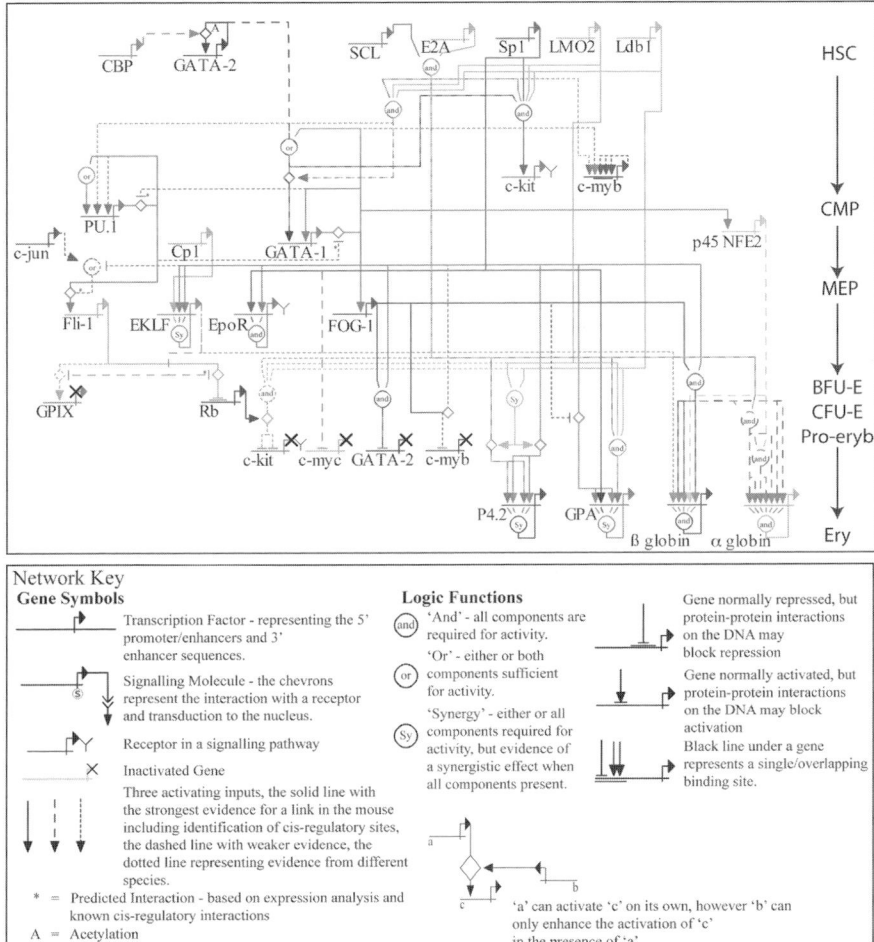

FIGURE 4. Genetic Regulatory Network derived from the published literature representing the interactions involved in erythroid commitment of a hematopoietic stem cell.

analyzing temporal expression profiles, while also quantifying the influence of one gene on another.[19]

Mammalian stem cell systems are more difficult to manipulate and therefore the quality of data seen in yeast cannot be achieved. This is in great part due to lack of synchrony and considerable cellular heterogeneity. Cellular heterogeneity is an intrinsic property of complex differentiation systems, such as hematopoiesis, and may arise through (*a*) heterogeneity in the initial population of multipotent cells, (*b*) a lack of cell synchrony either in the starting populations or introduced during differentiation, and (*c*) the extent to which cytokine-mediated differentiation is based on selective as opposed to instructive mechanisms. Resolving cellular heterogeneity is a key issue in building transcriptional networks to ensure that apparently linked events are indeed occurring within the same cell. It will therefore be important to gain an appreciation of the extent of heterogeneity at early time points as cells undergo unilineage commitment and differentiation decisions.

It has also been recognized that having many different sample types is beneficial to network inference. This was exploited in the case of network inference in B cells in which 336 samples of various B cell phenotypes were arrayed. Using a mutual information theory–based algorithm ARACNE,[20] the authors were able to generate a scale-free network[21] from the array data in which *Myc* was found to be one of the most highly connected transcription factors. The *Myc* binding site was overrepresented in the first neighbors of *Myc*, and CHiP assays were used to validate new targets. The outcome of this study was a very successful application of a computational approach to infer a network from a higher eukaryote, a subnetwork of which lead to a bench-testable hypothesis that was subsequently validated.

CONNECTIVITY VIA CIS-REGULATORY ELEMENTS

The control of metazoan development ultimately resides in cis-regulatory control of spatiotemporal gene expression.[22] Cis-regulatory information processing therefore is central to transcriptional network function. Studies in several model organisms have begun to decipher cis-regulatory codes that determine gene expression patterns in early embryos. However, comparatively much less is known about mammalian regulatory codes. To tackle this important issue, both computational and experimental approaches are being developed. Steps toward network validation can be taken at the DNA sequence level by asking whether a group of genes that have a common regulator possess conserved binding sites for the regulator in question. Many tools are available to do this, such as rVista,[23] TOUCAN,[24] and CONFAC[25] to name a few. One difficulty in such analysis is that transcription factors often have very simple binding motifs that will occur with regularity in DNA even when cross-species sequence conservation is used to increase the stringency of predictions. However, functional cis-regulatory elements typically consist of dense clusters of distinct binding sites. It therefore becomes more informative to look at the presence of a given site in context with other binding sites in close proximity to see whether they form part of a regulatory module that has consistent structure and conservation. This is illustrated by the +19 enhancer of SCL, which contains two conserved Ets sites and one Gata site within a 50-bp distance, forming a regulatory module to which a trimeric protein complex binds.[26] It was shown that Gata and Ets sites by themselves will occur more than 800,000 times in conserved regions, and 1555 times within a 50-bp range. By placing the constraints of spacing and orientation seen in the +19 enhancer, 67 genes were found of which 18 were transcription factors that contained the regulatory module. From the point of view of computationally derived networks, if two or more transcription factors are thought to be cooperatively regulating the same target gene—thus forming a putative regulatory module—a genome-wide scan for the occurrence of this module could be an effective way of discovering new targets that were not detected at the level of transcription.

Mathematical Modeling

The approaches so far described contribute to establishing the topology of transcriptional networks and the underlying logic of their regulatory connections. However, knowledge of network architecture (the topology and interaction logics) alone cannot describe the dynamics of these networks, which describes how gene expression patterns change over time and hence ultimately naps to cell fate. Insight into this crucial aspect will most likely come from mathematical modeling, which becomes possible once sufficient information on network structure is available. This requires the formulation and solving of, for example, ordinary differential equations that capture the regulatory relationships between the network components (i.e., the genes) as well as detailed quantitation of gene activity. These approaches are beginning to be applied to the analysis of differentiating mammalian systems. A well-examined network module is the interaction between the transcription factors GATA-1 and PU.1. These two transcription factors mutually inhibit each other and, in doing so, establish an either–or decision situation for the progenitor cell that can choose between the erythroid/megakaryocyte and the myeloid-monocytic lineages. This example provides a useful paradigm for modeling the process of lineage specification in hematopoietic cells.[27–29] These lineage-specific transcription factors also promote the expression of genes that implement the erythroid/megakaryocytic and myelomonocytic programs, respectively. Interestingly, in addition to cross-inhibition (reviewed in Ref. 30), both factors are able to autoregulate their own gene expression. This property appears—according to mathematical models—to stabilize the progenitor state and maintain bipotential property. This new insight that comes from mathematical modeling is in contrast to the traditional model of a graded "stoichiometric balance"[31] between these two factors that does not explain how the progenitor state is stabilized, nor how discrete cell fate decisions can be made and implemented.

The concept of "stable states" in network dynamics has come from early theoretical studies that have suggested that large complex gene networks—despite the vast space of potential gene expression patterns that are combinatorially possible—generate a relatively small number of stable states, known as attractors, in which expression of a large number of the genes is stationary. It has been proposed that these attractor states represent discrete cell types[32,33] or cell fates.[34] Such stable attractors were generated when the GATA-1/PU.1 interaction was modeled as a simple gene circuit involving autostimulation and mutual inhibition. The stable attractors correspond to the erythroid and myelomonocytic fates, with a metastable attractor located between them that is characterized by coexpression of both transcription factors and corresponds to the bipotent progenitor state.[29] This prediction of simultaneous expression at low levels of the two lineage-associated transcription factors in uncommitted progenitor cells provides a formal explanation for the phenomenon of multilineage gene priming.[35]

PERSPECTIVE

The conundrum that faces regulatory biology in the post genomic era is the following. With roughly 30,000 genes in mammalian genomes, there are a staggering 2 to the power 30,000 potential combinations of gene expression states even if, for simplicity, we assume just on–off states for each gene. In reality, gene expression is graded making the potential gene expression space available even larger. Although the vast majority of theoretically possible combinations are inevitably not compatible with life, it remains difficult to understand how such a large repertoire of potential states can be constrained such that only a few hundred cell types are generated, and furthermore that these remain relatively stable. Our view, as outlined earlier, is that this is largely achieved through the architecture and dynamic behavior of transcriptional networks, a deeper knowledge of which may ultimately allow the directed generation or reprogramming of cell fate for therapeutic or commercial benefit.

REFERENCES

1. WEISSMAN, I.L., D.J. ANDERSON & F. GAGE. 2001. Stem and progenitor cells: origins, phenotypes, lineage commitments, and transdifferentiations. Annu. Rev. Cell. Dev. Biol. **17:** 387–403.
2. ENVER, T. & M. GREAVES. 1998. Loops, lineage, and leukemia. Cell **94:** 9–12.
3. SHIVDASANI, R.A. & S.H. ORKIN. 1996. The transcriptional control of hematopoiesis. Blood **87:** 4025–4039.
4. ROTHENBERG, E.V. & M.K. ANDERSON. 2002. Elements of transcription factor network design for t-lineage specification. Dev. Biol. **246:** 29–44.
5. DAVIDSON, E.H. et al. 2002. A genomic regulatory network for development. Science **295:** 1669–1678.
6. LEVINE, M. & E.H. DAVIDSON. 2005. Gene regulatory networks for development. Proc. Natl. Acad. Sci. USA **102:** 4936–4942.
7. MILO, R. et al. 2002. Network motifs: simple building blocks of complex networks. Science **298:** 824–827.
8. SINGH, H. et al. 2005. Contingent gene regulatory networks and B cell fate specification. Proc. Natl. Acad. Sci. USA **102:** 4949–4953.

9. BASSO, K. et al. 2005. Reverse engineering of regulatory networks in human B cells. Nat. Genet. **37:** 382–390.
10. BRUNO, L. et al. 2004. Molecular signatures of self-renewal, differentiation, and lineage choice in multipotential hemopoietic progenitor cells in vitro. Mol. Cell. Biol. **24:** 741–756.
11. IVANOVA, N.B. et al. 2002. A stem cell molecular signature. Science **298:** 601–604.
12. VENEZIA, T. et al. 2004. Molecular signatures of proliferation and quiescence in hematopoietic stem cells. PLoS Biol. **2:** e301.
13. SWIERS, G. et al. 2006. Genetic regulatory networks programming hematopoietic stem cells and erythroid lineage specification. Dev. Biol. **294:** 525–540.
14. SPELLMAN, P.T. et al. 1998. Comprehensive identification of cell cycle-regulated genes of the yeast Saccharomyces cerevisiae by microarray hybridization. Mol. Biol. Cell. **9:** 3273–3297.
15. QIAN, J. et al. 2001. Beyond synexpression relationships: local clustering of time-shifted and inverted gene expression profiles identifies new, biologically relevant interactions. J. Mol. Biol. **314:** 1053–1066.
16. JI, L. & T. KIAN-LEE. 2004. Identifying time-lagged gene clusters using gene expression data. Bioinformatics **21:** 509–516.
17. FRIEDMAN, N. 2004. Inferring cellular networks using probabilistic graphical models. Science **303:** 799–805.
18. MURPHY, K. & S. MIAN. 1999. Modelling gene expression data using dynamic Bayesian networks. Technical Report. Available at http://www.cs.ubc.ca/~murphyk/papers/ismb99.pdf
19. SHMULEVICH, I. et al. 2002. Probabilistic boolean networks: a rule-based uncertainty model for gene regulatory networks. Bioinformatics **18:** 261–274.
20. MARGOLIN, A.A. et al. 2006. ARACNE: an algorithm for the reconstruction of gene regulatory networks in a mammalian cellular context. BMC Bioinformatics 7(Suppl 1): S7.
21. BARABSI, A.-L. & E. BONABEAU. 2003. Scale-free networks. Sci. Am. **288:** 60–69.
22. DAVIDSON, E.H. 2006. The Regulatory Genome. Academic Press. Burlington, MA.
23. LOOTS, G. & I. OVCHARENKO. 2004. rVista 2.0: evolutionary analysis of transcription factor binding sites. Nucleic Acids Res. **32:** W217–W221.
24. AERTS, S. et al. 2005. TOUCAN 2: the all-inclusive open source workbench for regulatory sequence analysis. Nucleic Acids Res. **33:** W393–W396.
25. KARANAM, S. & C.S. MORENO. 2004. CONFAC: automated application of comparative genomic promoter analysis to DNA microarray datasets. Nucleic Acids Res. **32:** W475–W484.
26. DONALDSON, I.J. et al. 2005. Genome-wide identification of cis-regulatory sequences controlling blood and endothelial development. Hum. Mol. Genet. **14:** 595–601.
27. CHICKARMANE, V. et al. 2006. Transcriptional dynamics of the embryonic stem cell switch. PLoS Comput. Biol. **2:** e123.
28. ROEDER, I. & I. GLAUCHE. 2006. Towards an understanding of lineage specification in hematopoietic stem cells: a mathematical model for the interaction of transcription factors GATA-1 and PU.1. J. Theor. Biol. **4:** 852–865.
29. HUANG, S. et al. 2007. Bifurcation dynamics in lineage-commitment in bipotent progenitor cells. Dev. Biol. **305(2):** 695–713.
30. GRAF, T. 2002. Differentiation plasticity of hematopoietic cells. Blood **99:** 3089–3101.

31. CANTOR, A.B. & S.H. ORKIN. 2001. Hematopoietic development: a balancing act. Curr. Opin. Genet. Dev. **11:** 513–519.
32. KAUFFMAN, S.A. 1969. Metabolic stability and epigenesis in randomly constructed genetic nets. J. Theor. Biol. **22:** 437–467.
33. KAUFFMAN, S.A. 1993. The Origins of Order. Oxford University Press. New York.
34. HUANG, S. 2002. Regulation of cellular states in mammalian cells from a genomewide view. *In* Gene Regulation and Metabolism: Post-Genomic Computational Approach. J. Collado-Vides & R. Hofestädt, Eds.: 181–220. MIT Press. Cambridge, MA.
35. HU, M. *et al.* 1997. Multilineage gene expression precedes commitment in the hemopoietic system. Genes. Dev. **11:** 774–785.

Maintenance of Quiescent Hematopoietic Stem Cells in the Osteoblastic Niche

FUMIO ARAI AND TOSHIO SUDA

Department of Cell Differentiation, The Sakaguchi Laboratory of Developmental Biology, School of Medicine, Keio University, Shinjuku-ku, 160-8582 Tokyo, Japan

ABSTRACT: Hematopoietic stem cells (HSCs) are responsible for blood cell production throughout an individual's lifetime. Interaction of HSCs with their specific microenvironments, known as stem cell niches, is critical for maintaining stem cell properties, including self-renewal capacity and the ability to differentiate into multiple lineages. During postnatal life, the bone marrow (BM) supports both self-renewal and differentiation of HSCs in specialized microenvironmental niches. In the adult BM, HSCs are located in the trabecular endosteum (osteoblastic niche) or sinusoidal perivascular (vascular niche) areas. Here we show that osteoblastic cells (OBs) are a critical component for sustaining slow-cycling or quiescent HSCs. Interaction of HSCs with OBs through signaling and cell adhesion molecules maintains the balance in HSCs between cell division/proliferation and quiescence. In particular, the quiescent state is thought to be an essential mechanism to protect HSCs from stress and to sustain long-term hematopoiesis.

KEYWORDS: hematopoietic stem cell; osteoblastic cells; quiescence; Tie2; angiopoietin-1; N-cadherin; ataxia telangiectasia mutated (ATM); reactive oxygen species (ROS); p38MAPK

INTRODUCTION

Stem cells are characterized by their ability to self-renew and produce numerous types of differentiated daughter cells; two properties that enable them to maintain tissues. Self-renewal activity and multi-lineage differentiation potential of tissue stem cells are required to supply functionally differentiated progenies as normal tissue turn over. It is now clear that the stem cell niche regulates stem cell-specific properties, including self-renewal, multipotentiality, and relative quiescence.[1–7] Interaction of stem cells with the niche is critical to sustain stem cell pools in tissues over long periods.

Address for correspondence: Fumio Arai, D.D.S., Ph.D., Department of Cell Differentiation, The Sakaguchi Laboratory of Developmental Biology, School of Medicine, Keio University, 35 Shinanomachi, Shinjuku-ku, Tokyo 160-8582, Japan. Voice: +81-3-5363-3475; fax: +81-3-5363-3474.
farai@sc.itc.keio.ac.jp

Hematopoietic stem cells (HSCs) are responsible for blood cell production throughout the lifetime of an individual. Bone marrow (BM) HSCs are the best-characterized stem cells, and small subsets of HSCs[8–10] and lymphoid and myeloid progenitor cells[11,12] can be isolated using cell-surface markers. Although HSCs have been identified, the localization of HSCs *in situ* or the structure of the HSC niche has not been solved. The balance between quiescence, self-renewal, and commitment of stem cells is controlled by a combination of cell-intrinsic and external regulatory mechanisms. Intrinsic cellular and molecular properties of stem cells have been characterized extensively, and recently, niches, or specific microenvironments, in which stem cells reside *in situ* have been characterized at the molecular level. The concept of the stem cell niche was first proposed for the human hematopoietic system in the 1970s.[13] A similar concept has also been postulated for stem cells of the epidermis, intestinal epithelium, nervous system, and gonads. The HSC niche is not as clearly defined as niches for *Drosophila* germline stem cells (GSCs)[14] or mammalian bulge cells in the skin.[15] Currently, the hematopoietic niche is conceptually divided into two parts: an endosteal surface (osteoblastic niche) and a sinusoidal endothelium (vascular niche). Specific properties of HSCs are controlled dynamically by signaling mediated by receptor/ligand and cell adhesion molecules produced by niche cells.[1–7] We describe here how HSC activity in the osteoblastic niche is regulated.

Quiescent HSCs in the Adult BM Niche

Key features of stem cells in a niche are that they are quiescent and adhere to surrounding niche cells.[1–7] Quiescence is critical for protecting the stem cell compartment.[16] When it is disrupted, as occurs with $p21^{Cip1}$ deficiency, HSCs cannot remain in G0 and $p21^{-/-}$ HSCs could not reconstitute lethally irradiated recipient mice in serial BM transplantation.[16] Indeed, it has been reported that HSCs are relatively quiescent compared to transient amplifying progenitor cells, and 75% of long-term, self-renewing HSCs (Lineage$^-$Sca-1$^+$c-Kit$^+$ (LSK)-Thy1^{low}) are in G0 phase.[17] Although a previous study also showed that 99% of LSK-Thy1^{low} cells divided approximately every 57 days,[18] it is not known whether all HSCs cycle *in vivo* or if a small proportion of cells remains in G0.

In this study, we analyzed side-population (SP) cells in LSK cells under 5-fluorouracil (5-FU)-induced myelosuppressive conditions and found that LSK-SP cells, but not LSK-non-SP cells, were resistant to BM suppression induced by 5-FU.[19] Since 5-FU induces apoptosis of actively cycling cells, our finding indicates that LSK-SP cells are quiescent and resistant to apoptosis. In addition, we investigated cell cycle status of SP cells by pyronin Y (PY) staining. It was previously reported that $PY^{low/-}$ and PY^+ cells are in G0 and G1 of the cell cycle, respectively (FIG. 1).[20] We confirmed that most LSK-SP cells are

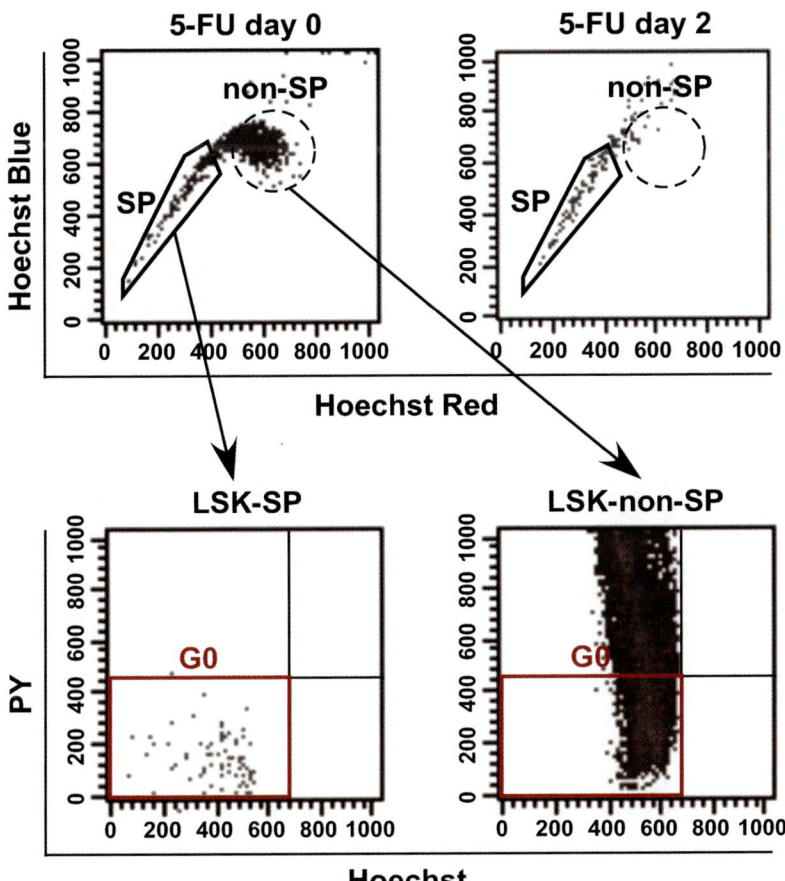

FIGURE 1. SP is a marker of quiescent HSCs. FACS analysis of SP cells in LSK fraction before (5-FU day 0) and after 2 days of 5-FU treatment (*upper panels*). Two days after 5-FU injection, most cells in the LSK fraction were SP, while numbers of non-SP cells were dramatically reduced. Since 5-FU treatment induces apoptosis in actively cycling cells, LSK-SP cells are thought to be in a noncycling, quiescent state. Hoechst and Pyronin Y (PY) staining emission pattern of LSK-SP and LSK-non-SP cells (*lower panels*). *Red boxes* indicate $PY^{low/-}$ G0 cells. LSK-SP cells were $PY^{low/-}$ fraction, but LSK-non-SP cells did not. These data suggest that LSK-SP cells are in G0. (Shown in color online.)

$PY^{low/-}$, suggesting that LSK-SP cells are in G0. In addition, a proportional increase in G0 cells in the LSK fraction was observed during mouse development, closely correlated with the relative increase in SP cells. Moreover, 5-FU-resistant LSK-Tie2$^+$ cells corresponded to SP cells. Therefore, in HSCs both the SP phenotype and Tie2 receptor tyrosine kinase are reliable markers of quiescent HSCs. Moreover, various forms of stresses can induce cell cycle progression of HSCs in order to expand the progenitor population and compensate for cell loss. It is also not clear whether HSCs divide within the niche or whether they exit the niche during cell division. We observed that HSCs shift from the SP to the non-SP fraction 6 days after 5-FU treatment and that hematopoietic cell clusters form adjacent to the endosteum and move toward the central marrow area, suggesting that cycling HSCs exit the niche and are mobilized to peripheral blood. Furthermore, when we transplanted the sorted non-SP fraction in HSCs mobilized by 5-FU treatment into irradiated mice, SP cells derived from mobilized LSK-non-SP cells were detected in recipients. Taken together, our observations support the notion that the niche is a microenvironment only for quiescent HSCs and that activated HSCs exit the niche.

Stem Cell Niche and Niche Cells in Adult BM

The niche is a term occasionally used to describe stem cells' location; however, the niche is actually composed of cellular components surrounding stem cells and signaling molecules provided by supporting cells.[1-7] Recently, it has been reported that long-term repopulating HSCs (LT-HSCs) are seen frequently at the trabecular bone surface in BM.[21,22]

Osteoblastic Niche

Bone-lining osteoblastic cells (OBs) critically regulate HSC pool size.[21,22] A long-term label retaining cell (LRC) study showed that 89% of $CD45^+Lin^-$ LRCs attached to the endosteal surface.[22] This observation suggests that HSCs remain in a quiescent state through close proximity to OBs. Genetic studies show that the number of LT-HSCs is a function of increases or decreases in the OB population in BM.[21-23] Increasing the number caused parallel increases in the HSC population, particularly LT-HSCs, without concomitant increases in other primitive progenitor cells. Such a specific increase in only the LT-HSC population suggested that a specific niche is functionally enhanced.[21,22] In contrast, OB depletion reduced BM cellularity.[23] These data suggest that OBs constitute part of the stromal cell support in BM. Taichman and Emerson also report that OBs are a part of the hematopoietic microenvironment after BM cavity formation.[24] Nilsson *et al.* report that HSCs are significantly enriched within the endosteal region after BM transplantation.[25]

To identify quiescent HSCs *in situ*, we analyzed Tie2 expression in 5-FU-treated BM and found that 5-FU-resistant $Tie2^+$ HSCs adhered to OBs at the endosteal surface, in agreement with previous findings.[21,22] In addition, angiopoietin-1 (Ang-1), a Tie2 ligand, is produced primarily by OBs, indicating that Tie2 and Ang-1 are expressed complementarily in the niche.[19] These observations support the idea that the niche is a microenvironment for quiescent

HSCs. HSCs thus may transit between niche and non-niche sites and/or between quiescence and active cell cycling *in vivo*.

Zhang *et al.* showed that N-cadherin-positive spindle-shaped osteoblasts (SNO cells) were localized to the endosteum of trabecular bone,[22] suggesting that some cells within the HSC niche constitute a very small subset of OBs expressing N-cadherin. It should now be determined whether hematopoietic activity is altered in osteoporosis or in osteopenia patients, who show decreased numbers of osteoblasts. In addition, the extent of the osteoblastic niche also should be defined. The distribution of SNO cells may differ between the trabecular and bone shaft regions, and thus quantitative analyses of the niche are required.

OBs also secrete several matrix proteins. It has been shown that osteopontin (Opn) negatively regulates HSC number in the BM niche. Two studies of Opn-null mice show that lack of Opn results in an increase in the number of HSCs.[26,27]

Adams *et al.* reported that levels of calcium ions in the HSC niche are critical for migration of HSCs to the endosteal surface. HSCs lacking calcium-sensing receptors can home to BM but are unable to remain in the endosteal niche.[28]

Vascular Niche

Hematopoietic and endothelial cells are derived from common progenitors: hemogenic endothelial cells or hemangioblasts.[29] Endothelial cells and HSCs coexpress CD31, CD34, CD133, vascular endothelial growth factor (VEGF) receptor 2, and Tie2.[30,31] In the mouse embryo, emergence of HSCs is closely associated with vascular endothelial cells. LT-HSCs develop from the intraembryonic mesodermal region, which contains para-aortic splanchnopleural mesoderm (P-Sp) as early as E7.5 in the absence of yolk sac circulation.[32] By E10.5, LTR-HSCs originate in the dorsal aorta, genital ridge/gonads, and pro/mesonephros region (aorta–gonad–mesonephros: AGM), the latter of which is related to the earlier P-Sp region and has been shown to harbor adult-type multipotent hematopoietic progenitors (MPPs) and pluripotent LT-HSCs.[32–35] Runx1 (runt-related transcription factor 1/acute myeloid leukemia 1 [AML1])-positive hemangiogenic endothelial cells can give rise to HSCs in the AGM region[36] and placenta labyrinth.[37,38] HSCs from this region are then presumed to colonize both the yolk sac and fetal liver, where they give rise to definitive hematopoietic precursors after E12.5.[39] During intraembryonic hematopoiesis, HSCs form clusters on endothelial cells. Thus, ligand–receptor signaling may occur between HSCs and endothelial cells.

It has been reported that expression of simple combinations of signaling lymphocyte activation molecule (SLAM) receptors ($CD150^+CD244^-CD48^-$ or $CD150^+CD48^-CD41^-$) can distinguish non-self-renewing MPPs ($CD244^+CD150^-CD48^-$).[40] Expression of CD150 in HSCs suggests that

LT-HSCs are localized not only to the endosteal surface but also to the perivascular area in both BM and extramedullary tissue. CD150$^+$ HSCs are associated with sinusoidal endothelial cells, thus establishing a vascular niche for HSC.[40,41] On the other hand, the vascular niche is thought to be the site where actively dividing stem or progenitor cells are located. Previous studies suggest that the vascular niche is where HSC differentiation and mobilization occur.[41,42] Further, sinusoidal endothelial walls consist of a single layer of endothelial cells, providing easy access of hematopoietic cells to the peripheral circulation. In addition, we have found that Tie2$^+$ cells adhering to BM endothelial cells are susceptible to myelosuppression and disappear following 5-FU treatment.[19] Thus, we propose that the osteoblastic niche is an environment promoting maintenance of quiescent HSCs and that the vascular niche supports mobilization and proliferation of activated HSCs. In extramedullary organs, such as spleen and liver, the vascular niche could serve as an alternative site for hematopoiesis under certain pathological conditions.[42,43]

Regulation of HSCs in the Osteoblastic Niche

Cell Adhesion Molecules

In *Drosophila* GSCs, niche cells produce factors inhibiting differentiation of stem cells and maintaining a stem cell state.[44] In the *Drosophila* ovary, DE-cadherin- and armadillo (the *Drosophila* beta-catenin homolog)-mediated cell adhesion promotes anchorage of GSCs to niche cells (cap cells) and is essential for their maintenance.[14] The importance of the adherens junction in GSC retention by the niche cap cells has been shown through mutation studies of adherens junction molecules, which are required for the niche to recruit and maintain GSCs.[14] In adult BM, cell adhesion molecules, such as cadherins and integrins, are also crucial for interactions between HSCs and the osteoblastic niche. Among members of the classical cadherin family, only N-cadherin is expressed in both quiescent mouse HSCs and OBs (FIG. 2).[19,22] Not only mouse HSCs but human CD34$^+$ HSCs express N-cadherin, and treatment of CD34$^+$ cells with an N-cadherin neutralizing antibody reduces colony formation.[45] These expression patterns suggest that an adherens junction between HSCs and osteoblasts is created via N-cadherin. Indeed, ectopic expression of N-cadherin by OP9 stromal cells substantially increases their ability to maintain immature phenotypes of HSCs *in vitro*.[19] In addition, enforced N-cadherin expression in both of HSCs and stromal cells enhances adhesion and inhibits cell division of HSCs *in vitro,* suggesting that N-cadherin-mediated adhesion mediates slow cell cycling of HSCs and may keep HSCs quiescent (our unpublished data).

In addition, since active bone remodeling takes place at the endosteum, extracellular calcium ion concentrations in the endosteum are likely higher than

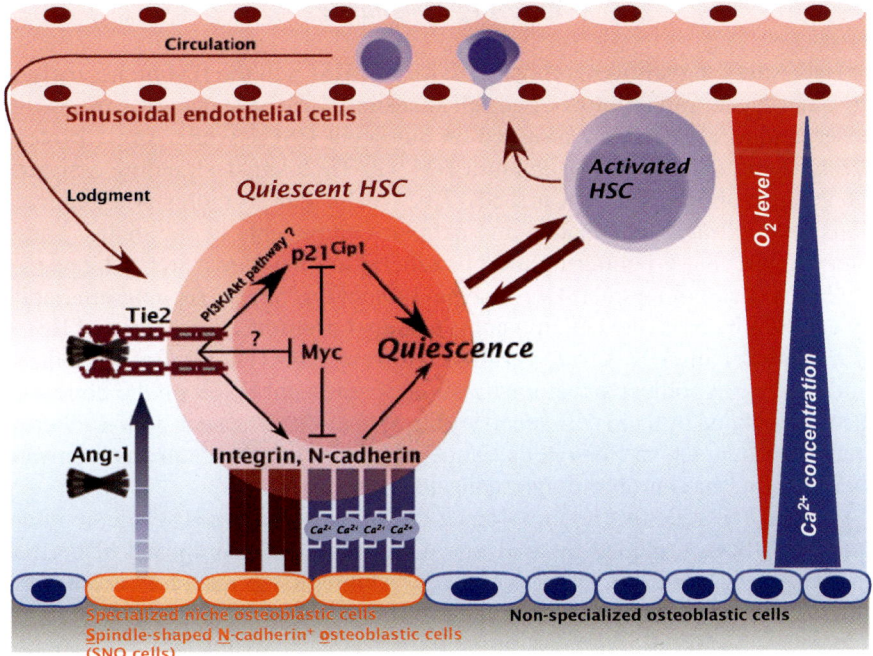

FIGURE 2. Regulation of HSC quiescence and adhesion in the osteoblastic niche. In the endosteal niche, HSCs adhere to spindle-shaped N-cadherin$^+$ OBs (SNO cells) and maintain their quiescence. In contrast to quiescent HSCs, activated HSCs may reside in the sinusoidal perivascular area. HSCs can also shift between the endosteal and perivascular niches, and their cell cycle status may depend on the nature of the interacting niche. Tie2/Ang-1 signaling between HSCs and osteoblastic niche cells enhances β1-integrin and N-cadherin-mediated cell adhesion and may activate cyclin-dependent kinase inhibitors, such as p21^{Cip1}. Both cell adhesion and CDKI activation contribute to HSC quiescence. In addition, both cell adhesion molecules and p21^{Cip1} transcription are negatively regulated by c-Myc. Therefore, c-Myc may be a negative mediator downstream of Tie2. The endosteal niche is low in oxygen and high in calcium ions. Hypoxic conditions inhibit HSC exposure to ROS, while high Ca^{2+} density is important for lodgment of HSCs and N-cadherin-mediated cell adhesion.

in the central marrow region.[46] Since calcium ions are absolutely essential for cadherins to form adherens junctions between HSCs and the niche, a high concentration of calcium ions in the endosteum facilitates N-cadherin-mediated interactions of HSCs with the osteoblastic niche (FIG. 2).

It has been reported that c-Myc-deficient HSCs upregulate cell adhesion molecules, such as N-cadherin and integrins (FIG. 2). In contrast, forced c-Myc expression in HSCs represses N-cadherin and integrins leading to loss of self-renewal activity.[47,48] These observations suggest that release of HSCs from the niche requires c-Myc activity.

Tie2/Ang-1 Signaling

Tie2/Ang-1 signaling regulates activities of HSCs in the BM niche, including quiescence, antiapoptosis, and cell adhesion.[19] We have made the following observations. (1) Tie2 is expressed on quiescent HSCs, while Ang-1 protein is mainly produced by OBs in adult BM. (2) Tie2/Ang-1 signaling activates β1-integrin and N-cadherin in LSK-Tie2$^+$ cells and promotes HSC interactions with extracellular matrix and cellular components of the niche. These data suggest that β1-integrin and N-cadherin are key downstream targets of Tie2/Ang-1 signaling in HSCs. (3) Ang-1 maintains *in vivo* repopulating ability of HSCs by blocking cell division. (4) Ang-1 promotes quiescence of HSCs *in vivo*, protecting HSCs from various cellular stresses (FIG. 2). Based on these findings, we hypothesize that localization of quiescent HSCs on the bone surface is regulated by stem cell-specific adhesion molecules, such as N-cadherin. Once HSCs attach to SNO cells, Ang-1 produced by the latter may activate Tie2 on HSCs and promote tight adhesion in the niche.

Genetic evidence for the importance of Tie2 signaling in HSC–niche interaction has been provided by analysis of chimeric mice composed of normal and Tie1/Tie2-deficient embryonic cells. Although Tie receptors are not required for fetal hematopoiesis, including emergence of definitive HSCs, or for relocation to and differentiation in fetal liver, HSCs lacking Tie1/Tie2 are not maintained in the adult BM microenvironment.[49] Tie1-deficient cells, which express normal levels of Tie2, contribute to hematopoiesis.[50,51] These findings indicate that Tie2 is required for postnatal BM hematopoiesis but not for embryonic hematopoiesis. In addition, analysis of chimeric mice that included Tie1 and Tie2 receptors double knockout donors and Rag2-deficient hosts, which do not produce mature lymphocytes, showed that Tie1/Tie2-deficient cells contribute to lymphopoiesis in the absence of competing host cells.[49] These findings strongly suggest that Tie2 is critical for the maintenance and survival of HSCs in adult BM and that Tie2-deficient cells cannot occupy the adult BM niche when competing with wild-type cells.

Ang-1/Tie2 signaling activates the phosphatidylinositol 3-kinase (PI3-K)/Akt signaling pathway.[52] PI3-K/Akt signaling also regulates several cell cycle regulators, such as the CDK inhibitor, p21^{Cip1}. Thus, Tie2/Ang-1 signaling may also regulate p21^{Cip1} protein via PI3-K/Akt activation, resulting in HSC quiescence. Further studies may reveal other molecules or pathways required for adhesion and cell cycle regulation mediated by niche factors. Understanding the nature of these factors should lead to the development of new strategies for regenerative medicine.

Reactive Oxygen Species (ROS)

Recently, we found that ataxia telangiectasia mutated (ATM) regulates LTR-activity of HSCs but not proliferation or differentiation.[53] ATM is a cell cycle checkpoint regulator activated after DNA damage. ATM also regulates oxidant levels, which in its absence, rise.[54] $Atm^{-/-}$ mice over the age of 24 weeks show progressive BM failure due to a defect in HSC function associated with elevated levels of reactive oxygen species (ROS). Elevation of ROS in HSCs upregulates the cyclin-dependent kinase (CDK) inhibitor p16^{Ink4a} and the retinoblastoma (*Rb*) gene in $Atm^{-/-}$ HSCs.[53] In addition, oxidative stress in HSCs specifically activates the p38 MAPK pathway, a signaling pathway responding to diverse cellular stresses.[55] Activation of p38 MAPK induces increases of *p16*^{Ink4a} and *p19*^{Arf}. p38 MAPK is activated through upstream MAPK kinases (MKK3 and MKK6) and MAPK kinase kinase (apoptosis signal-regulatory kinase 1, ASK1, encoded by *MAP3k5*). H_2O_2-induced activation of p38 MAPK is suppressed in *MAP3k5*$^{-/-}$ cells.[56] Inactivation of *MAP3K5* by retroviral infection of HSCs with a construct encoding a dominant-negative form of *MAP3K5* also restores defective repopulating capacity in serial BM transplantation experiments. Such redox-dependent activation is selective for HSCs and not observed in more differentiated hematopoietic progenitors, indicating that murine HSCs are exquisitely sensitive to oxidative stress.

Atm deficiency or ROS elevation stimulates p38 MAPK in HSCs leading to defective cell cycle quiescence.[53,55] Indeed, the proportion of SP cells and PY$^{low/-}$ cells in $Atm^{-/-}$ mice compared to wild-type mice, suggesting that $Atm^{-/-}$ HSCs are not quiescent. Treatment with antioxidative reagents, N-Acetyl Cysteine (NAC) or with a MAPK inhibitor restores reconstitutive capacity and quiescence of $Atm^{-/-}$ HSCs.[53,55]

The ROS-p38 MAPK pathway also functions in HSC exhaustion in normal aging as well as in $Atm^{-/-}$ mice. Levels of intracellular ROS in LSK cells from aging mice (24 months old) were higher than those seen in young mice (8 weeks old), suggesting that ROS levels increase in HSCs as mice age.[53] To investigate HSC life span, we undertook serial transplantation assays. ROS levels and p38 MAPK activity in HSCs gradually and markedly increased with each transplantation.[55] Also as ROS was elevated, repopulation capacity of HSCs was reduced in recipient mice after the third or fourth BMT.[55] Treatment with an antioxidant or a p38 MAPK inhibitor reversed these "aging" effects. Long-term treatment *in vivo* with NAC blocked p38 MAPK activation and p16^{Ink4a} and p19^{Arf} upregulation associated with serial transplantation. These results suggest that increases in intracellular ROS in HSCs—caused by genetic defects, such as *Atm* deficiency or by natural aging—activate the p38 MAPK-p16^{Ink4a}-Rb pathway and result in cell cycle progression of quiescent HSCs. Lack of quiescence appears to contribute ultimately to stem cell exhaustion and BM failure (FIG. 3).

Together, these data suggest that regulation of oxidative stress is critical to maintain HSC quiescence. Although it is still unclear how oxidative stress affects maintenance of HSC–niche interactions, ROS elevation in HSCs may affect interaction of HSCs with the niche and lead to the loss of quiescence. BM is a very low oxygen tension environment, and mesenchymal progenitors

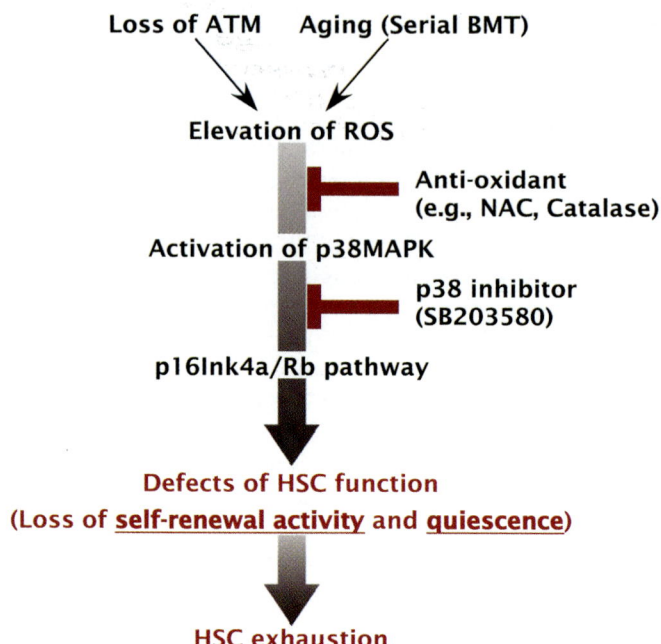

FIGURE 3. Effects of oxidative stress in HSC function. Loss of ATM or serial BMT-induced HSC aging elevates intracellular ROS in HSCs, causing defects in self-renewal activity and HSC quiescence through activation of the p38 MAPK/p16Ink4a/Rb pathway, resulting in HSC exhaustion. Antioxidants and p38 inhibitors can inhibit this process.

generate OBs efficiently in hypoxic conditions.[57] Therefore, we hypothesize that niches or niche cells required to keep stem cells quiescent maintain hypoxic conditions in BM (FIG. 2). The function of the BM niche is to provide an environment of low oxygen tension that would protect cells from exposure to ROS.

ACKNOWLEDGMENT

This work was supported by a grant-in-aid for Young Scientists to F.A. from the Ministry of Education, Science, Sports and Culture, Japan.

REFERENCES

1. LIN, H. 2002. The stem-cell niche theory: lessons from flies. Nat. Rev. Genet. **3:** 931–940.
2. FUCHS, E., T. TUMBAR & G. GUASCH. 2004. Socializing with the neighbors: stem cells and their niche. Cell **116:** 769–778.

3. LI, L. & T. XIE. 2005. Stem cell niche: structure and function. Annu. Rev. Cell. Dev. Biol. **21:** 605–631.
4. MOORE, K.A. & I.R. LEMISCHKA. 2006. Stem cells and their niches. Science **311:** 1880–1885.
5. SUDA, T., F. ARAI & A. HIRAO. 2005. Hematopoietic stem cells and their niche. Trends Immunol. **26:** 426–433.
6. ADAMS, G.B. & D.T. SCADDEN. 2006. The hematopoietic stem cell in its place. Nat. Immunol. **7:** 333–337.
7. WILSON, A. & A. TRUMPP. 2006. Bone-marrow haematopoietic-stem-cell niches. Nat. Rev. Immunol. **6:** 93–106.
8. SPANGRUDE, G.J., S. HEIMFELD & I.L. WEISSMAN. 1988. Purification and characterization of mouse hematopoietic stem cells. Science **241:** 58–62.
9. OSAWA, M. et al. 1996. Long-term lymphohematopoietic reconstitution by a single CD34-low/negative hematopoietic stem cell. Science **273:** 242–245.
10. ADOLFSSON, J. et al. 2001. Upregulation of Flt3 expression within the bone marrow Lin(-)Sca1(+)c-kit(+) stem cell compartment is accompanied by loss of self-renewal capacity. Immunity **15:** 659–669.
11. KONDO, M., I.L. WEISSMAN & K. AKASHI. 1997. Identification of clonogenic common lymphoid progenitors in mouse bone marrow. Cell **91:** 661–672.
12. AKASHI, K. et al. 2000. A clonogenic common myeloid progenitor that gives rise to all myeloid lineages. Nature **404:** 193–197.
13. SCHOFIELD, R. 1978. The relationship between the spleen colony-forming cell and the haemopoietic stem cell. Blood Cells **4:** 7–25.
14. SONG, X. et al. 2002. Germline stem cells anchored by adherens junctions in the Drosophila ovary niches. Science **296:** 1855–1857.
15. TUMBAR, T. et al. 2004. Defining the epithelial stem cell niche in skin. Science **303:** 359–363.
16. CHENG, T. et al. 2000. Hematopoietic stem cell quiescence maintained by p21cip1/waf1. Science **287:** 1804–1808.
17. CHESHIER, S.H. et al. 1999. In vivo proliferation and cell cycle kinetics of long-term self-renewing hematopoietic stem cells. Proc. Natl. Acad. Sci. USA **96:** 3120–3125.
18. ALLSOPP, R.C., S. CHESHIER & I.L. WEISSMAN. 2001. Telomere shortening accompanies increased cell cycle activity during serial transplantation of hematopoietic stem cells. J. Exp. Med. **193:** 917–924.
19. ARAI, F. et al. 2004. Tie2/Angiopoietin-1 signaling regulates hematopoietic stem cell quiescence in the bone marrow niche. Cell **118:** 149–161.
20. HUTTMANN, A. et al. 2001. Functional heterogeneity within rhodamine123(lo) Hoechst33342(lo/sp) primitive hemopoietic stem cells revealed by pyronin Y. Exp. Hematol. **29:** 1109–1116.
21. CALVI, L.M. et al. 2003. Osteoblastic cells regulate the haematopoietic stem cell niche. Nature **425:** 841–846.
22. ZHANG, J. et al. 2003. Identification of the haematopoietic stem cell niche and control of the niche size. Nature **425:** 836–841.
23. VISNJIC, D. et al. 2004. Hematopoiesis is severely altered in mice with an induced osteoblast deficiency. Blood **103:** 3258–3264.
24. TAICHMAN, R.S. & S.G. EMERSON. 1998. The role of osteoblasts in the hematopoietic microenvironment. Stem Cells **16:** 7–15.

25. NILSSON, S.K., H.M. JOHNSTON & J.A. COVERDALE. 2001. Spatial localization of transplanted hemopoietic stem cells: inferences for the localization of stem cell niches. Blood **97:** 2293–2299.
26. NILSSON, S.K. *et al.* 2005. Osteopontin, a key component of the hematopoietic stem cell niche and regulator of primitive hematopoietic progenitor cells. Blood **106:** 1232–1239.
27. STIER, S. *et al.* 2005. Osteopontin is a hematopoietic stem cell niche component that negatively regulates stem cell pool size. J. Exp. Med. **201:** 1781–1791.
28. ADAMS, G.B. *et al.* 2006. Stem cell engraftment at the endosteal niche is specified by the calcium-sensing receptor. Nature **439:** 599–603.
29. CHOI, K. *et al.* 1998. A common precursor for hematopoietic and endothelial cells. Development **125:** 725–732.
30. RAFII, S. *et al.* 2002. Vascular and haematopoietic stem cells: novel targets for anti-angiogenesis therapy? Nat. Rev. Cancer **2:** 826–835.
31. TAKAKURA, N. *et al.* 1998. Critical role of the TIE2 endothelial cell receptor in the development of definitive hematopoiesis. Immunity **9:** 677–686.
32. CUMANO, A., F. DIETERLEN-LIEVRE & I. GODIN. 1996. Lymphoid potential, probed before circulation in mouse, is restricted to caudal intraembryonic splanchnopleura. Cell **86:** 907–916.
33. MEDVINSKY, A.L. *et al.* 1993. An early pre-liver intraembryonic source of CFU-S in the developing mouse. Nature **364:** 64–67.
34. MULLER, A.M. *et al.* 1994. Development of hematopoietic stem cell activity in the mouse embryo. Immunity **1:** 291–301.
35. MEDVINSKY, A. & E. DZIERZAK. 1996. Definitive hematopoiesis is autonomously initiated by the AGM region. Cell **86:** 897–906.
36. NORTH, T.E. *et al.* 2002. Runx1 expression marks long-term repopulating hematopoietic stem cells in the midgestation mouse embryo. Immunity **16:** 661–672.
37. GEKAS, C. *et al.* 2005. The placenta is a niche for hematopoietic stem cells. Dev. Cell. **8:** 365–375.
38. OTTERSBACH, K. & E. DZIERZAK. 2005. The murine placenta contains hematopoietic stem cells within the vascular labyrinth region. Dev. Cell **8:** 377–387.
39. DZIERZAK, E. & A. MEDVINSKY. 1995. Mouse embryonic hematopoiesis. Trends Genet. **11:** 359–366.
40. KIEL, M.J. *et al.* 2005. SLAM family receptors distinguish hematopoietic stem and progenitor cells and reveal endothelial niches for stem cells. Cell **121:** 1109–1121.
41. KOPP, H.G. *et al.* 2005. The bone marrow vascular niche: home of HSC differentiation and mobilization. Physiology (Bethesda) **20:** 349–356.
42. AVECILLA, S.T. *et al.* 2004. Chemokine-mediated interaction of hematopoietic progenitors with the bone marrow vascular niche is required for thrombopoiesis. Nat. Med. **10:** 64–71.
43. HEISSIG, B. *et al.* 2002. Recruitment of stem and progenitor cells from the bone marrow niche requires MMP-9 mediated release of kit-ligand. Cell **109:** 625–637.
44. SPRADLING, A., D. DRUMMOND-BARBOSA & T. KAI. 2001. Stem cells find their niche. Nature **414:** 98–104.
45. PUCH, S. *et al.* 2001. N-cadherin is developmentally regulated and functionally involved in early hematopoietic cell differentiation. J. Cell Sci. **114:** 1567–1577.

46. SILVER, I.A., R.J. MURRILLS & D.J. ETHERINGTON. 1988. Microelectrode studies on the acid microenvironment beneath adherent macrophages and osteoclasts. Exp. Cell Res. **175:** 266–276.
47. WILSON, A. *et al.* 2004. c-Myc controls the balance between hematopoietic stem cell self-renewal and differentiation. Genes Dev. **18:** 2747–2763.
48. MURPHY, M.J., A. WILSON & A. TRUMPP. 2005. More than just proliferation: myc function in stem cells. Trends Cell. Biol. **15:** 128–137.
49. PURI, M.C. & A. BERNSTEIN. 2003. Requirement for the TIE family of receptor tyrosine kinases in adult but not fetal hematopoiesis. Proc. Natl. Acad. Sci. USA **100:** 12753–12758.
50. PARTANEN, J. *et al.* 1996. Cell autonomous functions of the receptor tyrosine kinase TIE in a late phase of angiogenic capillary growth and endothelial cell survival during murine development. Development **122:** 3013–3021.
51. RODEWALD, H.R. & T.N. SATO. 1996. Tie1, a receptor tyrosine kinase essential for vascular endothelial cell integrity, is not critical for the development of hematopoietic cells. Oncogene **12:** 397–404.
52. SHIOJIMA, I. & K. WALSH. 2002. Role of Akt signaling in vascular homeostasis and angiogenesis. Circ. Res. **90:** 1243–1250.
53. ITO, K, *et al.* 2004. Regulation of oxidative stress by ATM is required for self-renewal of haematopoietic stem cells. Nature **431:** 997–1002.
54. ROTMAN, G. & Y. SHILOH. 1997. Ataxia-telangiectasia: is ATM a sensor of oxidative damage and stress? Bioessays **19:** 911–917.
55. ITO, K. *et al.* 2006. Reactive oxygen species act through p38 MAPK to limit the lifespan of hematopoietic stem cells. Nat. Med. **12:** 446–451.
56. TOBIUME, K. *et al.* 2001. ASK1 is required for sustained activations of JNK/p38 MAP kinases and apoptosis. EMBO Rep. **2:** 222–228.
57. LENNON, D.P., J.M. EDMISON & A.I. CAPLAN. 2001. Cultivation of rat marrow-derived mesenchymal stem cells in reduced oxygen tension: effects on *in vitro* and *in vivo* osteochondrogenesis. J. Cell. Physiol. **187:** 345–355.

Cytokine Signaling, Lipid Raft Clustering, and HSC Hibernation

SATOSHI YAMAZAKI,[a,b] ATSUSHI IWAMA,[a,c] YOHEI MORITA,[a] KOJI ETO,[a] HIDEO EMA,[a] AND HIROMITSU NAKAUCHI[a]

[a]*Laboratory of Stem Cell Therapy, Center for Experimental Medicine, The Institute of Medical Science, University of Tokyo, Tokyo 108-8639, Japan*

[b]*ReproCELL Inc, Tokyo 100-0011, Japan*

[c]*Department of Cellular and Molecular Medicine, Graduate School of Medicine, Chiba University, Chiba 260-8670, Japan*

ABSTRACT: Hematopoietic stem cells (HSCs) reside in the bone marrow (BM) niche in a noncycling state and enter the cell cycle at long intervals. This unique property of HSCs is reminiscent of hibernation in mammals. However, little is known about inter- and intracellular signaling mechanisms underlying this unique property of HSCs. This is largely due to the paucity of HSCs making application of traditional signal transduction assays difficult. To address these issues, we have developed a novel assay based on in-droplet single-cell staining and quantitative fluorescence imaging analysis. Using this assay system, we demonstrate that freshly isolated HSCs from the BM niche lack lipid raft clustering, exhibit repression of the AKT-FOXO signaling pathway, and express abundant $p57^{Kip2}$ cyclin-dependent kinase inhibitor. Lipid raft clustering induced by cytokines was essential for HSC re-entry into the cell cycle. Conversely, inhibition of lipid raft clustering caused sustained nuclear accumulation of FOXO transcription factors and induced HSC hibernation *ex vivo*. Among niche signals examined, transforming growth factor-β (TGF-β) efficiently inhibited lipid raft clustering and induced $p57^{Kip2}$ expression, leading to HSC hibernation. These data uncover a critical role for lipid rafts in HSC fate decision and establish the role of TGF-β as a niche signal in control of HSC hibernation in the BM niche.

KEYWORDS: HSCs; hibernation; lipid raft clustering; FOXO; p57

PROSPECTIVE ISOLATION AND SINGLE-CELL IMAGING ANALYSIS OF HEMATOPOIETIC STEM CELLS

Hematopoietic stem cells (HSCs) are clonogenic cells with self-renewal and multilineage differentiation capabilities. In adult mouse bone marrow (BM),

Address for correspondence: Hiromitsu Nakauchi, Laboratory of Stem Cell Therapy, Center for Experimental Medicine, The Institute of Medical Science, University of Tokyo, Tokyo 108-8639, Japan. Voice: +81-3-5449-5330; fax: +81-3-5449-5451.

nakauchi@ims.u-tokyo.ac.jp

most HSCs are in nondividing quiescent state in the so-called stem-cell niche in close contact with supporting cells like osteoblasts and vascular endothelium.[1-3] The capacity to enter and to leave a hibernation-like state is one of the properties necessary for "stemness" and is of critical biological importance in preventing premature HSC exhaustion under conditions of hematopoietic stress.[4]

We have established a very efficient method for HSC purification using fluorescence-activated cell sorting and have prospectively identified HSCs in adult mouse BM. $CD34^-$ $c\text{-}Kit^+$ $Sca\text{-}1^+$ lineage marker-negative cells ($CD34^-KSL$) from mouse BM, a population representing 0.004% of BM mononuclear cells, are exclusively enriched in HSCs, whereas $CD34^+$ $c\text{-}Kit^+$ $Sca\text{-}1^+$ lineage marker-negative cells ($CD34^+KSL$) are progenitors with short-term repopulating capacity.[5-8] However, the more we purify HSCs, the fewer cells we can obtain. This has made the application of conventional biochemical analysis to HSCs almost impossible. Therefore, little is known about the intracellular signaling events in HSCs. To address these issues, we have developed an immunostaining method that can efficiently examine very small numbers of cells. This method enables quantitative measurement of the immunofluorescent molecules at the single-cell level (FIG. 1).[9,10] Not only the expression of proteins but also the intracellular localization/mobilization of various proteins can be studied. Moreover, signaling can be quantified by measuring fluorescent intensity on individual HSCs using anitphosphoprotein antibodies.[11] Using this assay system, we attempted to clarify the mechanisms that regulate cell cycle status of HSCs.

CELL CYCLE ANALYSIS AND FOXO3 EXPRESSION IN HSCs

To evaluate the cell cycle status of BM HSCs, Ki-67 expression and Pyronin-Y staining were directly analyzed in $CD34^-KSL$ HSCs and $CD34^+KSL$ progenitor cells. As expected, most $CD34^-KSL$ cells (>94%) were negative for Ki-67 and Pyronin-Y staining, whereas most $CD34^+KSL$ progenitor cells were positive (FIG. 2). These data confirmed dormancy in HSCs as previously demonstrated by long-term bromoxyuridine incorporation analyses and Pyronin-Y staining.[12,13] Interestingly, HSCs are not always in G_0. It has been demonstrated that HSCs are recruited into the cell cycle at long intervals, on average every 1–2 months, in analyses using long-term bromodeoxyuridine incorporation.[12,14] Thus, HSC quiescence in the BM niche is reminiscent of mammalian entry into hibernation and of *Caenorhabditis elegans* dauer formation. This similarity prompted us to examine the PI3K-Akt-FOXO signaling pathway that regulates these unique states of adaptation to harsh environmental conditions. FOXO1 (FKHR), FOXO3a (FKHRL), and FOXO4 (AFX) are mammalian orthologues of *C. elegans* DAF-16 forkhead transcription factor. In freshly isolated $CD34^-KSL$ HSCs, phosphorylated Akt, an active form of

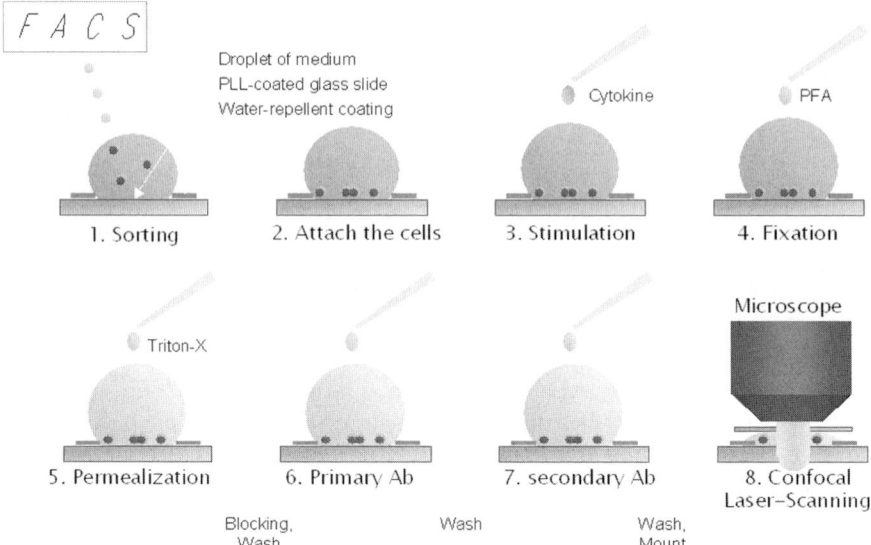

FIGURE 1. In-droplet single-cell immunostaining. Fifty CD34⁻KSL cells are directly sorted into a droplet of medium on a poly-L-lysine–coated glass slide. They were stimulated with cytokines for 30 min. After fixation on PLL-coated glass slides, cells were stained with antibodies and then analyzed with a confocal laser-scanning microscope. All fluorescence intensities of individual cells were computationally quantified by using ImageJ 1.33 software (http://rsb.info.nih.gov/if), and were normalized against the mean intensity of unstimulated cells.

Akt, was not detected at all, and one of its downstream targets, FOXO3a, accumulated in the nucleus. In contrast, in most CD34$^+$KSL progenitor cells, Akt was highly activated, whereas FOXO3a was largely excluded from the nucleus and restricted to the cytoplasm (FIG. 3). Another orthologue of DAF-16, FOXO1, behaved like FOXO3a (data not shown). Signals mediated by receptor tyrosine kinases and cytokine receptors activate the PI3K-Akt pathway, and thereby negatively regulate FOXO activity.[15]

LIPID RAFT CLUSTERING IN WAKING UP HSCs

To evaluate the degree of cytokine-signal activation in HSCs, we next examined the localization of c-Kit, an HSC cell-membrane receptor tyrosine kinase, relative to lipid rafts. Lipid raft distribution was assessed by using cholera toxin subunit B (CTxB) to label endogenous GM1 ganglioside, a component of lipid rafts (FIG. 4). On most (92%) of the freshly isolated CD34⁻KSL HSCs, both c-Kit and lipid rafts were diffusely distributed. On CD34$^+$KSL progenitor cells, both c-Kit and lipid rafts were polarized; in addition, they were largely colocalized with one another, indicating that c-Kit was condensed in the lipid raft

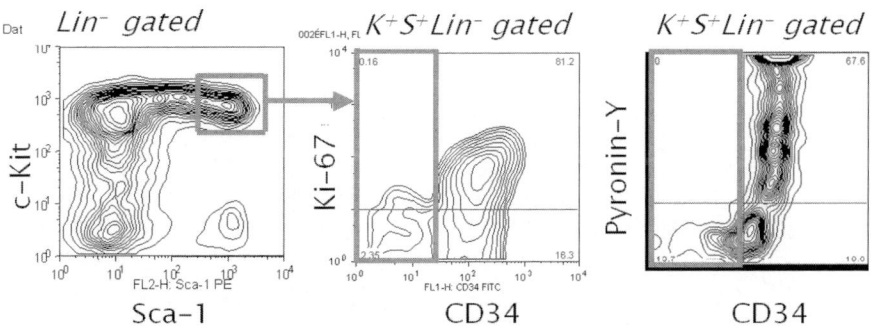

FIGURE 2. Cell cycle status of BM HSCs and progenitor cells. Ki-67 expression and Pyronin-Y staining were directly analyzed in CD34⁻KSL and CD34⁺KSL cells. Most HSCs in BM niche are in G_0, whereas great majority of CD34⁺ progenitor cells are proliferating.

cluster. Stimulation of HSCs by stem cell factor (SCF) and thrombopoietin (TPO), either independently or in combination, induced clustering within agglomerated lipid rafts of both c-Kit and c-mpl, the receptors for SCF and TPO, respectively. These data indicate that cytokine signals concentrate activated receptors together with inactive ones for efficient and versatile transduction of signals using lipid rafts. As observed in freshly isolated CD34⁺KSL progenitor cells, cytokine stimulation of CD34⁻KSL HSCs induced activation of Akt. It also induced restriction of FOXO3a to the cytoplasm and exclusion of FOXO3a from the nucleus (FIG. 3). These data demonstrate a tight correlation of cell cycle status with lipid raft status as well as with activity of the PI3K-Akt-FOXO pathway.

INHIBITION OF LIPID RAFT CLUSTERING MAINTAINS HSCs IN HIBERNATING STATE

The striking contrast in lipid raft and cell cycle status between CD34⁻KSL HSCs and CD34⁺KSL progenitor cells evoked the possibility that lipid rafts regulate the cell cycle status of HSCs by modulating cytokine signals. To address this, we inhibited lipid raft clustering in response to cytokine stimulation by depleting plasma membrane cholesterol with methyl-β-cyclodextrin (MβCD). Pretreatment of freshly isolated CD34⁻KSL HSCs with MβCD inhibited lipid raft clustering in 81.2% of the cells tested. The downstream signals that activate Akt also stayed repressed, leaving FOXO3a in the nucleus. Even in hibernating HSCs, cyclin D1, one of the D-type G_1 cyclins, is abundantly expressed. Intriguingly cyclin D1, localized exclusively in the cytoplasm in HSCs, was immediately translocated into the nucleus after cytokine stimulation (FIG. 5). Pretreatment of freshly isolated CD34⁻KSL HSCs with MβCD again inhibited cytokine-induced nuclear translocation of cyclin D1 (FIG. 5).

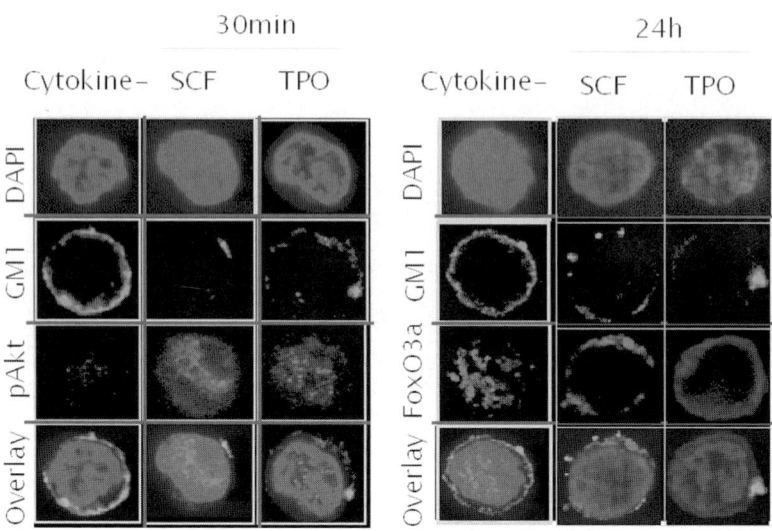

FIGURE 3. Activation of the Akt-FOXO signaling pathway in HSCs. Freshly isolated CD34⁻KSL cells were incubated for 30 min in the presence or absence of cytokines (SCF or TPO) and were stained with DAPI (blue), CTB (green), and antiphospho-Akt (red), or an anti-FOXO3a antibody (red). (In color in Annals online.)

On the other hand, cyclin D1 in CD34$^+$KSL progenitor cells was present largely in the nucleus.

p57 APPEARS TO BE RESPONSIBLE FOR CELL CYCLE REGULATION IN HSCs

Among *Cip/Kip* gene products (p21, p27, and p57), p21, a late G_1-phase cyclin-dependent kinase inhibitors (CDKIs), is of critical biological importance in preventing premature HSC exhaustion,[16] whereas p27, another late G_1-phase CDKI, regulates proliferation and pool size of hematopoietic progenitor cells rather than of HSCs.[17] No role for p57, another G_1-phase CDKI, has been identified in HSCs. RT-PCR analysis of primitive hematopoietic cell fractions unexpectedly unveiled specific expression of *p57* in CD34⁻KSL HSCs (FIG. 5). Immunocytological analysis confirmed abundant expression of p57 in CD34⁻KSL HSCs and drastic downregulation of p57 expression in CD34$^+$KSL progenitor cells. The vast majority of p57 lay in the cytoplasm, like cyclin D1. After cytokine stimulation, p57 immediately disappeared from HSCs. In the presence of MβCD, however, p57 remained in HSC cytoplasm, as did cyclin D1 (FIG. 5). Although p21 and p27 are known direct transcriptional targets for FOXO1, FOXO3a, and FOXO4,[18,19] expression of *p27* in CD34⁻KSL HSCs was lower than that in lineage marker-negative (Lin⁻)

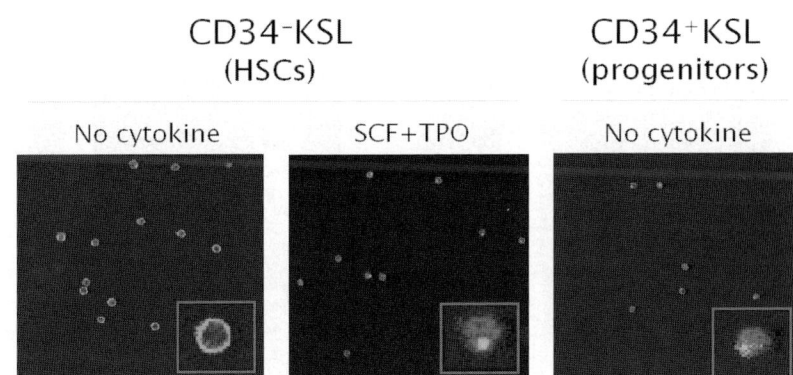

FIGURE 4. Formation of lipid raft clustering in progenitor cells and cytokine stimulated HSCs, but not in freshly isolated HSCs. Freshly isolated CD34$^-$KSL HSCs and CD34$^+$KSL progenitor cells were incubated for 30 min in the presence or absence of cytokines (20 ng/mL SCF and 50 ng/mL TPO), and were stained with DAPI (blue) and CTB (green). (In color in Annals online.)

progenitor fraction cells, and immunocytochemical analyses detected only weak p27 expression in CD34$^-$KSL HSC cytoplasm (data not shown). Nor was expression of *p21* in CD34$^-$KSL HSCs significant. These data indicate a major role of p57 as a specific CDKI that in HSCs binding to the cyclin D/CDK complex and suppresses its activity. After dissociating from the cyclin D/CDK complex, p27 is rapidly degraded through the ubiquitin-proteasome pathway.[20] If p57 behaves similarly, its degradation at the G_0 to G_1 transition may be critical in promoting HSC cell cycle progression from G_0 to G_1.

INDUCTION OF HIBERNATION STATE IN HSCs BY LIPID RAFT CLUSTERING INHIBITORS *EX VIVO*

To understand the biological meaning of lipid raft status for HSC proliferation, MβCD was added to single-cell CD34$^-$KSL HSC cultures in the presence of SCF and TPO. An optimal dose (1 mM) of MβCD efficiently inhibited division of single HSCs with minimal apoptotic death; single HSCs stayed alive for up to 5–10 days. During this observation period, 93% of single HSCs did not undergo cell division and they remained single cells, without apoptosis. By contrast, CD34$^+$KSL progenitor cells underwent apoptotic death by day 2 under the same conditions. Without cytokines, both CD34$^-$KSL HSCs and CD34$^+$KSL progenitors died within 24 h. These data indicate that lipid raft clustering in response to cytokine stimulation is crucial for the proliferation of both HSCs and progenitors. They also indicate that cytokine signals muted by nonclustering of lipid rafts keep HSCs dormant without inducing apoptosis. Such signals, however, cannot secure survival of progenitor cells.

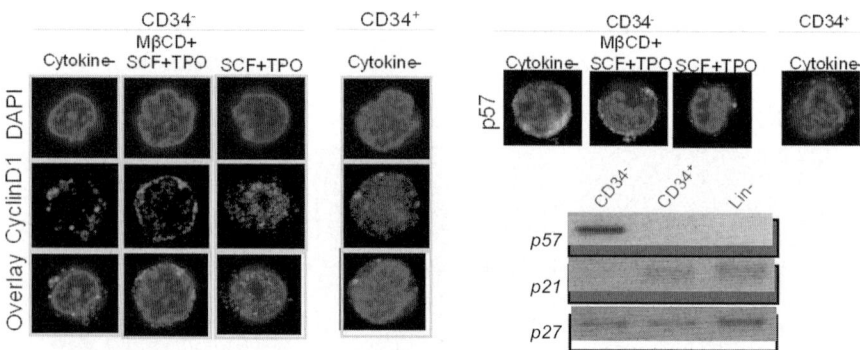

FIGURE 5. To examine subcellular localization of cyclin D1 in HSCs, freshly isolated CD34⁻KSL HSCs (CD34⁻) and CD34⁺KSL hematopoietic progenitor cells (CD34⁺) were incubated for 30 min in the presence or absence of SCF and TPO or were preincubated with MβCD for 30 min and further incubated in the presence of SCF and TPO for another 30 min. The cells were stained with DAPI (blue) and with an anti-cyclin D1 antibody (red). To analyze subcellular localization of p57 in HSCs, freshly isolated CD34⁻KSL HSCs (CD34⁻) were preincubated with and without MβCD for 30 min and further incubated in the presence of SCF and TPO for 12 h. The incubated cells, together with freshly isolated CD34⁻KSL HSCs (CD34⁻) and CD34⁺KSL hematopoietic progenitor cells (CD34⁺), were stained with DAPI (blue) and with an anti-p57 antibody (red) (*upper panel*). mRNA expression for mouse *Cip/Kip* genes is presented (*lower panel*). Cells analyzed are BM CD34⁻KSL HSCs (CD34⁻), CD34⁺KSL progenitors (CD34⁺), and Lineage marker cells (Lin⁻). (In color in Annals online.)

We then tried to examine whether or not such HSCs maintained *in vitro* in the presence of cytokines and an lipid raft clustering inhibitor can re-enter the cell cycle and retain a full range of HSC functions. After varying culture periods in the presence of SCF, TPO, and 1 mM MβCD, surviving single HSCs that had not divided were allowed to form colonies by replacing culture medium with an optimal medium supplemented with SCF, TPO, IL-3, and EPO. Surprisingly, almost half of single HSCs survived up to 5 days in the presence of SCF, TPO, and MβCD. After the change in culture medium, 66% of single HSCs gave rise to colonies (data not shown), and 57% of colonies were neutrophil/macrophage/Erythroblast/Megakaryocyte (nmEM) colonies, derived from colony-forming units-nmEM (CFU-nmEM) with multipotency, that is, a full range of differentiation capacity along myeloid lineages. Similarly, *in vivo* competitive repopulation analysis also revealed that those *in vitro* cultured HSCs retain long-term multilineage repopulation capacity comparable to freshly isolated hibernating HSCs. All these data strongly support the proposition that inhibition of lipid raft clustering modulates cytokine signals and induces hibernation in HSCs *ex vivo* without affecting HSC capacity to self-renew and to differentiate into a full range of hematopoietic cell lineages.

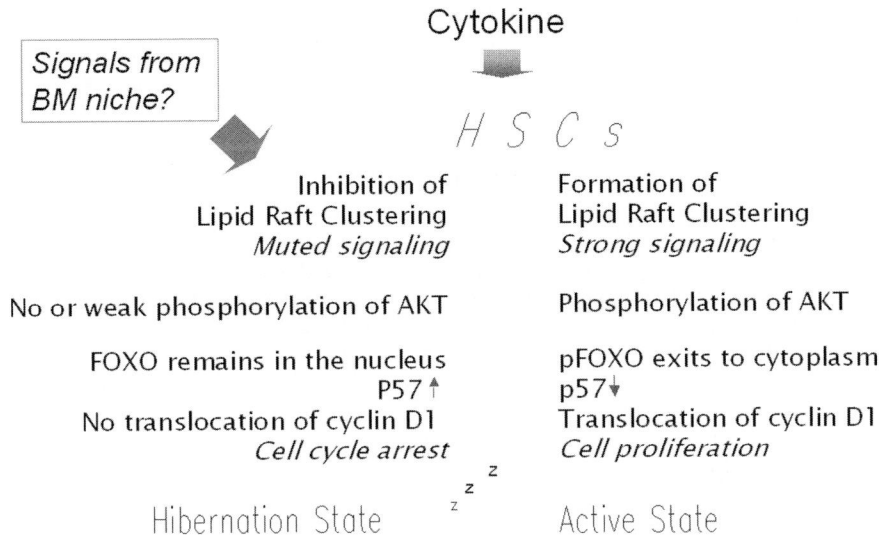

FIGURE 6. Inhibition of lipid raft clustering induces HSC hibernation. On cytokine stimulation, lipid rafts are clustered and give strong signals in side the HSCs that activate AKT-FOXO pathway inducing translocation of cyclin D1 and cell proliferation. We speculate that signals from niche somehow inhibit the formation of lipid raft clustering that attenuate the cytokine signaling leading to survival of HSCs without activating AKT-FOXO pathway and cell proliferation.

INHIBITION OF LIPID RAFT CLUSTERING MAY BE A ROLE OF THE BM NICHE SIGNAL

We have shown that inhibition of lipid raft clustering by MβCD maintains HSCs in a nondividing state even in the presence of stimulating cytokines *ex vivo* at least for a week. Our findings clearly demonstrate that lipid raft clustering is tightly inhibited in BM niche HSCs (FIG. 2). Thus, we may postulate that a role of the "niche signal" is to inhibit lipid raft clustering to maintain hibernation state in the HSCs (FIG. 6). On the basis of this hypothesis, we screened several putative "niche factors." Among those, Ang-1 partially inhibited clustering but not enough to inhibit activation of the Akt-FOXO pathway and to induce cell cycle arrest (data not shown). TGF-β strikingly inhibited lipid raft clustering in response to SCF and TPO, implying that it may be one of the niche signals necessary to induce hibernation in HSCs (Yamazaki *et al.*, unpublished data).

It is surprising that the PI3K-Akt-FOXO pathway involved in *C. elegans* dauer formation also operates in mammalian hibernation and in dormant HSCs in the BM niche. Hibernation, an evolutionary adaptation to harsh environmental conditions, may be an effective means to protect HSCs, a long-lived cell population that supplies all blood cells throughout life, from environmental hazards. The identification of niche signal and understanding of regulatory mechanisms of hibernating and waking HSC is the next critical step.

REFERENCES

1. ZHANG, J., C. NIU, L. YE, et al. 2003. Identification of the haematopoietic stem cell niche and control of the niche size. Nature **425:** 836–841.
2. CALVI, L.M., G.B. ADAMS, K.W. WEIBRECHT, et al. 2003. Osteoblastic cells regulate the haematopoietic stem cell niche. Nature **425:** 841–846.
3. KIEL, M.J., O.H. YILMAZ, T. IWASHITA, et al. 2005. SLAM family receptors distinguish hematopoietic stem and progenitor cells and reveal endothelial niches stem cells. Cell **121:** 1109–1121.
4. CHENG, T. & D.T. SCADDEN. 2002. Cell cycle entry of hematopoietic stem and progenitor cells controlled by distinct cyclin-dependent kinase inhibitors. Int. J. Hematol. **75:** 460–465.
5. OSAWA, M., K.-I. HANADA, H. HAMADA, et al. 1996. Long-term lymphohematopoietic reconstitution by a single CD34-low/negative hematopoietic stem cells. Science **273:** 242–245.
6. EMA, H., H. TAKANO, K. SUDO, et al. 2000. In vitro self-renewal division of hematopoietic stem cells. J. Exp. Med. **192:** 1281–1288.
7. TAKANO, H., H. EMA, SUDO, H. TAKANO, et al. 2004. Asymmetric division and lineage commitment at the level of hematopoietic stem cells: inference from differentiation in daughter cell and granddaughter cell pairs. J. Exp. Med. **199:** 295–302.
8. EMA, H., K. SUDO, J. SEITA, et al. 2005. Quantification of self-renewal capacity in single hematopoietic stem cells from normal and Lnk-deficient mice. Dev. Cell **8:** 907–914.
9. YAMAZAKI, S., A. IWAMA, S. TAKAYANAGI, et al. 2006. Cytokine signals modulated via lipid rafts mimic niche signals and induce hibernation in hematopoietic stem cells. EMBO J. **25:** 3515–3523.
10. EMA, H., Y. MORITA, S. YAMAZAKI, et al. 2007. Adult mouse hematopoietic stem cells: purification and single-cell assays. Nat. Protoc. **1:** 2979–2987.
11. SEITA, J., H. EMA, J. OOEHARA, et al. 2007. Lnk negatively regulates self-renewal of hematopoietic stem cells by modifying thrombopoietin-mediated signal transduction. Proc. Natl. Acad. Sci. USA **104:** 2349–2354.
12. SUDO, K., H. EMA, Y. MORITA, et al. 2000. Age-associated characteristics of murine hematopoietic stem cells. J. Exp. Med. **192:** 1273–1280.
13. ARAI, F., A. HIRAO, M. OHMURA, et al. 2004. Tie2/angiopoietin-1 signaling regulates hematopoietic stem cell quiescence in the bone marrow niche. Cell **118:** 149–161.
14. CHESHIER, S.H., S.J. MORRISON, X. LIAO, et al. 1999. In vivo proliferation and cell cycle kinetics of long-term self-renewing hematopoietic stem cells. Proc. Natl. Acad. Sci. USA **96:** 3120–3125.
15. BURGERING, B.M. & G.J. KOPS. 2002. Cell cycle and death control: long live forkheads. Trends Biochem. Sci. **27:** 352–360.
16. CHENG, T., N. RODORIGUES, H. SHEN et al. 2000. Hematopoietic stem cell quiescence maintained by $p21^{cip1/waf1}$. Science **287:** 1804–1808.

17. CHENG, T., N. RODRIGUES, D. DOMBKOWSKI et al. 2000. Stem cell repopulation efficiency but not pool size is governed by p27(kip1). Nat. Med. **6:** 1235–1240.
18. MEDEMA, R.H., G.J. KOPS, J.L. BOS, et al. 2000. AFX-like forkhead transcription factors mediate cell-cycle regulation by Ras and PKB through $p27^{kip1}$. Nature **404:** 782–787.
19. SEOANE, J., H.V. LE, L. SHEN, et al. 2004. Integration of Smad and forkhead pathways in the control of neuroepithelial and glioblastoma cell proliferation. Cell **117:** 211–223.
20. PAGANO, M., S.W. TAM, A.M. THEODRUS, et al. 1995. Role of the ubiquitin-proteasome pathway in regulating abundance of the cyclin-dependent kinase inhibitor p27. Science **269:** 682–685.

Dormant and Self-Renewing Hematopoietic Stem Cells and Their Niches

ANNE WILSON,[a] GABRIELA M. OSER,[b] MAIKE JAWORSKI,[b,c] WILLIAM E. BLANCO-BOSE,[b] ELISA LAURENTI,[b,c] CHRISTELLE ADOLPHE,[b] MARIEKE A. ESSERS,[b] H. ROBSON MACDONALD,[a] AND ANDREAS TRUMPP[b,c]

[a] *Ludwig Institute for Cancer Research Lausanne Branch, University of Lausanne, 1066 Epalinges, Switzerland*

[b] *Genetics and Stem Cell Laboratory, Swiss Institute for Experimental Cancer Research (ISREC), CH-1066 Epalinges, Switzerland*

[c] *Ecole Polytechnique Fédérale de Lausanne (EPFL), School of Life Sciences, CH-1015 Lausanne, Switzerland*

ABSTRACT: In the mouse, over the last 20 years, a set of cell-surface markers and activities have been identified, enabling the isolation of bone marrow (BM) populations highly enriched in hematopoietic stem cells (HSCs). These HSCs have the ability to generate multiple lineages and are capable of long-term self-renewal activity such that they are able to reconstitute and maintain a functional hematopoietic system after transplantation into lethally irradiated recipients. Using single-cell reconstitution assays, various marker combinations can be used to achieve a functional HSC purity of almost 50%. Here we have used the differential expression of six of these markers (Sca1, c-Kit, CD135, CD48, CD150, and CD34) on lineage-depleted BM to refine cell hierarchies within the HSC population. At the top of the hierarchy, we propose a dormant HSC population (Lin$^-$Sca1$^+$c-Kit$^+$ CD48$^-$CD150$^+$CD34$^-$) that gives rise to an active self-renewing CD34$^+$ HSC population. HSC dormancy, as well as the balance between self-renewal and differentiation activity, is at least, in part, controlled by the stem cell niches individual HSCs are attached to. Here we review the current knowledge about HSC niches and propose that dormant HSCs are located in niches at the endosteum, whereas activated HSCs are in close contact to sinusoids of the BM microvasculature.

KEYWORDS: hematopoietic stem cell (HSC); niche; dormancy; endosteum; osteoblast; self-renewal

---

Address for correspondence: Andreas Trumpp, ISREC/EPFL, Boveresses 155, CH-1066 Epalinges, Switzerland. Voice: +41 21 692 5817; fax: +41 21 652 6933.
Andreas.Trumpp@isrec.ch

A POSSIBLE HSC-PROGENITOR HIERARCHY

Somatic stem cells are present in most self-renewing tissues, including the skin, the intestinal epithelium, the mammary gland, and the hematopoietic system. On a single-cell basis, they have the capacity to both produce more stem cells by self-renewal and to give rise to a defined set of differentiated progeny critical for the maintenance and repair of their host tissue.[1–4] The best-characterized adult stem cell is the mouse hematopoietic stem cell (HSC), which is located in the bone marrow (BM) during homeostasis but can be mobilized to the spleen and liver after BM injury.[5,6] Since HSCs were first identified, advances in technology have made it possible to purify functional adult mouse HSCs close to homogeneity. Several groups have achieved long-term reconstitution of the hematopoietic system of a lethally irradiated recipient by transplantation of a single-purified BM HSC, providing functional proof for the existence of bona fide stem cells (TABLE 1). In the adult mouse, functional HSCs are found in a subset of the BM population that does not express cell-surface markers normally present on lineage (Lin)-committed hematopoietic cells but expresses high levels of the stem-cell antigen-1 (Sca-1) and the c-Kit receptor. Although all functional HSC activity is found in the $Lin^- Sca1^+ c\text{-}Kit^+$ (LSK) population, less than 1 out of 10 of these LSK-cells has repopulation capacity, suggesting substantial functional heterogeneity. On further subdivision of this population using differential surface expression of the antiadhesive sialomucin CD34,[7,8] the signal lymphocyte activating molecule (SLAM) receptors (CD150 and CD48),[9] or by the ability to efflux DNA-binding dyes such as Rhodamine-123[10] or Hoechst 33342 via side population (SP) activity,[11] single-cell reconstitution efficiencies of between 18% and 47% have been achieved (TABLE 1).[12] The results of these functional studies suggest that the cell-surface markers that seem to be most critical for defining repopulating HSCs within the LSK population are the SLAM receptors (CD150 and CD48) and CD34. Because all HSC activity is retained in the $LSKCD34^-$ population,[7] we examined the expression of CD34 in the $CD48^-CD150^+LSK$ population that contains 47% repopulating HSCs.[9] To do this, six-color fluorescence-activated cell sorting (FACS) was performed on lineage-depleted BM. An example of how the LSK population can be subdivided on the basis of these markers is shown in FIGURE 1A. The LSK fraction can be split into three subsets by expression of CD48 and CD150. Around 0.0055% of total BM is $CD150^+CD48^-LSK$, whereas $CD150^+CD48^+LSK$ and $CD150^-CD48^+LSK$ are 0.0066% and 0.039%, respectively. Each of these three subsets can be further divided on the basis of CD34 and CD135 (FLT3 receptor) expression. Most interestingly, only around 23% of the $CD150^+CD48^-$ subset of LSK cells do not express CD34. On the basis of single-cell reconstitution studies, this should represent an HSC subset with a repopulating activity superior to those using the SLAM or CD34 markers alone (TABLE 1).[7,9] The $CD34^-CD150^+CD48^-LSK$ population comprises

TABLE 1. Summary of single-cell BM reconstitution data

Study no.	Phenotype of transferred cell	Frequency of reconstitution (%)	Reference
1	LSK, CD150$^+$CD48$^-$	47	Kiel et al. (2005)[9]
2	CD41$^-$CD150$^+$CD48$^-$	45	Kiel et al. (2005)[9]
3	LSK, SP^{lo}, Thy1.1^{lo}, CD34$^-$, CD135$^-$	35	Camargo et al. (2006)[11]
4	Lin^{neg}, SP, Rho^{lo}	33	Uchida et al. (2003)[10]
5	LSK, CD34$^{lo/-}$	22	Osawa et al. (1996)[7]
6	CD150$^+$CD48$^-$	21	Kiel et al. (2005)[9]
7	LSK, Thy1.1^{lo}	18	Wagers et al. (2002)[12]

LSK = Lin$^-$Sca1$^+$c-Kit$^+$; SP = side population (can efflux Hoechst 33342); SP^{lo} = TIP of SP; Rho: Rhodamine.

only 0.00125% (12 cells/million) of total BM (FIG. 1B; TABLE 2), and likely represents the most dormant HSC population based on BrdU label retaining assays (Wilson and Trumpp, unpublished data). Strikingly, CD34$^-$ cells were only found in the CD150$^+$CD48$^-$LSK subset, suggesting that acquisition of CD34 expression on the cell surface is one of the earliest events during activation of HSCs from their dormant state (FIG. 1B). Interestingly, actively expanding HSCs, such as the earliest definitive HSCs generated in the embryo and later on present in the fetal liver as well as G-CSF mobilized adult HSCs, are all CD34$^+$.[13–15] We therefore propose that the acquisition of CD34 expression in the CD150$^+$CD48$^-$LSKs represents the transition from dormant to actively self-renewing adult HSCs (FIG. 1B).

The first step in initiating the differentiation of HSCs is likely to be the appearance of CD48 cell-surface expression (multipotent progenitor 1 [MPP1]). This would be followed by the loss of CD150 (MPP2) and gain of CD135 expression (MPP3), which has been functionally associated with differentiation toward a lymphoid primed multilineage progenitor.[16] However, this is still controversial,[17] and the exact branch points (MPP1–3) at which lineage commitment occurs still remains to be elucidated (FIG. 1B).[18] Taken together, we propose a scheme in which a dormant/quiescent CD34$^-$ HSC is on top of the hierarchy (FIG. 1B and TABLE 2). This HSC probably does not contribute to the day-to-day production of progenitors, but rather serves as a reserve pool of stem cells that are only activated on injury. Self-renewing HSCs are CD34$^+$ andare those from which progenitors arise during normal homeostasis. For unknown reasons, these activated HSCs are less efficient in functional repopulation assays that typically introduce HSCs by intravenous (i.v.) injection into conditioned hosts.[7] However, this observed deficiency may be due to technical aspects of this assay, whereby homing, survival, extravasation, and lodging of HSCs into niches are all required to achieve reconstitution, but are not essential features of endogenous (i.e., nontransplanted) HSCs. Intrafemoral injections

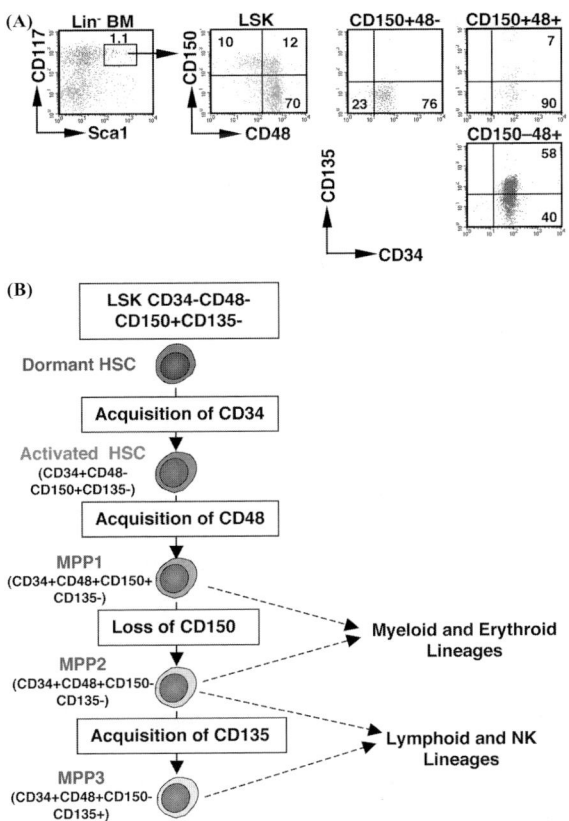

FIGURE 1. (A) Isolation of hematopoietic stem and progenitor subsets from mouse bone marrow. Mouse bone marrow cells were depleted of all lineage (lin) positive cells with a cocktail of monoclonal antibodies (CD3, CD4, CD8, CD11b, CD19, CD161, Gr1, Ter119) by negative selection on goat anti-rat Ig magnetic beads. The resultant population (lin$^-$) was then stained with fluorescent conjugates to CD117 (c-Kit), Sca1, CD48, CD150, CD34, and CD135 (FLT3R) and analyzed on a FACS Canto flow cytometer (Becton Dickinson, San Jose, CA). After gating on the Lin$^-$Sca1$^+$CD117/c-Kit$^+$ (*LSK*) population (*left panel*), the CD48 versus CD150 profile is shown (*second panel from left*). The resulting three subsets can be further subdivided on the basis of CD34 and CD135 staining (*right panels*). The relative percentages of each subset within each two-parameter histogram are shown and are representative of several independent experiments. Total cell numbers per mouse of each of these populations are shown in TABLE 2. **(B)** Proposed hierarchy of adult bone marrow stem cell subsets in the mouse. A "dormant" hematopoietic stem cell (HSC) with the phenotype LSK CD34$^-$CD48$^-$CD150$^+$CD135$^-$ is placed at the top of this hierarchy. On acquisition of CD34 (consistent with activation), a dormant HSC becomes an "activated" HSC. This activated HSC sequentially loses surface expression of first CD48, and then CD150, and then upregulates CD135, giving rises to a series of multipotent progenitor (MPP) populations, here termed MPP1, 2, and 3. MPPs, in turn give rise to mature cells of the myeloid, erythroid, lymphoid, and NK lineages.

TABLE 2. Relative abundance and total number of BM HSC subsets per mouse (fore and hind legs, vertebrae, sternum and skull included, assuming around 300×10^6 cells)

Subset	Phenotype	Percentage of total BM	Relative abundance	No. of total BM
Dormant HSC	$CD34^-CD48^-$ $CD135^-CD150^+LSK$	0.00125	1	3,750
Activated HSC	$CD34^+CD48^-$ $CD135^-CD150^+LSK$	0.00425	3.4	12,750
MPP1	$CD34^+CD48^+$ $CD135^-CD150^+LSK$	0.0065	5.2	19,500
MPP2	$CD34^+CD48^+$ $CD135^-CD150^-LSK$	0.016	12.8	48,000
MPP3	$CD34^+CD48^+$ $CD135^+CD150^-LSK$	0.0225	18	67,500

BM = bone marrow; HSC = hematopoietic stem cell; MPP = multipotent progenitor; LSK = $Lin^-Sca1^+c-Kit^+$.

have been shown to be more efficient for assaying human HSC activity,[19] thus it will be interesting to see whether activated $CD34^+$ mouse HSCs would show long-term reconstitution activity if injected directly into the BM rather than intravenously.

BONE MARROW STEM CELL NICHES

Although during homeostasis, a small minority of HSCs are present in the circulation,[20] the vast majority of HSCs are found in the BM and are thought to be localized in specific microenvironments called "stem-cell niches."[21] In general, a stem-cell niche can be defined as a three-dimensional spatial structure in which one or several stem cells are housed. Several different types of niche may exist. The most obvious function of a stem cell niche is maintenance of stem cell numbers by allowing self-renewal in the absence of differentiation. This type of niche may be thought of as a "self-renewing" or "homeostatic" niche.[21] It is widely assumed that regulation of this balance between HSC self-renewal and differentiation is, at least in part, dependent on the interactions between HSCs and the cells comprising the niches. Moreover, the maintenance of long-term dormant HSCs, which may serve as a reserve pool of stem cells to be activated in response to injury signals, is thought to occur in "dormant" niches[21,22] (FIG. 2). We are just beginning to understand the basic principles of HSC–niche interactions as well as the molecular and cellular basis of stem cell niches.

As so often occurs, progress was initially made in invertebrate models such as *Drosophila*, where it has been shown that germline stem cell production and function in the testis and ovary are controlled by specific interactions between stem and niche cells, and involve cell adhesion and secreted short range

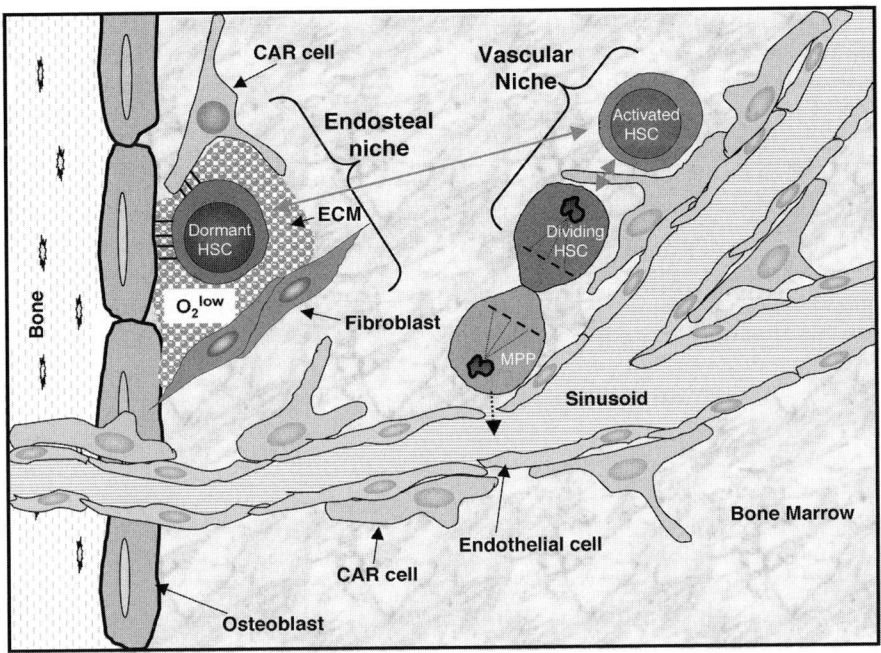

FIGURE 2. Model showing the two different locations of hematopoietic stem cells (HSCs) in the mouse bone marrow. Dormant HSCs are in direct contact with specialized osteoblasts lining the endosteal surface of the bone. These cells are also in contact with CXCL12-abundant reticular (CAR) cells that are found at the endosteum, but are more frequently associated with sinusoids. In the "endosteal niche" or "dormant niche," CAR cells and osteoblasts together with stromal fibroblasts and potentially other cell types such as osteoclasts (not shown) generate a hypoxic (O_2^{low}) environment with a dense extracellular matrix (ECM) that retains the dormancy of HSCs. Dormant HSCs are activated stochastically or in response to injury signals. Activated HSCs are located adjacent to perivascular CAR cells near sinusoids forming the vascular niche. On a self-renewing division, asymmetry is generated, which leads to the generation of two daughter cells: a multipotent progenitor (MPP) and an activated HSC. MPPs and/or their progeny enter the circulation through the fenestrated endothelium of sinusoidal microvasculature.

signaling molecules.[23] In mammals, the best-characterized mammalian niche is the bulge region adjacent to the hair follicle of the skin harboring epidermal stem cells and melanocyte stem cells.[24–26] In contrast to what is known about the bulge niche, the localization and cellular composition of niches harboring HSCs is still poorly understood. The main reason for this paucity of information is that unlike most other stem cells, HSCs are not spatially embedded in a solid tissue structure, but are migratory and hidden inside bones. Nevertheless, significant progress has been made in the last years and a number of techniques that have been used to localize HSCs *in situ* are summarized below. According to these studies, putative nonhematopoietic cell types found to be in close contact to HSCs, and thus likely to be part of the HSC niche have been identified. These include specialized osteoblasts and CXCL12-abundant reticular (CAR) cells often associated with the microvasculature.[21,27,28] In the BM two types of niches have been identified. First, the endosteal niche located at the endosteum (inner cellular lining of the bone cavities interfacing with the BM) of trabecular bone and comprising specialized osteoblasts, CAR cells osteoclasts, and stromal fibroblasts.[21,26,28,29] And second, the vascular niche where HSCs are found next to CAR cells adjacent to BM sinusoids, which are low-pressure vessels with a fenestrated endothelium.[9,27] Below we review the key techniques used to identify these niches and present a model suggesting a distinct function for both BM microenvironments.

HSC LOCALIZATION USING HOMING ASSAYS

This assay takes advantage of the known capacity of HSCs to home to and engraft into BM niches after injection into the circulation of conditioned (i.e., irradiated) or unconditioned recipients. To visualize the HSCs and the sites to which they home, FACS-enriched HSC/progenitor cells are labeled with fluorescence dyes (i.e., CFSE) and injected i.v. into recipient mice. Transplanted HSCs can be identified by their CFSE fluorescence on histological bone sections about 15 h after transfer. These studies have revealed a significant enrichment of CFSE-labeled stem/progenitor cells at, or near, the endosteum of trabecular bone.[30] HSC-homing assays have also been combined with staining for markers expressed on osteoblasts. These studies have revealed that homed HSC/progenitors at the endosteum are in close contact to osteopontin-, N-cadherin-, and BMP-receptor 1-expressing osteoblasts,[31] further supporting the concept that specialized osteoblasts are part of the endosteal niche. A further advantage of this type of assay is that it is relatively easy, can be combined with markers expressed on putative niche stromal cells, and is rapid compared with classical, functional stem cell assays. Nevertheless, it is important to note that homing of stem cells after i.v. transfer may not reflect endogenous niche location, given that irradiation damage may alter the BM vasculature, thereby facilitating or altering the homing process; thus the localization of HSCs using this technique may not always be specific.

HSC IDENTIFICATION BY IMMUNOLOCALIZATION

Although it is possible to highly purify HSCs by multicolor flow cytometry, the combination of multiple fluorescence dyes typically used (see above) is too complex for localization of HSCs by immunostainings on bone sections using fluorescence microscopy. However, the recent discovery that HSCs can also be highly purified on the basis of the expression of two SLAM family receptors CD150 and CD48 ($CD150^+CD48^-$) has made immunolocalization more feasible. Moreover, further exclusion of CD41 expressing megakaryocytic lineage cells reveals a population of cells that gives 45% long-term multilineage reconstitution if transplanted as single cells.[9] Using these markers (CD150, CD48, and CD41), it has become possible to identify cells very likely to be functional HSCs by immunofluorescence. Surprisingly, about two-thirds of $CD150^+CD48^-CD41^-$ HSCs were found adjacent to sinusoids (vascular niche) with most of the others present at the endosteum as expected.[9] In a subsequent study, the cell type that $CD150^+CD48^-CD41^-$ HSCs are in contact with at the vascular niche were identified as CAR cells.[27] CAR cells express high levels of the CXCL-12 (previously called stromal-derived factor 1, SDF-1) chemokine that activates the G protein–coupled receptor CXCR4 expressed on HSCs. Interestingly, the CXCL12-CXCR4 pathway controls a number of HSC functions including migration during homing and mobilization, as well as adhesion to the niche. In addition, CXCR4 itself is required for long-term maintenance of HSC activity.[6,27] Not only does the highest transcriptional activity for CXCL12 in the BM occur in CAR cells, but they are also the cell type with which $CD150^+CD48^-CD41^-$ HSCs are most frequently associated.[27] Interestingly, although CAR cells were mostly found near the sinusoids, a significant number were also located near to, and at the endosteum, raising the possibility that CAR cells may provide a cellular and possible molecular link between vascular and endosteal niches. However, as CAR cells also serve as a niche for BM B cells,[32] it is unlikely that these cells alone are sufficient to define HSC niches. This raises the possibility that BM niches comprise several different cell types, including osteoblasts, CAR cells, and possibly osteoclasts, stromal fibroblasts, and endothelial cells. Although the use of complex immunostaining has provided immense insight into the localization of putative HSCs *in situ*, this technique, like any other, has limitations. The identification of HSCs using a complex set of markers by immunofluorescence relies on both positive (CD150) and negative signals (CD48 and CD41), which are critical to exclude $CD150^+$ cells that are not stem cells. Although negatively stained cells can readily be discriminated by FACS, identification of negative cells is very problematic by nonquantitative immunofluorescence on fixed and embedded tissue sections, questioning the reliability of identifying HSCs by negative markers.

HSC IDENTIFICATION BY LABEL-RETAINING ASSAYS

A classical method that has been used recently to identify the "bulge" niche in the skin epidermis, and which was initially used to localize putative intestinal epithelial stem cells is the so-called label retaining assay.[25,33] This assay takes advantage of the fact that most adult stem cells divide rather infrequently and can be quiescent or dormant for weeks or even months. The DNA of adult stem cells is first labeled *in vivo* by nucleotide analogues such as ^{3}H-thymidine or BrdU or by the histone H2B–enhanced green-fluorescent-protein (H2B-GFP) fusion protein.[25] After the cell-labeling phase, the animals are "chased" for several months in the absence of the DNA labeling agent. Dividing cells rapidly dilute out the label, which is therefore lost. In contrast, long-lived quiescent cells such as stem cells retain the label for months and can be identified as "label retaining cells" (LRCs) by immunohistochemistry.[25,33,34] Although LRCs can also be identified in the BM, this tissue, in contrast to the skin epidermis, also harbors a number of long-lived quiescent cell types. Thus, the majority of BM LRCs are in fact memory T cells and NK cells and not HSCs (Wilson *et al.*, unpublished data). We have combined this technique with six-color flow cytometry and found that only about 6% of BrdU$^+$LRCs present in the BM have an HSC phenotype. Therefore, anti-BrdU staining needs to be combined with other markers not expressed on memory T and NK cells, (for example c-Kit), to reliably identify HSCs on histological bone sections (Wilson *et al.*, unpublished data). One study in particular used Lin$^-$BrdU$^+$LRCs to show that putative HSCs are much more frequently (sixfold) located at the endosteum compared with the center of the BM.[28] The above-mentioned label-retaining assays both confirm and extend previous classical studies in which morphological criteria and colony-forming assays were used to show that HSCs are enriched in the endosteal region compared with the central part of the BM, which is rather populated by progenitors and more differentiated cell types.[35,36] In addition, it was shown that BrdU$^+$LRCs expressing c-Kit could be in direct contact with osteoblasts at the endosteum, and that both of these cell types express the cell adhesion anchor N-cadherin.[28] This led to the hypothesis that HSCs are anchored to niche osteoblasts via homotypic N-cadherin interactions. This hypothesis is strengthened by the findings that Angipoietin-Tie2 signaling, as well as c-*Myc*, regulate both N-cadherin expression and HSC–niche interactions.[31,37] However, genetic data are still required to prove that N-cadherin alone serves as the HSC niche anchor. Nevertheless, it is more likely that N-cadherin is an important component of a cell adhesion network that also includes other cell adhesion molecules such as integrins and selectins, which together control HSC adhesion to the niche.

GENETIC EVIDENCE FOR THE ROLE OF THE NICHE

Although some data suggest that a substantial number of HSCs are present in peri-vascular niches next to sinusoids, most of the genetic data to date support the hypothesis that osteoblasts at the endosteum are a crucial component of the BM niche (reviewed in Ref. 21). For example, the increase in osteoblasts observed on conditional deletion of BMPR1a[28] correlates with an increased number of HSCs and the elimination of osteoblasts by using an *in vivo* cell-ablation technique causes HSCs to exit the BM and induces extramedullary hematopoiesis.[38] In addition, activation of the parathyroid hormone receptor on osteoblasts increases HSC number, potentially mediated by increased Notch signaling.[39,40] Furthermore, the high calcium concentrations found next to the endosteum are important for HSC localization to the endosteal niche,[41] and osteoblasts lacking membrane bound stem cell factor, which activates the c-Kit receptor on HSCs, are unable to maintain functional HSCs *in vivo*.[42,43]

CONCLUDING REMARKS

There is overwhelming evidence supporting an important role for specific osteoblasts in the formation of the endosteal BM HSC niche. However, other cell types are certainly involved as well. These include CAR cells,[27] osteoclasts,[29] and endothelial cells. In addition, it is highly likely that several types of niche, each serving different roles, may exist in the BM (FIG. 2). For example, self-renewing HSCs, potentially $CD34^+CD150^+CD48^-$LSKs (FIG. 1B), that are actively involved in the daily production of progenitors may be present in the vascular niches in close contact to the circulation. In contrast, long-term dormant $CD34^-CD150^+CD48^-$LSKs may represent a reserve pool of HSCs that are "activatable" on injury may rather be located at the endosteum (FIG. 2). This would also be in agreement with the hypothesis that dormant niches exist in hypoxic areas (distant from the vessels) because HSCs are especially sensitive to reactive oxygen species.[44] As CAR cells are a part of both endosteal and vascular niches, they may serve as a cellular link between both these niches (FIG. 2). The new field of stem cell niches is developing fast, and a more comprehensive cellular and molecular understanding of niches will become essential to enable clinical influencing of stem cell function by modifying niche activities, a strategy for which first examples have just been reported.[40]

ACKNOWLEDGMENTS

This work was supported in part by postdoctoral fellowships from EMBO (M.E.) and Roche (C.A.), and grants to A.T. from the Swiss National Science Foundation and the Swiss Cancer League.

REFERENCES

1. WEISSMAN, I.L. 2000. Stem cells: units of development, units of regeneration, and units in evolution. Cell **100:** 157–168.
2. FUCHS, E., T. TUMBAR & G. GUASCH. 2004. Socializing with the neighbors: stem cells and their niche. Cell **116:** 769–778.
3. JOSEPH, N.M. & S.J. MORRISON. 2005. Toward an understanding of the physiological function of Mammalian stem cells. Dev. Cell. **9:** 173–183.
4. MURPHY, M.J., A. WILSON & A. TRUMPP. 2005. More than just proliferation: Myc function in stem cells. Trends Cell. Biol. **15:** 128–137.

5. SHIZURU, J.A., R.S. NEGRIN & I.L. WEISSMAN. 2005. Hematopoietic stem and progenitor cells: clinical and preclinical regeneration of the hematolymphoid system. Annu. Rev. Med. **56:** 509–538.
6. KOLLET, O., A. DAR & T. LAPIDOT. 2007. The multiple roles of osteoclasts in host defense: bone remodeling and hematopoietic stem cell mobilization. Annu. Rev. Immunol. **25:** 51–69.
7. OSAWA, M. *et al.* 1996. Long-term lymphohematopoietic reconstitution by a single CD34- low/negative hematopoietic stem cell. Science **273:** 242–245.
8. DREW, E. *et al.* 2005. CD34 and CD43 inhibit mast cell adhesion and are required for optimal mast cell reconstitution. Immunity **22:** 43–57.
9. KIEL, M.J. *et al.* 2005. SLAM family receptors distinguish hematopoietic stem and progenitor cells and reveal endothelial niches for stem cells. Cell **121:** 1109–1121.
10. UCHIDA, N. *et al.* 2003. Different *in vivo* repopulating activities of purified hematopoietic stem cells before and after being stimulated to divide *in vitro* with the same kinetics. Exp. Hematol. **31:** 1338–1347.
11. CAMARGO, F.D. *et al.* 2006. Hematopoietic stem cells do not engraft with absolute efficiencies. Blood **107:** 501–507.
12. WAGERS, A.J. *et al.* 2002. Little evidence for developmental plasticity of adult hematopoietic stem cells. Science **297:** 2256–2259.
13. SANCHEZ, M.J. *et al.* 1996. Characterization of the first definitive hematopoietic stem cells in the AGM and liver of the mouse embryo. Immunity **5:** 513–525.
14. OGAWA, M. *et al.* 2001. CD34 expression by murine hematopoietic stem cells. Developmental changes and kinetic alterations. Ann. N. Y. Acad. Sci. **938:** 139–145.
15. FURNESS, S.G. & K. MCNAGNY. 2006. Beyond mere markers: functions for CD34 family of sialomucins in hematopoiesis. Immunol. Res. **34:** 13–32.
16. ADOLFSSON, J. *et al.* 2005. Identification of Flt3+ lympho-myeloid stem cells lacking erythro-megakaryocytic potential a revised road map for adult blood lineage commitment. Cell **121:** 295–306.
17. FORSBERG, E.C. *et al.* 2006. New evidence supporting megakaryocyte-erythrocyte potential of flk2/flt3+ multipotent hematopoietic progenitors. Cell **126:** 415–426.
18. AKASHI, K. 2007. Cartography of hematopoietic stem cell commitment dependent upon a reporter for transcription factor activation. Ann. N. Y. Acad. Sci. [Epub ahead of print]
19. MAZURIER, F. *et al.* 2003. Rapid myeloerythroid repopulation after intrafemoral transplantation of NOD-SCID mice reveals a new class of human stem cells. Nat. Med. **9:** 959–963.
20. WRIGHT, D.E. *et al.* 2001. Physiological migration of hematopoietic stem and progenitor cells. Science **294:** 1933–1936.
21. WILSON, A. & A. TRUMPP. 2006. Bone-marrow haematopoietic-stem-cell niches. Nat. Rev. Immunol. **6:** 93–106.
22. ADAMS, G.B. & D.T. SCADDEN. 2006. The hematopoietic stem cell in its place. Nat. Immunol. **7:** 333–337.
23. OHLSTEIN, B. *et al.* 2004. The stem cell niche: theme and variations. Curr. Opin. Cell. Biol. **16:** 693–699.
24. NISHIMURA, E.K. *et al.* 2002. Dominant role of the niche in melanocyte stem-cell fate determination. Nature **416:** 854–860.

25. TUMBAR, T. et al. 2004. Defining the epithelial stem cell niche in skin. Science **303:** 359–363.
26. MOORE, K.A. & I.R. LEMISCHKA. 2006. Stem cells and their niches. Science **311:** 1880–1885.
27. SUGIYAMA, T. et al. 2006. Maintenance of the hematopoietic stem cell pool by CXCL12-CXCR4 chemokine signaling in bone marrow stromal cell niches. Immunity **25:** 977–988.
28. ZHANG, J. et al. 2003. Identification of the haematopoietic stem cell niche and control of the niche size. Nature **425:** 836–841.
29. KOLLET, O. et al. 2006. Osteoclasts degrade endosteal components and promote mobilization of hematopoietic progenitor cells. Nat. Med. **12:** 657–664.
30. NILSSON, S.K., H.M. JOHNSTON & J.A. COVERDALE. 2001. Spatial localization of transplanted hemopoietic stem cells: inferences for the localization of stem cell niches. Blood **97:** 2293–2299.
31. WILSON, A. et al. 2004. c-Myc controls the balance between hematopoietic stem cell self-renewal and differentiation. Genes Dev. **18:** 2747–2763.
32. TOKOYODA, K. et al. 2004. Cellular niches controlling B lymphocyte behavior within bone marrow during development. Immunity **20:** 707–718.
33. COTSARELIS, G., T.T. SUN & R.M. LAVKER. 1990. Label-retaining cells reside in the bulge area of pilosebaceous unit: implications for follicular stem cells, hair cycle, and skin carcinogenesis. Cell **61:** 1329–1337.
34. POTTEN, C.S. & M. LOEFFLER. 1990. Stem cells: attributes, cycles, spirals, pitfalls and uncertainties. Lessons for and from the crypt. Development **110:** 1001–1020.
35. LORD, B.I., N.G. TESTA & J.H. HENDRY. 1975. The relative spatial distributions of CFUs and CFUc in the normal mouse femur. Blood **46:** 65–72.
36. GONG, J.K. 1978. Endosteal marrow: a rich source of hematopoietic stem cells. Science **199:** 1443–1445.
37. ARAI, F. et al. 2004. Tie2/angiopoietin-1 signaling regulates hematopoietic stem cell quiescence in the bone marrow niche. Cell **118:** 149–161.
38. VISNJIC, D. et al. 2004. Hematopoiesis is severely altered in mice with an induced osteoblast deficiency. Blood **103:** 3258–3264.
39. CALVI, L.M. et al. 2003. Osteoblastic cells regulate the haematopoietic stem cell niche. Nature **425:** 841–846.
40. ADAMS, G.B. et al. 2007. Therapeutic targeting of a stem cell niche. Nat. Biotechnol. **25:** 238–243.
41. ADAMS, G.B. et al. 2006. Stem cell engraftment at the endosteal niche is specified by the calcium-sensing receptor. Nature **439:** 599–603.
42. BARKER, J.E. 1994. Sl/Sld hematopoietic progenitors are deficient *in situ*. Exp. Hematol. **22:** 174–177.
43. BARKER, J.E. 1997. Early transplantation to a normal microenvironment prevents the development of Steel hematopoietic stem cell defects. Exp. Hematol. **25:** 542–547.
44. ITO, K. et al. 2006. Reactive oxygen species act through p38 MAPK to limit the lifespan of hematopoietic stem cells. Nat. Med. **12:** 446–451.

Cartography of Hematopoietic Stem Cell Commitment Dependent upon a Reporter for Transcription Factor Activation

KOICHI AKASHI

Department of Cancer Immunology and AIDS, Dana-Farber Cancer Institute, Boston, Massachusetts 02115 USA; and Center for Cellular and Molecular Medicine, Kyushu University Hospital, 812-8582 Fukuoka, Japan

ABSTRACT: A hierarchical hematopoietic developmental tree has been proposed based on the result of prospective purification of lineage-restricted progenitors. For more detailed mapping for hematopoietic stem cell (HSC) commitment, we tracked the expression of PU.1, a major granulocyte/monocyte (GM)- and lymphoid-related transcription factor, from the HSC to the myelolymphoid progenitor stages by using a mouse line harboring a knockin reporter for PU.1. This approach enabled us to find a new progenitor population committed to GM and lymphoid lineages within the HSC fraction. This result suggests that there should be another developmental pathway independent of the conventional one with myeloid versus lymphoid bifurcation, represented by common myeloid progenitors and common lymphoid progenitors, respectively. The utilization of the transcription factor expression as a functional marker might be useful to obtain cartography of the hematopoietic development at a higher resolution.

KEYWORDS: hematopoietic stem cell; commitment; transcription factor

INTRODUCTION

Prospective purification of population with limited lineage potential is critical to understand the hierarchical development of hematopoiesis initiating from hematopoietic stem cells (HSCs). HSCs were first identified within the lineage antigen negative (Lin$^-$) Sca-1$^+$Thy-1^{lo} fraction in the mouse bone marrow by using a multicolor FACS.[1] Thereafter, HSC subclasses and their downstream progenitors have been identified by adding new cell-surface markers.[2,3] The

---

Address for correspondence: Koichi Akashi, Smith 770, 44 Binney St., Boston, MA 02115. Voice: 617-632-3595; fax: 617-632-3809.
koichi_akashi@dfci.harvard.edu

hematopoietic differentiation map, which is based on these prospectively isolated stem and progenitor cells, has greatly contributed toward the understanding of hematopoietic development.

There is a general agreement that the most primitive HSC with long-term (LT) self-renewal potential can be purified within the $Lin^-Sca\text{-}1^+c\text{-}Kit^+$ (LSK) fraction of the $Thy1^{lo}$ or $CD34^-$ profile[4] (FIG. 1A). These cells can reconstitute hematopoiesis at the single cell level after transplantation for >6 months,[5] and therefore are called as LT-HSCs.[2] In contrast, $Thy1^-$ or $CD34^+$ LSK cells are termed as short-term (ST)-HSCs or multipotential progenitors since they are capable of reconstituting multilineages only for ~3 months.[2,5] Lineage commitment has been considered to occur after the ST-HSC stage. In the myeloid pathway, common myeloid progenitors (CMPs), granulocyte/monocyte progenitors (GMPs), and megakaryocyte/erythrocyte progenitors (MEPs) have been purified within the $Lin^-Sca\text{-}1^-c\text{-}Kit^+$ fraction[6] (FIG. 1A). In the lymphoid pathway, common lymphoid progenitors (CLPs) have been identified as the $IL\text{-}7R\alpha^+Lin^-Sca\text{-}1^{lo}c\text{-}Kit^{lo}$ population.[7] Based on the existence of these prospectively isolated progenitor populations, the lymphoid and myeloid developmental pathways have been proposed to be generally independent.[4] These progenitors exist in the bone marrow as a result of hematopoietic lineage commitment, and therefore, the differentiation scheme established by arraying these progenitors should be revised if a new progenitor population could be isolated.

CONTROVERSY IN EARLY LINEAGE COMMITMENT WITHIN THE HSC FRACTION

Several studies, however, have shown that lineage commitment could already have initiated at the ST-HSC stage. The ST-HSCs might consist of

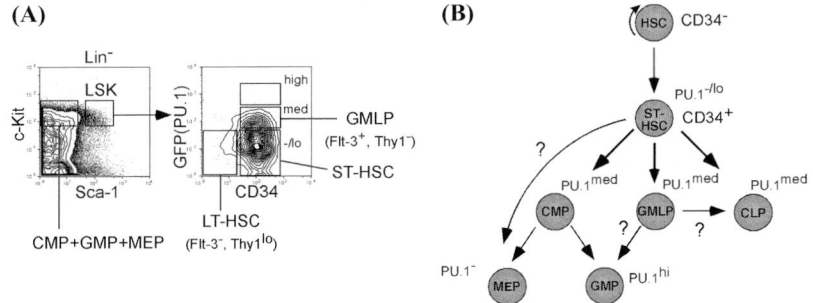

FIGURE 1. (**A**) The $Lin^-Sca\text{-}1^+c\text{-}Kit^+$ (LSK) population was subdivided into three subpopulations by using additional markers, such as CD34 and PU.1-GFP. The progenitor population with restricted differentiation activity to granulocyte/monocyte/lymphoid lineages (GMLP) was isolated as $CD34^+PU.1^{med}$ LSK cells. (**B**) The proposed lineage commitment scheme by tracking PU.1-GFP expression.

heterogeneous progenitor populations. Igarashi et al.[8] reported that the CD34⁻ LSK fraction contained a small population activating RAG-1, a lymphoid-specific gene. The RAG-1⁺ LSK population differentiated mainly into the lymphoid lineage but retrained a weak GM potential, and was termed as the earliest lymphoid progenitor (ELP).[8] Recently, two independent groups have reported that Flt-3, a tyrosine kinase receptor, is expressed in a fraction of ST-HSCs but not in LT-HSCs.[9,10] Each group further studied the detailed differentiation activity of Flt-3⁺ LSK cells, and reached different conclusions. Adolfsson et al.[11] subdivided CD34⁺ LSK cells into Flt-3⁺ and Flt-3⁻ fractions, and reported that although Flt-3⁻CD34⁺ LSK cells were multipotent, Flt-3⁺CD34⁺ LSK population maintains the GM and the T/B lymphoid but not the megakaryocyte/erythrocyte (MegE) potential.[11] The loss of MegE potential in the Flt-3⁺CD34⁺ LSK population was demonstrated by both *in vitro* and *in vivo* studies: Flt-3⁺CD34⁺ LSK cells gave rise to GM but rarely formed MegE colonies, and MegE progeny from the Flt-3⁺CD34⁺ LSK cells was not seen on 7–9 days post transplantation. According to this result, they proposed that in addition to the lymphoid versus myeloid developmental pathway represented by CLPs and CMPs, respectively, there is a critical developmental stage common to GM, T, and B lymphoid cells.[11] We have also found that RAG-1-activating ELPs are Flt-3⁺, constituting a fraction of the Flt-3⁺CD34⁺ LSK population (unpublished data). The other group has shown that Flt-3⁻ LSK cells (LT-HSCs) are entirely Thy1ˡᵒ, whereas the Flt-3⁺ LSK cell population contains both Thy1⁻ and Thy1ˡᵒ cells. Flt-3⁺Thy1ˡᵒ LSK cells are multipotent ST-HSCs. In contrast to Adolfsson's report, Flt-3⁺Thy1⁻ LSK cells could generate MegE colonies in the spleen, and substantial numbers of circulating platelets were detected when analyzed 12–16 days after transplantation, claiming that the stage common to GM/lymphoid lineages does not constitute a major pathway for hematopoietic development.[12]

Such discrepancies could be simply due to the difference in timing of evaluation for MegE readouts after transplantation in these studies as Forsberg et al. claim.[12] It is also possible that this reflects the functional difference between the Flt-3⁺CD34⁺ and the Flt-3⁺Thy1⁻ LSK populations. These populations should be substantially overlapped, but it is difficult to precisely compare populations purified based on different combinations of antibodies in independent labs.[3] These previous reports urged us to further subdivide the LSK population by using transcription factor expression as a functional marker.

A HIGH-RESOLUTION MAPPING OF HSC COMMITMENT BY A REPORTER FOR PU.1 ACTIVATION

We are currently further subdividing the LSK fraction by using mouse lines having a reporter for RAG-1,[13] Lysozyme M,[14] PU.1,[15] or GATA-1.[16] Here, we briefly summarize the data of the PU.1 reporter mouse. The Ets family transcription factor PU.1 (Spi-1) is one of the most important regulators of GM and lymphoid lineage development. PU.1 transactivates a number of GM (i.e., myeloperoxidase, lysozyme M, G-CSFR, and GM-CSFRα), and lymphoid-related genes (i.e., IL-7Rα).[17-20] Conditional disruption of PU.1 from hematopoiesis has revealed that PU.1 is necessary for development of early GM and lymphoid progenitors, such as CMPs, GMPs, and CLPs in adult hematopoiesis, but not for the MEP development.[15] We thus analyzed a mouse line possessing EGFP knocked into the PU.1 locus[15] in which the EGFP level should represent the endogenous activation of PU.1 transcription.

In the PU.1-EGFP knockin mice, GMPs expressed a high level of PU.1-EGFP, and MEPs were negative for PU.1-EGFP. The CD34$^-$ LSK (LT-HSC) population expressed negative to low levels of PU.1-GFP (PU.1$^{-/lo}$), while CD34$^+$ LSK cells expressed a wide range of PU.1-GFP (FIG. 1A). Almost 20% of CD34$^+$ LSK cells expressed a medium level of PU.1-GFP (PU.1^{med}). A fraction of PU.1$^{-/lo}$CD34$^+$ LSK cells expressed a low level of Thy1. The PU.1$^{-/lo}$CD34$^+$ LSK population consisted of Flt-3$^+$ and Flt-3$^-$ cells. PU.1^{med} CD34$^+$ LSK cells were all Flt-3$^+$ but did not express Thy-1. Thus, the PU.1^{med}CD34$^+$ LSK cells should constitute a minor population within either the Flt-3$^+$CD34$^+$ LSK[11] or the Flt-3$^+$Thy1$^-$ LSK population.[12]

PU.1$^{-/lo}$CD34$^+$ LSK cells formed a number of mixed myeloerythroid colonies, and displayed multilineage reconstitution, which lasts only for ~3 months. PU.1$^{-/lo}$CD34$^+$Flt-3$^+$ LSK cells formed abundant GM colonies, but were also able to form a few percent of MegE colonies. In contrast, the PU.1^{med}CD34$^+$ (Flt-3$^+$Thy-1$^-$) LSK cells did not form MegE colonies in the presence of myeloid cytokine cocktail containing stem cell factor (SCF), interleukin-3 (IL-3), granulocyte/macrophage colony-stimulating factor (GM-CSF), erythropoietin (Epo), and thrombopoietin (TPO). PU.1^{med}CD34$^+$ LSK cells possessed potent T and B cell potential: 1 in 20 PU.1^{med}CD34$^+$ LSK cells could readout T and B cell differentiation on the OP9-δ and the OP9 stromal layers, respectively.

PU.1^{med}CD34$^+$ LSK cells lacked MegE potential even *in vivo*. These cells did not form day 8 or day 12 CFU-S that is mainly composed of MegE lineage cells.[21] We then crossed the PU.1-EGFP knockin with the GPIIb-EYFP knockin mouse line[22] to mark their platelets with EYFP. PU.1$^{-/lo}$ but not PU.1^{med}CD34$^+$ LSK cells generated EYFP$^+$ platelets on day 16, indicating that the PU.1^{med}CD34$^+$ LSK fraction cannot contribute to the platelet production even *in vivo*. These data clearly demonstrate that at least a fraction of CD34$^+$Flt-3$^+$ LSK cells expressing a high level of PU.1 have lost MegE potential while retaining GM/lymphoid differentiation activity.

If the GM/lymphoid progenitor (GMLP) stage is the alternative pathway for GMPs and CLPs, it is difficult to draw a simple developmental tree integrating CMPs and MEPs. Adolfsson *et al*.[11] proposed that there might be a pathway to MEPs directly from the CD34$^+$ ST-HSC stage (FIG. 1B). CMPs coexpressing both GM and MegE lineage genes, but not lymphoid ones are isolatable downstream of ST-HSCs.[6,14] CMPs may also be able to develop directly from the ST-HSC fraction. Does the complexity mean that the commitment process is at random or stochastic, rather than sequential or ordered? Our data suggest that reporters for transcription factors should be useful for isolation of functional populations within early hematopoiesis. As we have successfully isolated GMLPs within the LSK fraction by using PU.1 activation as a marker, the mouse line having reporters for other transcription factors should be used to isolate committed progenitors within the ST-HSC fraction. We should hold the "stochastic" or "deterministic" debate until we could separate all committed progenitors and clarify their lineal relationships.

Thus, the use of a functional PU.1 activation as a cell isolation marker revealed the existence of progenitor population that has lost MegE but retains GM/lymphoid potential, within the $CD34^+$ or $Thy\text{-}1^-$ LSK multipotential ST-HSC or progenitors. This raises a possibility that lineage commitment has already initiated within the "ST-HSC" LSK fraction, and suggests that other unknown committed progenitors may also be isolatable by using additional functional markers. Of course, the commitment process might occur in a continuum manner rather than abruptly. In fact, in visualizing stem and progenitor cells on multicolor FACS, each population continuously exists within the dot plot. In order to purify a "functionally homogenous" population, one need to arbitrarily draw a line to set a gate that encompasses a certain range of levels for antigen expression in each experiment. Nonetheless, it should be important to separate progenitor populations with different lineage potentials to understand the underlying mechanisms of lineage commitment in the stem cell system. Our data suggest that we are still in the middle of drawing a high-resolution map for hematopoietic development. The map should be continuously improved by using new functional markers.

REFERENCES

1. SPANGRUDE, G.J., S. HEIMFELD & I.L. WEISSMAN. 1988. Purification and characterization of mouse hematopoietic stem cells. Science **241**: 58–62.
2. MORRISON, S.J. & I.L. WEISSMAN. 1994. The long-term repopulating subset of hematopoietic stem cells is deterministic and isolatable by phenotype. Immunity **1**: 661–673.
3. AKASHI, K., D. TRAVER & L.I. ZON. 2005. The complex cartography of stem cell commitment. Cell **121**: 160–162.
4. TRAVER, D. & K. AKASHI. 2004. Lineage commitment and developmental plasticity in early lymphoid progenitor subsets. Adv. Immunol. **83**: 1–54.
5. OSAWA, M., K. HANADA, H. HAMADA & H. NAKAUCHI. 1996. Long-term lymphohematopoietic reconstitution by a single CD34- low/negative hematopoietic stem cell. Science **273**: 242–245.

6. AKASHI, K., D. TRAVER, T. MIYAMOTO & I.L. WEISSMAN. 2000. A clonogenic common myeloid progenitor that gives rise to all myeloid lineages. Nature **404**: 193–197.
7. KONDO, M., I.L. WEISSMAN & K. AKASHI. 1997. Identification of clonogenic common lymphoid progenitors in mouse bone marrow. Cell **91**: 661–672.
8. IGARASHI, H., S.C. GREGORY, T. YOKOTA, et al. 2002. Transcription from the RAG1 locus marks the earliest lymphocyte progenitors in bone marrow. Immunity **17**: 117–130.
9. ADOLFSSON, J., O.J. BORGE, D. BRYDER, et al. 2001. Upregulation of Flt3 expression within the bone marrow Lin(-)Sca1(+)c-kit(+) stem cell compartment is accompanied by loss of self-renewal capacity. Immunity **15**: 659–669.
10. CHRISTENSEN, J.L. & I.L. WEISSMAN. 2001. Flk-2 is a marker in hematopoietic stem cell differentiation: a simple method to isolate long-term stem cells. Proc. Natl. Acad. Sci. USA **98**: 14541–14546.
11. ADOLFSSON, J., R. MANSSON, N. BUZA-VIDAS, et al. 2005. Identification of flt3(+) lympho-myeloid stem cells lacking erythromegakaryocytic potential a revised road map for adult blood lineage commitment. Cell **121**: 295–306.
12. FORSBERG, E.C., T. SERWOLD, S. KOGAN, et al. 2006. New evidence supporting megakaryocyte-erythrocyte potential of flk2/flt3 +multipotent hematopoietic progenitors. Cell **126**: 415–426.
13. SHIGEMATSU, H., B. REIZIS, H. IWASAKI, et al. 2004. Plasmacytoid dendritic cells activate lymphoid-specific genetic programs irrespective of their cellular origin. Immunity **21**: 43–53.
14. MIYAMOTO, T., H. IWASAKI, B. REIZIS, et al. 2002. Myeloid or lymphoid promiscuity as a critical step in hematopoietic lineage commitment. Dev. Cell. **3**: 137–147.
15. IWASAKI, H., C. SOMOZA, H. SHIGEMATSU, et al. 2005. Distinctive and indispensable roles of PU.1 in maintenance of hematopoietic stem cells and their differentiation. Blood **106**: 1590–1600.
16. IWASAKI, H., S.I. MIZUNO, R. MAYFIELD, et al. 2005. Identification of eosinophil lineage-committed progenitors in the murine bone marrow. J. Exp. Med. **201**: 1891–1897.
17. ZHANG, D.E., C.J. HETHERINGTON, S. MEYERS, et al. 1996. CCAAT enhancer-binding protein (C/EBP) and AML1 (CBF alpha2) synergistically activate the macrophage colony-stimulating factor receptor promoter. Mol. Cell. Biol. **16**: 1231–1240.
18. IWAMA, A., P. ZHANG, G.J. DARLINGTON, et al. 1998. Use of RDA analysis of knockout mice to identify myeloid genes regulated in vivo by PU.1 and C/EBPalpha. Nucleic Acids Res. **26**: 3034–3043.
19. DEKOTER, R.P., J.C. WALSH & H. SINGH. 1998. PU.1 regulates both cytokine-dependent proliferation and differentiation of granulocyte/macrophage progenitors. EMBO J. **17**: 4456–4468.
20. DEKOTER, R.P., H.-J. LEE & H. SINGH. 2002. PU.1 regulates expression of the interleukin-7 receptor in lymphoid progenitors. Immunity **16**: 297–309.
21. NA NAKORN, T., D. TRAVER, I.L. WEISSMAN & K. AKASHI. 2002. Myeloerythroid-restricted progenitors are sufficient to confer radioprotection and provide the majority of day 8 CFU-S. J. Clin. Invest. **109**: 1579–1585.
22. SCHULZE, H., M. KORPAL, J. HUROV, et al. 2006. Characterization of the megakaryocyte demarcation membrane system and its role in thrombopoiesis. Blood **107**: 3868–3875.

Gradients of Antigen Expression and Developmental Potential in Hematopoiesis

JAE-YONG KWAK,[a] SCOTT CHO,[b] AND GERALD J. SPANGRUDE[b]

[a]*Division of Hematology/Oncology, Department of Internal Medicine, Chonbuk National University Medical School, Jeonju 561-712, Korea*

[b]*Division of Hematology, Department of Internal Medicine, and Department of Pathology, University of Utah School of Medicine, Salt Lake City, Utah 84132-2408, USA*

> ABSTRACT: **Prospective isolation of hematopoietic stem and progenitor cell subsets depends upon the premise that expression of combinations of surface antigens reflects developmental potential. During the process of differentiation, however, the loss of antigens associated with stem cells and the concomitant gain of those associated with progenitor cells often occurs as a continuum rather than by discrete binary steps. Coupled with the fact that assay conditions can profoundly influence the developmental fates of prospectively isolated cells, gradients of antigen expression during differentiation have led to a variety of interpretations of lineage commitment in hematopoiesis.**
>
> KEYWORDS: fluorescence-activated cell sorting; lymphoid progenitor cells; erythropoiesis; *in vitro* assays

INTRODUCTION

Transcriptional programs that drive hematopoietic differentiation are well described,[1–4] as are the discrete stages of development in the various hematopoietic lineages.[5–7] A number of lines of evidence point to an early segregation of the megakaryocyte and erythroid lineages from the remainder of the myeloid lineages, including targeted mutation studies[8,9] as well as prospective isolation of cell populations representing intermediate stages of hematopoietic development.[10,11] The degree to which megakaryocyte and erythroid lineages segregate from the lymphoid lineage early in development has been controversial, with one group showing near-absolute separation[12] while

Address for correspondence: Gerald J. Spangrude, Ph.D., Division of Hematology, University of Utah, RM4C416, 30 N 1900 East, Salt Lake City, UT 84132-2408. Voice: 801-585-5544; fax: 801-585-3178.
 gspangrude@mac.com

Ann. N.Y. Acad. Sci. 1106: 82–88 (2007). © 2007 New York Academy of Sciences.
doi: 10.1196/annals.1392.002

another reports much more overlap at the progenitor stage.[13] We show here that at least part of this controversy can be accounted for by a lack of discrete phenotypic stages that correspond to functionally committed progenitor cells. Due in part to a continuum in expression levels of antigens used for prospective isolation, differing arbitrary cut off limits in studies reported by different laboratories have resulted in interpretations that are in disagreement. An additional difference between studies is that unique assays with variable sensitivities are used to detect erythroid and platelet lineage potentials. Using a highly sensitive assay for erythroid lineage development, we show here that residual erythroid potential is maintained even very late during the commitment to lymphoid development. We conclude that in the absence of standardization in the field, differing interpretations of experimental data seem inevitable.

MATERIALS AND METHODS

Isolation of Bone Marrow Cell Populations

Bone marrow cells were isolated from young adult C57BL/Ka-Thy-1.1 mice. After lysis of erythrocytes using ammonium chloride, the cells were incubated in a solution containing monoclonal antibodies specific for lineage-specific antigens followed by depletion using magnetic beads. The lineage-depleted (Lin^{neg}) population was then reacted with labeled monoclonal antibodies specific for Thy-1.1 (fluorescein), Sca-1 (phycoerythrin), Flt3 (biotinylated antibody detected using streptavidin-ECD), and c-kit (allophycocyanin). Dead cells were excluded based on DAPI staining. Cell subsets were isolated by aseptic cell sorting and placed into culture.

Cell Culture Studies

Cell populations isolated by cell sorting were plated in αMEM-based methylcellulose containing 10% fetal calf serum, 10% deionized bovine serum albumen, antibiotics, glutamine, and 2-mercaptoethanol. Colony growth and differentiation were stimulated with recombinant cytokines, including steel factor (100 ng/mL), G-CSF, Flt3L, IL-3, IL-6, IL-7 (all at 10 ng/mL), and erythropoietin (4 U/mL). After the time intervals indicated in the figures, quadruplicate plates were stained using benzidine hydrochloride and scored for total colonies and for colonies containing hemoglobinized erythroid lineage cells colonies based on benzidine staining.[14] Results are reported as mean number of colonies per 100 cells plated, and error bars represent standard error values. Statistical analysis used a one-tailed, unpaired t-test with equal variance.

RESULTS AND DISCUSSION

All studies reported here focused on the subset of mouse bone marrow cells characterized by expression of c-kit and Sca-1 in the absence of lineage antigens. This population is referred to as KLS cells. Within this population, progenitor cells biased for lymphoid differentiation can be identified by Flt3 expression, while stem cells and multipotent progenitor (MPP) cells can be identified by Thy-1.1 expression. One report in the literature shows that Flt3$^+$ lymphoid progenitors lack erythroid and platelet potential,[12] while another report demonstrates potential for both lineages.[13] When methylcellulose cultures established with either KLS Thy-1.1$^+$ MPP or KLS Flt3$^+$ progenitors were evaluated for erythroid lineage potential, we observed the results shown in FIGURE 1. The total colony-forming potential of each population was similar in this experiment, but the proportion of colonies containing erythroid lineage cells was quite distinct. Approximately 50% of the colonies formed by the Thy-1.1$^+$ MPP population included erythroid lineage cells, and this proportion remained fairly constant throughout the observation period of 6–12 days. In contrast, while a large fraction of colonies formed by Flt3$^+$ progenitors contained erythroid lineage cells at day 6 of culture, this proportion decreased steadily so then by day 12 only about 10% of colonies included hemoglobin as detected by benzidine staining. The experiment shown in FIGURE 1 emphasizes the importance of assay parameters in the interpretation of this experiment, since analysis at early versus late times of culture resulted in profound differences in the frequency of colonies that include erythroid lineage cells.

Several recent publications by Kondo and colleagues have shown that the Flt3$^+$ subset of KLS cells can be further segregated based on co-expression of the VCAM-1 antigen, with the brightest expression of Flt3 correlating with decreased expression of VCAM-1.[15,16] This group has shown that erythroid potential within the Thy-1.1^{neg} KLS population, as detected in a methylcellulose assay, was exclusively associated with cells expressing low, but not high, levels of surface Flt3. To further explore this observation in the context of the data shown in FIGURE 1, we segregated Thy-1.1^{neg} KLS cells into Flt3^{low} and Flt3^{high} subsets as previously described by Lai and Kondo.[16] We also isolated MPP cells as Thy-1.1$^+$ Flt3$^+$ KLS cells (FIG. 2, top panel). These three populations were cultured in methylcellulose as described for FIGURE 1 and evaluated for erythroid differentiation by benzidine staining. Consistent with the data shown in FIGURE 1, all three populations generated colonies containing erythroid lineage cells when evaluated after 6 days of culture. On subsequent days, however, the frequency of colonies containing benzidine-positive erythroid cells declined dramatically in cultures derived from either the Flt3^{low} or Flt3^{high} subsets. On day 10 of culture, cultures seeded with Flt3^{high} cells contained significantly fewer benzidine-positive colonies relative to either the Flt3^{low} or the MPP cultures (FIG. 2, *lower panel*), and

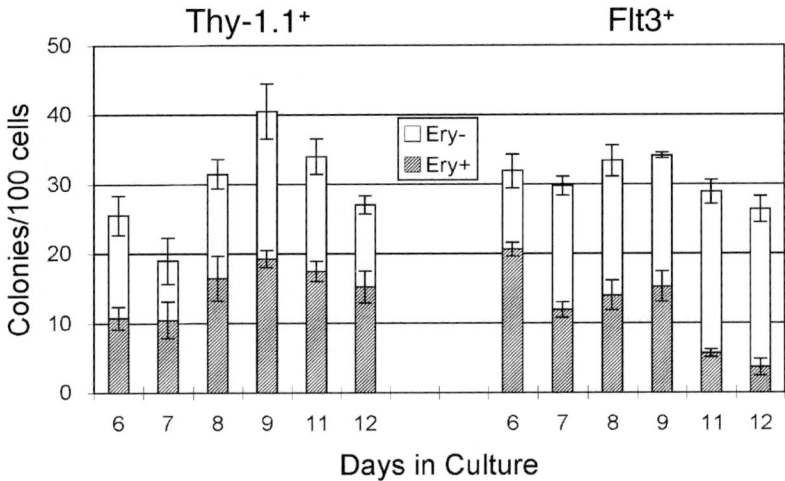

FIGURE 1. KLS cells isolated from mouse bone marrow were separated into Thy-1.1+ and Flt3+ subsets and cultured in methylcellulose in the presence of cytokines as described in "Materials and Methods" section. After the indicated days of culture, replicate cultures were stained with benzidine hydrochloride and colonies were scored as benzidine positive (Ery+) or benzidine negative (Ery–). Data are shown as the mean ± SEM of quadruplicate cultures.

on day 13 both the Flt3^{low} and Flt3^{high} cultures contained significantly fewer benzidine-positive colonies compared to MPP cultures. Apart from quantitative differences that are presumably due to the increased sensitivity of the benzidine detection method for scoring erythroid lineage-containing colonies, the results shown in FIGURE 2 are consistent with the studies reported by Lai and Kondo[16] in that erythroid potential is lost concomitantly with decreasing expression of Thy-1.1 and increasing expression of Flt3. However, discrete expression levels of these two antigens are not observed, and as a result the selection of subsets for functional analysis, as shown in FIGURE 2, is somewhat arbitrary. It is likely that some of the conflicting reports in the literature are due to differences in this arbitrary selection of cell populations. In addition, experimental design and assay sensitivity will have a major impact on the ability to detect differentiation into specific lineages.

FIGURE 3 depicts the general concept that the transitions between some stages of development in early hematopoiesis should be considered as gradients rather than as the discrete steps that are implied by classical tree diagrams of lineage commitment. This view reflects the concept that lineage decisions in early hematopoiesis are likely driven by relative levels of multiple transcriptional regulators and are characterized by quantitative as well as combinatorial aspects. During transitional states, such as the multipotent cell to lymphoid progenitor transition shown in FIGURE 3, extrinsic signals provided by the environment

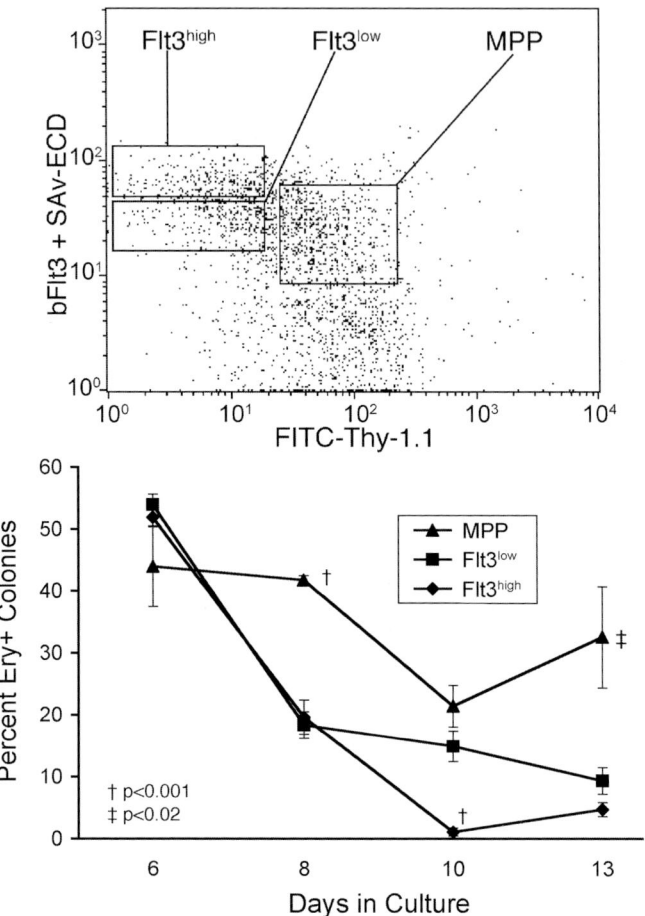

FIGURE 2. KLS cells isolated from mouse bone marrow were separated into three subsets as indicated in the upper panel and cultured in methylcellulose in the presence of cytokines as described in "Materials and Methods" section. After the indicated days of culture, replicate cultures were stained with benzidine hydrochloride and colonies were scored as benzidine positive (Ery+) or benzidine negative (Ery–). Data are shown as the mean percentage of Ery+ colonies ± SEM of quadruplicate cultures. The total number of colonies per 100 cells plated, averaged over the entire experiment ± SEM, was 64 ± 4 (MPP), 44 ± 2 (Flt3^{low}), and 26 ± 1 (Flt3^{high}) with $n = 16$ in all cases. Symbols adjacent to specific data points indicate the probability that the value is significantly different from the values of the other two groups assayed on the same day.

are likely able to shift developmental decisions in an instructive manner, even though the overall process has stochastic characteristics. The transition between early lymphoid progenitors and the committed precursors for the T and B lymphocyte lineages is thus depicted as a "gray zone" in FIGURE 3, meaning

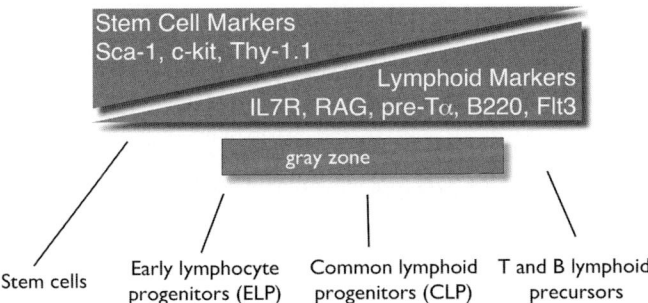

FIGURE 3. Gradients of antigen expression and developmental potential during lymphoid commitment. Mouse stem and progenitor cells can be identified by coexpression of Sca-1, c-kit, and Thy-1.1. These antigens are downregulated and others are upregulated, including IL-7R, RAG (as detected by an eGFP knockin reporter),[7] pre-Tα (as detected by an eGFP transgene reporter),[17] B220, and Flt3, as lymphoid lineage fates are specified. The "gray zone" indicates stages of development that are not resolved well by surface antigen expression and are thus difficult to study by prospective isolation approaches.

that absolute and reproducible physical separations of cells representing the transitional stages in this process of lineage commitment may not be possible. Furthermore, the conditions chosen by experimentalists to evaluate lineage potentials can greatly influence the outcome and interpretation of prospective isolation studies, and lead to conflicting and controversial findings.

While it may be convenient to conceptualize lineage commitments in hematopoietic development as the discrete stages depicted in classical tree diagrams, our efforts to understand mechanistic regulation of hematopoiesis are probably undermined by this notion of binary states. Recognizing that the gradients of antigen expression that are apparent during some lineage transitions in hematopoiesis are likely reflective of similar gradients in developmental potential will likely provide relevant insight into the mechanistic underpinnings of these developmental processes.

REFERENCES

1. KIRITO, K. & K. KAUSHANSKY. 2006. Transcriptional regulation of megakaryopoiesis: thrombopoietin signaling and nuclear factors. Curr. Opin. Hematol. **13:** 151–156.
2. KEE, B.L. 2005. Helix-loop-helix proteins in lymphocyte lineage determination. Curr. Top. Microbiol. Immunol. **290:** 15–27.
3. MIYAMOTO, T. & K. AKASHI. 2005. Lineage promiscuous expression of transcription factors in normal hematopoiesis. Int. J. Hematol. **81:** 361–367.
4. ORKIN, S.H. 2001. Transcription factors that regulate lineage decisions: the molecular basis of blood diseases. *In* The Molecular Basis of Blood Diseases, 3rd ed. G. Stamatoyannopoulos, P.W. Majerus, R. Perlmutter, H. Varmus, Eds.: 80–102. W.B. Saunders. Philadelphia, PA.

5. SPANGRUDE, G.J., S. HEIMFELD & I.L. WEISSMAN. 1988. Purification and characterization of mouse hematopoietic stem cells. Science **241**: 58–62.
6. MORRISON, S.J. & I.L. WEISSMAN. 1994. The long-term repopulating subset of hematopoietic stem cells is deterministic and isolatable by phenotype. Immunity **1**: 661–673.
7. IGARASHI, H., S.C. GREGORY, T., YOKOTA *et al*. 2002. Transcription from the RAG1 locus marks the earliest lymphocyte progenitors in bone marrow. Immunity **17**: 117–130.
8. SALEQUE, S., S. CAMERON & S.H. ORKIN. 2002. The zinc-finger proto-oncogene Gfi-1b is essential for development of the erythroid and megakaryocytic lineages. Genes Dev. **16**: 301–306.
9. HALL, M.A., D.J. CURTIS, D. METCALF, *et al*. 2003. The critical regulator of embryonic hematopoiesis, SCL, is vital in the adult for megakaryopoiesis, erythropoiesis, and lineage choice in CFU-S12. Proc. Natl. Acad. Sci. USA **100**: 992–997.
10. TRAVER, D., T. MIYAMOTO, J. CHRISTENSEN, *et al*. 2001. Fetal liver myelopoiesis occurs through distinct, prospectively isolatable progenitor subsets. Blood **98**: 627–635.
11. EDVARDSSON, L., J. DYKES & T. OLOFSSON. 2006. Isolation and characterization of human myeloid progenitor populations–TpoR as discriminator between common myeloid and megakaryocyte/erythroid progenitors. Exp. Hematol. **34**: 599–609.
12. ADOLFSSON, J., R. MANSSON, N. BUZA-VIDAS, *et al*. 2005. Identification of Flt3+ lympho-myeloid stem cells lacking erythro-megakaryocytic potential: a revised road map for adult blood lineage commitment. Cell **121**: 295–306.
13. FORSBERG, E.C., T. SERWOLD, S. KOGAN, *et al*. 2006. New evidence supporting megakaryocyte-erythrocyte potential of flk2/flt3+ multipotent hematopoietic progenitors. Cell **126**: 415–426.
14. SLAYTON, W.B., M.P. MOJICA, L.J. PIERCE & G.J. SPANGRUDE. 2001. Observations of residual differentiation potential during lineage commitment. Ann. N. Y. Acad. Sci. **938**: 157–165.
15. LAI, A.Y., S.M. LIN & M. KONDO. 2005. Heterogeneity of Flt3-expressing multipotent progenitors in mouse bone marrow. J. Immunol. **175**: 5016–5023.
16. LAI, A.Y. & M. KONDO. 2006. Asymmetrical lymphoid and myeloid lineage commitment in multipotent hematopoietic progenitors. J. Exp. Med. **203**: 1867–1873.
17. GOUNARI, F., I. AIFANTIS, C. MARTIN, *et al*. 2002. Tracing lymphopoiesis with the aid of a pTalpha-controlled reporter gene. Nat. Immunol. **3**: 489–496.

Biological and Molecular Evidence for Existence of Lymphoid-Primed Multipotent Progenitors

SIDINH LUC,[a,*] NATALIJA BUZA-VIDAS,[b,*] AND STEN EIRIK W. JACOBSEN[a,b]

[a]*Hematopoietic Stem Cell Laboratory, Lund Strategic Research Center for Stem Cell Biology and Cell Therapy, Lund University, 221 84 Lund, Sweden*

[b]*Haematopoietic Stem Cell Laboratory, Weatherall Institute of Molecular Medicine, John Radcliffe Hospital, University of Oxford, Headington, Oxford, OX3 9DS, United Kingdom*

ABSTRACT: Studies from our and other laboratories have over the last 2 years implicated the existence of multipotent progenitors (MPPs) with combined granulocyte–macrophage, B cell, and T cell potential, but little or no megakaryocyte–erythroid (MkE) potential in the adult bone marrow Lineage$^-$SCA-1$^+$KIT$^+$ (LSK) compartment of multipotent stem and progenitor cells. The evidence for the existence of LSKCD34$^+$FLT3^{hi} lymphoid-primed MPPs (LMPPs) implicates that a strict separation into common myeloid and lymphoid pathways might not be the first lineage commitment step of hematopoietic stem cells (HSCs). Together with the evidence for existence of common myeloid and common lymphoid progenitors (CMPs and CLPs, respectively), the identification of LMPPs also suggests that at least the granulocyte–macrophage lineage can be generated through alternative pathways. However, the existence of LMPPs has recently been questioned, as there is evidence that at least a fraction of LSKCD34$^+$FLT3^{hi} cells sustains MkE potential. Thus, in more recent studies we have in more detail compared the molecular signature of adult LMPPs to populations of LSK cells enriched for cells with pluripotent HSC activity. Notably, we have found at the global as well as single-cell level that LMPPs when compared with pluripotent HSCs downregulate the transcriptional priming of genes typically expressed in cells of the MkE lineage, while upregulating early lymphoid genes. Although other studies have suggested that the earliest HSC commitment steps might differ in fetal and adult hematopoiesis, we have also obtained evidence suggesting that the LMPP is defined already during fetal development.

KEYWORDS: hematopoiesis; hematopoietic stem cell; lineage commitment

Address for correspondence: Sten Eirik W. Jacobsen, Haematopoietic Stem Cell Laboratory, Weatherall Institute of Molecular Medicine, John Radcliffe Hospital, University of Oxford, Headington, Oxford, OX3 9DS, UK. Voice: +44-0-1865-222425; fax: +44-0-1865-222501.
sten.jacobsen@imm.ox.ac.uk

*These authors contributed equally to this work.

Ann. N.Y. Acad. Sci. 1106: 89–94 (2007). © 2007 New York Academy of Sciences.
doi: 10.1196/annals.1392.023

INTRODUCTION

Parallel to major efforts and progress in understanding the molecular regulation of development of the diverse blood cell lineages originating from hematopoietic stem cells (HSCs), considerable progress has been achieved toward delineating the cellular pathways by which HSCs undergo lineage commitment.[1,2] Through the seminal identification and prospective purification of common myeloid and common lymphoid progenitors (CMPs and CLPs, respectively),[3,4] it was implicated that the first lineage commitment step of HSCs results in a strict separation of myelopoiesis and lymphopoiesis.[2] Although the evidence for a CMP–CLP pathway for lineage commitment has been strong, this does not exclude that lineage commitment might also occur through alternative pathways. In fact, using different markers and approaches, three different groups have recently provided evidence for the existence, within the adult bone marrow (BM) Lineage$^-$SCA-1$^+$KIT$^+$ (LSK) compartment of multipotent progenitors (MPPs) with combined granulocyte–macrophage (GM), B cell, and T cell potential at the single-cell level, but little or no megakaryocyte-erythroid (MkE) potential.[5-9] The identification of these lymphoid-primed MPPs (LMPPs) suggests that the first lineage commitment step of HSCs might not be a binary decision separating myelopoiesis and lymphopoiesis. However, because of their residual MkE potential suggested through clonal *in vitro* studies to be limited to 1–4% of the phenotypically purified cell populations in each of these studies,[6,8,9] and in more recent *in vivo* studies to be more extensive,[10] it remains possible that LMPPs might have an intrinsic MkE potential, but that the probability to uncover this through available *in vitro* studies is very low. On the other side, as the available *in vivo* studies did not allow the frequency of LSKCD34$^+$FLT3^{hi} cells with *in vivo* MkE potential to be precisely established,[10] it remains important to pursue the further purification, as well as biological and molecular characterization of LMPPs, and their positioning in the hematopoietic hierarchy.

MOLECULAR EVIDENCE FOR DISTINCT PATTERNS OF MULTILINEAGE PRIMING IN ADULT AND FETAL HSCs AND LMPPs

In contrast to lineage-restricted progenitors, it has been demonstrated that pluripotent HSCs and MPPs (at the single-cell level) are transcriptionally primed for expression of genes otherwise restricted in a lineage-specific pattern.[11-14] Although the multilineage priming was initially thought to reflect the pluripotentiality of HSCs, later studies suggested that despite of having lymphoid potential, the multilineage priming of HSCs might be largely restricted to genes of multiple myeloid lineages, and that HSCs might not yet be transcriptionally primed for lymphopoiesis.[11] The original studies identifying candidate LMPPs indicated that the LMPPs might be the first lymphoid-primed progenitors by demonstrating at the population level downregulation of some MkE-related genes and upregulation of lymphoid-associated genes.[6,8,9] However, being restricted to a few investigated genes and performed on whole cell populations, neither of these studies could establish the frequency of cells within the isolated LMPP populations, which were in fact lymphoid-primed, or to what degree the lymphoid genes were coexpressed with genes of the GM and MkE lineages, in these or any other cells in the LSK HSC-MPP hierarchy. Therefore, it could also not be ruled out that the observed lymphoid priming might be restricted to a small population of cells that might not simultaneously be transcriptionally primed for myeloid genes, and thus potentially rather represent more lymphoid-restricted progenitor cells.

In more recent studies, we have examined in detail the transcriptional multilineage priming of hierarchically related LSK subsets in adult BM: LSKCD34$^-$FLT3$^-$ LT-HSCs, LSKCD34$^+$FLT3$^-$ ST-HSCs, and LSKCD34$^+$FLT3^{hi} LMPPs.[15] Comparing the gene expression of each of these populations with published patterns of gene expression of the FDCP mix cell line differentiated along the Mk, E, and GM lineages,[16] and B cell lines,[17] we obtained evidence for a distinct hierarchical organization of lineage programs within the hierarchy of LSK cells, with a gradual downregulation of MkE transcriptional priming from LT-HSCs to LMPPs, sustained GM priming, and a distinct upregulation of lymphoid genes in the LMPPs, whereas expression of genes of the GM lineage showed no distinct differences between the three investigated LSK populations.[15] This pattern was confirmed by quantitative PCR analysis, and of further interest most of the typical lymphoid genes found to be upregulated in LSKCD34$^+$FLT3^{hi} LMPPs are those expressed early in lymphoid development, such as *Il7r*, *Rag1*, and sterile IgH transcript, whereas expression of investigated genes otherwise specific for each of the B cell and T cell lineages were not yet expressed. Most important, when extended to the single-cell level, we found typical coexpression of MkE (*Gata1*, *Vwf*, and/or *Epor*) and GM (*Mpo* and/or *Csf3r*) genes in LSKCD34$^-$FLT3$^-$ LT-HSCs and LSKCD34$^+$FLT3$^-$ ST-HSCs, whereas virtually all of the approximately 30% LSKCD34$^+$FLT3^{hi} LMPPs expressing early lymphoid genes (*Rag1*, sterile IgH transcript, and/or *Il7r*) sustained expression of GM but not MkE genes.[15] In fact, throughout the adult LSK hierarchy, the single-cell expression of the investigated MkE and lymphoid genes was almost mutually exclusive.

Although it has been implicated that adult and fetal HSC lineage commitment pathways might largely differ,[2,18] some of the proposed differences might at least in part reflect that quite different approaches have been used to provide the experimental evidence for each of these models.[2,18] In more recent studies, we have identified in fetal liver (E14.5), LSKCD34$^+$FLT3^{hi} LMPPs that similar to their adult counterpart sustain combined B cell, T cell, and GM potential, but little or no MkE potential,[15] suggesting that at least this early lineage commitment step might be conserved between definitive fetal and adult

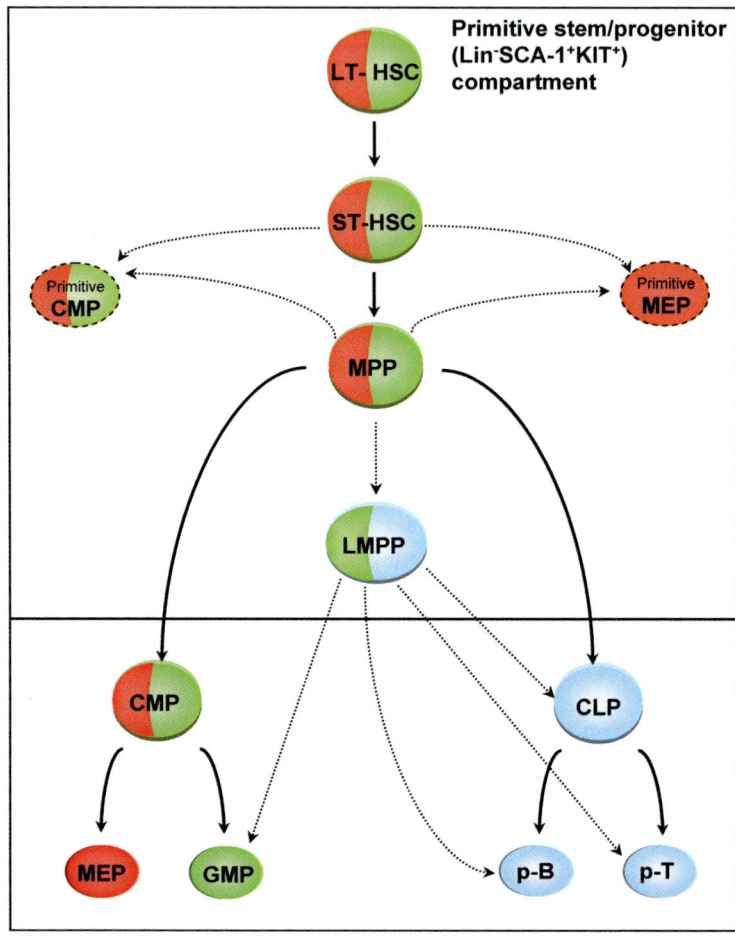

FIGURE 1. Positioning of LMPPs into the CMP–CLP model for hematopoietic lineage commitment. Figure shows the proposed cellular pathways and changes in transcriptional lineage priming in the CMP–CLP model, with the recently identified LMPP incorporated into the model, showing the combined lymphoid (blue) and GM (green) priming, in contrast to LT-HSCs and ST-HSCs with combined GM and MkE (red) priming. Solid arrows indicate lineage commitment pathways according to the classical CMP–CLP model. Broken arrows indicate possible (yet to be established) connections between the LMPP and its predecessors and progeny. Although not indicated, along each line intermediate progenitors might be present. Also indicated is the possibility that ST-HSCs and/or MPPs prior or simultaneously to the generation of LMPPs could give rise to subsets of more primitive (Lin$^-$SCA-1$^+$KIT$^+$) CMPs and/or MEPs. Abbreviations: LT-HSC = long-term HSC; ST-HSC = short-term HSC; CMP = common lymphoid progenitor; CLP = common lymphoid progenitor; HSC = hematopoietic stem cell; MPP = multipotent progenitor; LMPP = lymphoid-primed MPP; MEP = megakaryocyte/erythroid progenitor; GMP = granulocyte/macrophage progenitor; p-B = B cell progenitor; p-T = T cell progenitor.

hematopoiesis. Notably, we found the multilineage priming (for all lineages) to be much more extensive in fetal than adult LSK populations, although importantly without changing the hierarchical pattern observed between different lineage programs and distinct populations.[15] Thus, specifically more than 50% of single fetal LSKCD34$^+$FLT3^{hi} LMPPs were simultaneously primed for lymphoid and GM genes, whereas no cells at any level in the LSK hierarchy showed combined MkE and lymphoid priming without also coexpressing GM genes. However, a few LSKCD34$^+$FLT3^{hi} LMPP cells coexpressed genes of the GM, MkE, as well as lymphoid lineages, but notably only one of the investigated MkE genes were expressed in these rare cells.[15]

Although our and other recent studies have provided biological and molecular evidence for the existence of alternative early pathways for HSC lineage commitment (FIG. 1), through the identification of LMPPs with combined GM, B cell, and T cell, but reduced MkE potential,[5,6,8,9,15] this remains controversial,[10] highlighting the need for the development of improved tools and more uniform standards to trace and establish the lineage potentials of multipotent but lineage-restricted progenitor populations. Such a development will facilitate the identification of the alternative cellular pathways of lineage commitment and thereby also a better identification and understanding of the molecular regulation of the lineage commitment process. However, the process of establishing the complete roadmap for hematopoietic lineage commitment has probably just begun.

ACKNOWLEDGMENTS

These studies were generously supported by grants from the Swedish Research Council, Swedish Foundation for Strategic Research, and the EU project LHSB-CT-2003-503005 (EUROSTEMCELL). The Lund Stem Cell Center is supported by a center of excellence grant from the Swedish Foundation for Strategic Research. The authors declare no competing financial interests.

REFERENCES

1. LAIOSA, C.V., M. STADTFELD & T. GRAF. 2006. Determinants of lymphoid-myeloid lineage diversification. Annu. Rev. Immunol. **24:** 705–738.
2. REYA, T. *et al.* 2001. Stem cells, cancer, and cancer stem cells. Nature **414:** 105–111.
3. AKASHI, K. *et al.* 2000. A clonogenic common myeloid progenitor that gives rise to all myeloid lineages. Nature **404:** 193–197.
4. KONDO, M., I.L. WEISSMAN & K. AKASHI. 1997. Identification of clonogenic common lymphoid progenitors in mouse bone marrow. Cell **91:** 661–672.

5. ADOLFSSON, J. *et al*. 2001. Upregulation of Flt3 expression within the bone marrow Lin(-)Sca1(+)c-kit(+) stem cell compartment is accompanied by loss of self-renewal capacity. Immunity **15:** 659–669.
6. ADOLFSSON, J. *et al*. 2005. Identification of Flt3+ lympho-myeloid stem cells lacking erythro-megakaryocytic potential a revised road map for adult blood lineage commitment. Cell **121:** 295–306.
7. YANG, L. *et al*. 2005. Identification of Lin(-)Sca1(+)kit(+)CD34(+)Flt3- short-term hematopoietic stem cells capable of rapidly reconstituting and rescuing myeloablated transplant recipients. Blood **105:** 2717–2723.
8. LAI, A.Y. & M. KONDO. 2006. Asymmetrical lymphoid and myeloid lineage commitment in multipotent hematopoietic progenitors. J. Exp. Med. **203:** 1867–1873.
9. YOSHIDA, T. *et al*. 2006. Early hematopoietic lineage restrictions directed by Ikaros. Nat. Immunol. **7:** 382–391.
10. FORSBERG, E.C. *et al*. 2006. New evidence supporting megakaryocyte-erythrocyte potential of flk2/flt3+ multipotent hematopoietic progenitors. Cell **126:** 415–426.
11. AKASHI, K. *et al*. 2003. Transcriptional accessibility for genes of multiple tissues and hematopoietic lineages is hierarchically controlled during early hematopoiesis. Blood **101:** 383–389.
12. HU, M. *et al*. 1997. Multilineage gene expression precedes commitment in the hemopoietic system. Genes Dev. **11:** 774–785.
13. MIYAMOTO, T. *et al*. 2002. Myeloid or lymphoid promiscuity as a critical step in hematopoietic lineage commitment. Dev. Cell. **3:** 137–147.
14. YE, M. *et al*. 2003. Hematopoietic stem cells expressing the myeloid lysozyme gene retain long-term, multilineage repopulation potential. Immunity **19:** 689–699.
15. MANSSON, R. *et al*. 2007. Molecular evidence for hierarchical transcriptional lineage priming in fetal and adult stem cells and multipotent progenitors. Immunity **26:** 407–419.
16. BRUNO, L. *et al*. 2004. Molecular signatures of self-renewal, differentiation, and lineage choice in multipotential hemopoietic progenitor cells *in vitro*. Mol. Cell. Biol. **24:** 741–756.
17. TSAPOGAS, P. *et al*. 2003. RNA analysis of B cell lines arrested at defined stages of differentiation allows for an approximation of gene expression patterns during B cell development. J. Leukoc. Biol. **74:** 102–110.
18. KATSURA, Y. 2002. Redefinition of lymphoid progenitors. Nat. Rev. Immunol. **2:** 127–132.

Insertional Mutagenesis by Replication-Deficient Retroviral Vectors Encoding the Large T Oncogene

ZHIXIONG LI,[a] OLGA S. KUSTIKOVA,[a,b] KENJI KAMINO,[c]
THOMAS NEUMANN,[a] MATHIAS RHEIN,[a] ELKE GRASSMAN,[d]
BORIS FEHSE,[b] AND CHRISTOPHER BAUM[a,d]

[a]*Department of Experimental Hematology, Hannover Medical School, 30625 Hannover, Germany*

[b]*Clinic for Stem Cell Transplantation, University Medical Center Hamburg-Eppendorf, 20251 Hamburg, Germany*

[c]*Institute for Cell and Molecular Pathology, Hannover Medical School, 30625 Hannover, Germany*

[d]*Division of Experimental Hematology, Cincinnati Children's Hospital Medical Center, Cincinnati, Ohio 45229-3039, USA*

ABSTRACT: Insertion sites of replication-deficient retroviral vectors may trigger clonal dominance of hematopoietic cells *in vivo*. Here, we tested whether this would also be the case when using vectors that express powerful oncogenes, such as the large tumor antigen (TAg) of simian virus 40. TAg inactivates the tumor-suppressor proteins p53 and Rb by virtue of a chaperone-like activity. Primary hematopoietic stem/progenitor cells transduced with retroviral vectors encoding TAg-induced histiocytic sarcoma (HS) or myeloid leukemia (ML) in transplanted mice (average survival of 21 weeks). Retrovirally introducing TAg into pretransformed 32D cells generated a monocytic leukemia, with faster kinetics (~8 weeks). Leukemic clones showed retroviral insertions in genes contributing to all known TAg cooperation pathways, acting mitogenic and/or modulating apoptosis (such as *BclX, Crk, Pim2, Csfr1/Pdgfrb, Osm/Lif, Axl, Fli, Sema4b, Sox4*). 32D-derived monocytic leukemias showed hits in *Pim2* and *Max* proto-oncogenes, or the chaperone *Hspa4*, plus additional signaling genes. Vector-mediated insertional mutagenesis thus revealed a broad spectrum of potential TAg complementation genes. These findings have important implications for the use of retroviral transgenesis in cancer research, and the expression of signaling genes in somatic gene therapy.

Address for correspondence: Christopher Baum, M.D., Department of Experimental Hematology, OE6960, Hannover Medical School, Carl-Neuberg-Straße 1, 30625 Hannover, Germany. Voice: 49-511-532-6067; fax: +49-511-532-6068.
baum.christopher@mh-hannover.de

Ann. N.Y. Acad. Sci. 1106: 95–113 (2007). © 2007 New York Academy of Sciences.
doi: 10.1196/annals.1392.003

KEYWORDS: simian virus 40; histiocytic sarcoma; myeloid leukemia; retroviral vectors; insertional mutagenesis

INTRODUCTION

To examine the phenotypic and genetic consequences of ectopic (proto-) oncogene expression in primary cells or established cell lines, replication-deficient retroviral vectors are widely used. However, the semirandom insertion pattern of retroviruses generates a complex chimerism of the transduced population with respect to the retroviral insertion site(s). Our recent studies have demonstrated that insertional mutagenesis of crucial cellular alleles by retroviral vectors encoding relatively neutral genes may contribute to clonal dominance and leukemia induction in mice.[1,2] Accordingly, the mutagenic potential of retroviral vectors increases as a function of the copy number per cell.[2] Dose-escalated retroviral gene transfer may even lead to the combinatorial alteration of independent proto-oncogene loci,[2] similar to the situation encountered in tumors elicited by replication-competent retroviruses.[3,4]

Nevertheless, very few cases of murine and human leukemias induced by retroviral vectors that do not encode oncogenes have been reported to date.[2,5,6] Based on initial insights into the pathogenesis of these cases, we and others suggested that the oncogenic risk of retroviral gene transfer (and thus the spectrum of insertion sites selected for) greatly depends on the nature of the proteins encoded by the vector.[5,7–9] A recent study showed that replication-deficient vectors expressing the *Sox4* oncogene require insertional upregulation of cellular proto-oncogenes for transformation of hematopoietic cells.[9] The aim of this study was thus to test this concept in the context of retroviral vectors expressing the large tumor antigen (TAg) of the DNA tumor virus simian virus 40 (SV40).

TAg is a well-established oncogene that transforms cells and induces tumors in animals by inactivating the tumor-suppressor proteins p53 and Rb.[10,11] Although SV40 is a monkey virus, vaccines contaminated with infectious SV40 were administered worldwide to humans during the period 1955–1963.[12] Recent data demonstrated that SV40 is associated (although not necessarily causally linked) with some types of human tumors, and can also be found in normal blood.[13,14] As inactivation of p53 and Rb is not known to be sufficient for leukemogenesis and the spectrum of tumors potentially elicited by TAg is not fully understood, it is important to study the influence of the target cells and to identify cooperating proto-oncogenes.

A number of complementation studies have revealed that genes belonging to mitotic and antiapoptotic pathways are powerful cooperation partners in the clonal evolution of TAg-associated tumors. Well-studied examples are (proto-)oncogenic receptors, such as *Csfr1*=*c-Fms*,[15] signal transducers, such as *Ras*[15,16] or *Src*,[15] transcription factors, such as *Fli1*,[17] and apoptosis modulators, such as *Bcl2*.[18] Moreover, it has been shown that TAg acts as a chaperone, collaborating with cellular heat shock proteins (*Hsp*).[11]

Here, we found that retroviral vector-mediated expression of TAg in primary murine hematopoietic cells induces histiocytic sarcoma (HS) or myeloproliferative disease-like myeloid leukemia (MPD-ML), whereas acute myeloid leukemia (AML) developed following transduction of an established myeloid cell line (32D cells). In each tumor we discovered that the retroviral insertion sites marked genes or pathways that are well-known collaborators of TAg-mediated transformation.

MATERIALS AND METHODS

Animals

C57BL/6J mice and C3H/Hej mice, aged 12–16 weeks, were obtained from and kept in the animal laboratories of Hannover Medical School, Hannover, Germany. Animal experiments were approved by the local ethical committee.

Retroviral Vectors and Vector Production

SF91.IRES-EGFP.WPRE and SF11dLNGFR have been described.[19,20] SF91TAg (long name FMEV-lox-EGFP2A-SV40LT-tCD34) contains sequences of a loxP-site flanked cDNA of TAg, coexpressed with enhanced green fluorescent protein (EGFP) as a self-cleaving EGFP2A-TAg fusion protein.[21] Cell-free supernatants of retroviral vector preparations were produced by transient transfection in Phoenix-gp cells (kindly provided by G. Nolan) as described,[22] using a plasmid encoding the ecotropic envelope.

Bone Marrow Transduction and Transplantation

Lineage negative (Lin–) BM cells were transduced as described.[23] Briefly, Lin– cells were isolated from BM by magnetic sorting using lineage-specific antibodies (Gr1, CD11b, CD45R/B220, CD3e, TER-119; Pharmingen, Hamburg, Germany), and cultured in StemSpan HS2000 medium (CellSystems, St. Katharinen, Germany) supplemented with cytokines and penicillin/streptomycin. Cells were transduced on the following days (day 4 and 5). On day 6, transgene expression was measured and at least 5×10^4 Lin– cells per lethally (9.5 Gy) irradiated recipient were transplanted by tail vein injection.

32D Cell Culture, Retroviral Transduction, and Injection

32D cells were obtained from DSMZ (Braunschweig, Germany), and cultured in IMDM supplemented with 10% fetal calf serum and recombinant mIL-3 (2 ng/mL). Withdrawal of IL-3 led to cell death, mutation rates to IL-3 independence were $<10^{-7}$. Cells were retrovirally transduced using a single exposure at an MOI of 10. Polyclonal cultures were used for transplantation into C3H mice irradiated with 2.5 Gy, at a dose of 10^7 cells per recipient.

Tumor Phenotyping

Mice were humanely killed when moribund, or analyzed when found dead before onset of autolysis. Enlarged organs were weighed. Bone marrow, spleen, liver, kidney, lung, and thymus were fixed in a buffered 4% formalin solution and embedded in Paraplast plus (Kendall, Mansfield, MA). Sections (2 μm) were stained with hematoxylin and eosin, and examined by light microscopy. Cells from infiltrated organs and peripheral blood (PB) were analyzed by flow cytometry, using antibodies against Gr1, CD11b (myeloid cells), CD19 (B cells), Ter119 (erythroid cells), CD3, CD4, CD8 (T cells), Sca1, c-Kit (Pharmingen, Hamburg, Germany), or F4/80 (CALTAG, Burlingame, CA). Dead cells were excluded by propidium iodide staining. Leukocyte morphology was evaluated in Pappenheim-stained blood smears and cytospins of BM and spleen cells. Blood cell counts were obtained using an automatic analyzer (ABC Counter, Scil, Viernheim, Germany). Mice that did not develop leukemia were analyzed with this protocol after a maximal observation period of 37 weeks.

Southern Blot

Tumor DNA was prepared using a kit from Qiagen (Hilden, Germany). Ten micrograms of each DNA were digested with *Bgl*II (New England BioLabs, Frankfurt, Germany), size separated by agarose gel electrophoresis and blotted onto Pall Biodyne (Pall Corporation, Pensacola, FL) membranes. Hybridization conditions followed established protocols.[20] The radioactive probe was a 2158-bp fragment containing the coding sequence of TAg.

Ligation-Mediated Polymerase Chain Reaction (LM-PCR)

Our protocol followed published conditions.[1,2,24] Briefly, DNA was digested with 5U of restriction enzyme Tsp509 I (New England BioLabs) per microgram DNA for 2 h at 65°C. For primer extension, 0.25 pmol of biotinylated retroviral primer A1RV (5′-CTGGGGACCATCTGTTCTTGGCCTC-3′) was used. The first PCR (94°C for 2 min; 94°C for 15 s, 60°C for 30 s, 68°C for 1 min for 30 cycles; 68°C for 10 min) was performed using Extensor Hi-Fidelity PCR Master Mix (ABGENE, Hamburg, Germany), retroviral primer A2RV (5′-GCCCTTGATCTGAACTTCTC-3′), and linker-specific primer OCI.[24] The nested PCR was performed under identical conditions, but using retroviral primer A3RV (5′-CCATGCCTTGCAAAATGGC-3′) and linker-specific primer OCII.[24] PCR products were isolated after gel electrophoresis using QIA quick Gel Extraction Kit (Qiagen) and sequenced directly using the primer RAseq (5′-CTTGCAAAATGGCGTTAC-3′). Recovered sequences were screened using the NCBI mouse genome database.

RESULTS

Vector Construction and Evaluation

The retroviral vector SF91TAg coexpresses TAg with EGFP using the 2A-proteinase of foot and mouth disease virus (FIG. 1A). The EGFP2A-TAg coexpression cassette was flanked by loxP sites. The truncated CD34 (tCD34) cell surface marker[20] was cloned downstream of TAg such that is only expressed after Cre-mediated excision of TAg (confirmed by flow cytometry, data not shown). Additional vectors SF91.IRES-EGFP.WPRE and SF11dLNGFR

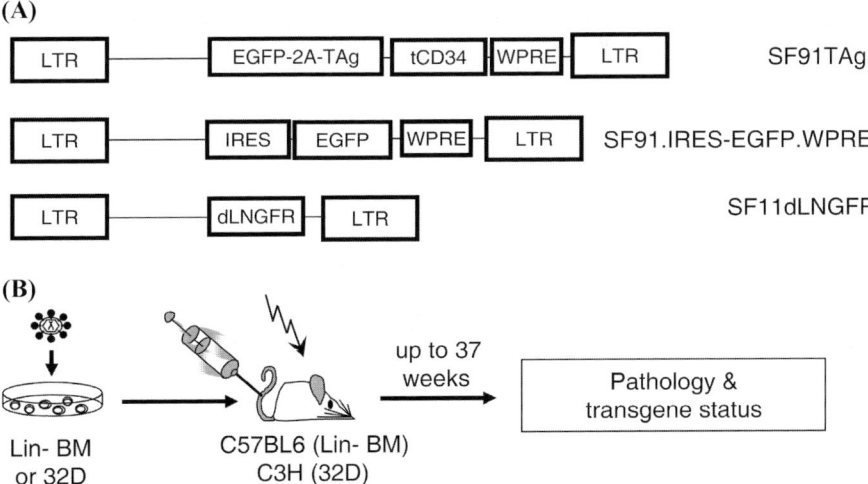

FIGURE 1. Vectors used in the present study and experimental scheme. (**A**) The gammaretroviral vector SF91TAg encodes the SV40 TAg from a self-cleaving EGFP fusion protein. The tCD34 reading frame is only expressed after Cre-mediated excision of EGFP2ATAg, which is flanked by loxP sites. (**B**) We transduced either lineage-depleted primary bone marrow cells or 32D cells and transplanted these into syngeneic hosts. Experimental groups are further described in TABLE 1.

encoded only EGFP and the deleted low affinity nerve growth factor receptor (dLNGFR), respectively (FIG. 1B). Ecotropic replication-defective vectors were produced by transient transfection into safety-modified packaging cells.[22] The oncogenic potential of SF91TAg was confirmed by inducing anchorage-independent colonies of RAT1 fibroblasts.[21]

Retroviral Gene Transfer of TAg into and Transplantation of Hematopoietic Cells

Primary hematopoietic stem cells and progenitor cells (HSC/HPC) of C57Bl6 were transduced with SF91TAg or cotransduced with SF91TAg and SF11dLNGFR, to increase the mutagenic risk of retroviral gene transfer without escalating the expression levels of TAg. A control population of HSC/HPC was transduced with SF91.IRES-EGFP.WPRE. Moreover, we transduced nontumorigenic IL-3-dependent 32D cells with SF91TAg or SF91.IRES-EGFP.WPRE. Two days after transduction, 30–50% of 32D cells expressed EGFP (data not shown). Expression of TAg in 32D cells was not sufficient to induce growth factor-independence (data not shown).

Transduced HSC/HPC were transplanted into syngeneic, lethally irradiated (9.5 Gy) C57BL6/J recipients. Four mice received HSC/HPC modified with SF91TAg only (group I), another 4 mice HSC/HPC modified with SF91TAg plus SF11dLNGFR (group II), and yet another 4 mice HSC/HPC modified with SF91.IRES-EGFP.WPRE only (group III). Polyclonal populations of 32D cells were injected into syngeneic, sublethally irradiated (2.5 Gy) C3H recipients via the tail veins,[25] resulting in experimental groups IV (32D: SF91TAg, $n = 6$) and V (32D: SF91.IRES-EGFP.WPRE, $n = 3$). A last group of C3H mice received 32D cells that were mock transduced (group VI, $n = 3$). TABLE 1 gives an overview.

Induction of HS and Leukemia by Retroviral Vectors Encoding TAg

Disease phenotypes associated with retroviral expression of TAg in hematopoietic cells were consistent but depended on the target cell type. Among all recipients of TAg-modified HSC/HPC (groups I and II), the latency to induction of overt disease was 17–23 weeks (TABLE 1). The first death occurred in week 17 (recipient 421) with the coexistence of MPD-ML and HS. Three animals of the same group (group I) developed HS without signs of ML, as revealed by the typical histopathology in hematopoietic organs and characteristic infiltrates in nonhematopoietic organs (FIG. 2A). Blood smear analyses (data not shown) showed that HS was associated with pancytopenia in the PB. One mouse (recipient 420) showed MPD-ML without evidence of HS (data not shown). In this case, peripheral-marked leukocytosis (19,000 leukocytes/μL), anemia (Hb of 8 g/dL), and thrombocytopenia (40,000/μL) were observed.

TABLE 1. Experimental groups and outcome

Group	Vector	Cells	No. of recipients (strain)	Outcome per animal (week of final analysis)				
				Histiocytic sarcoma	MPD-like myeloid leukemia	Acute myeloid leukemia	Death of unknown cause	Healthy
I	SF91TAg	Lin⁻ BM	4 (C57BL/6)	194 (23) 195 (23) 421 (17)	420 (18)	—	—	—
II	SF91TAg + SF11dLNGFR	Lin⁻ BM	4 (C57BL/6)	196 (23) 197 (23)	—	—	199 (21)	198 (23)
III	SF91.IRES-EGFP.WPRE	Lin⁻ BM	4 (C57BL/6)	—	—	—	—	414 (37) 415 (37) 416 (24) 417 (24)
IV	SF91TAg	32D	6 (C3H)	—	—	379 (8) 380 (10) 381 (8)	388 (7) 389 (7) 390 (7)	—
V	SF91.IRES-EGFP.WPRE	32D	3 (C3H)	—	—	—	—	393 (23) 394 (23) 395 (23)
VI	Mock	32D	3 (C3H)	—	—	—	—	396 (23) 397 (23) 398 (23)

Animals with death of unknown cause were found dead with signs of autolysis, precluding histopathology and DNA analysis. Mouse 421 had both HS and MPD-like myeloid leukemia.
MPD-like, myeloproliferative disorder-like.

HS was transplantable. Cells from liver and bone marrow isolated from animals 195 and 196 were intravenously injected into lethally irradiated recipients. The animals developed HS after 6 weeks (FIG. 3).

HS was also observed in 2 of the 4 animals that received cells cotransduced with TAg and dLNGFR (group II). While one animal of this group died

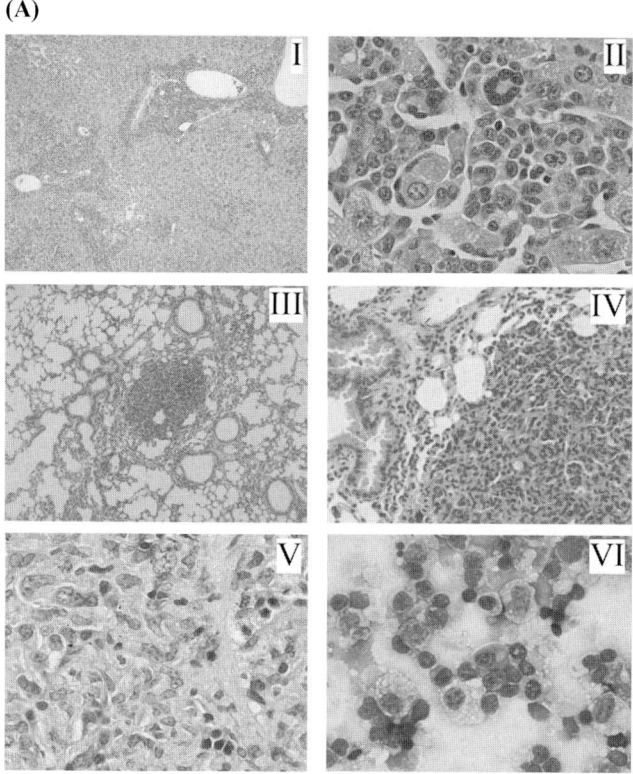

FIGURE 2. Histopathology of TAg-associated hematopoietic tumors. (**A**) HS (I–II, case 194; III–IV, case 195; V–VI, case 197): Predominantly perivascular infiltration of tumor cells in the liver (I, ×100) with a typical multinucleated giant cell (II, ×1000). Lung with perivascular and peribronchial sarcoma infiltration (III, ×100) consisting of oval cells with pleomorphic nuclei, eosinophilic abundant cytoplasm, and an atypical mitosis (IV, ×630). Tumor cells in the spleen with marked nuclear pleomorphism (V, ×1000). Cytospin of splenocytes showing numerous HS cells with polymorphic nuclei, coarse chromatin structure, and abundant partly phagocytizing cytoplasm (VI, ×1000). (**B**) AML of the monocytic subtype (case 380) with organ infiltrations: in the liver marked infiltration of tumor cells (I, ×100) consisting of immature myeloid cells with pleomorphic nuclei and scanty cytoplasm (II, ×1000). Unrecognizable structure of the spleen (III, ×100) that is infiltrated by pleomorphic leukemic cells with some mitotic figures (IV, ×1000). Blood smear (V, ×1000) and cytospin of splenocytes (VI, ×1000) showing numerous monocytes, partly with phagocytosis.

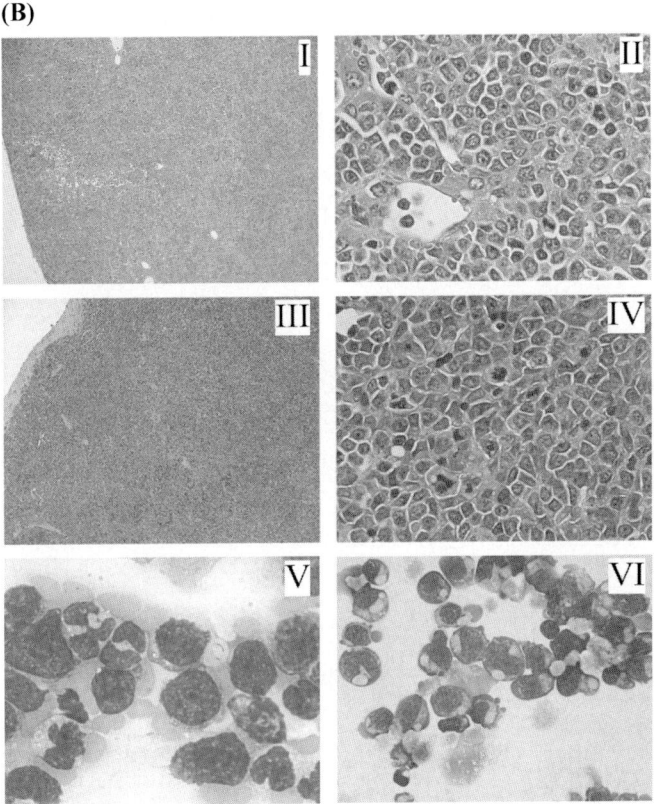

FIGURE 2. Continued

unexpectedly (199, week 22) and was not available for further analyses, another (198) remained healthy and showed only expression of SF11dLNGFR in detailed analyses of hematopoietic organs and PB (data not shown).

Interestingly, according to the coexpressed EGFP, TAg cells were reduced to almost undetectable level in the PB several weeks before the development of overt malignancy, in contrast to the durable engraftment with EGFP-marked cells in the control group (data not shown). This suggested that either TAg expression in primary HSC/HPC produced apoptosis in the majority of transplanted cells or that TAg expression changed the organ distribution of hematopoietic cells.

Retroviral transfer of SF91TAg into 32D cells gave rise to a very rapid monocytic AML (latency <9 weeks, TABLE 1). Three animals died suddenly in week 7, precluding detailed molecular and phenotypic studies. The other mice showed a profound leukocytosis (up to 52,000 leukocytes/μL) with predominance of blasts in the PB (FIG. 2B). *In vitro*, 32D-derived leukemic cells were

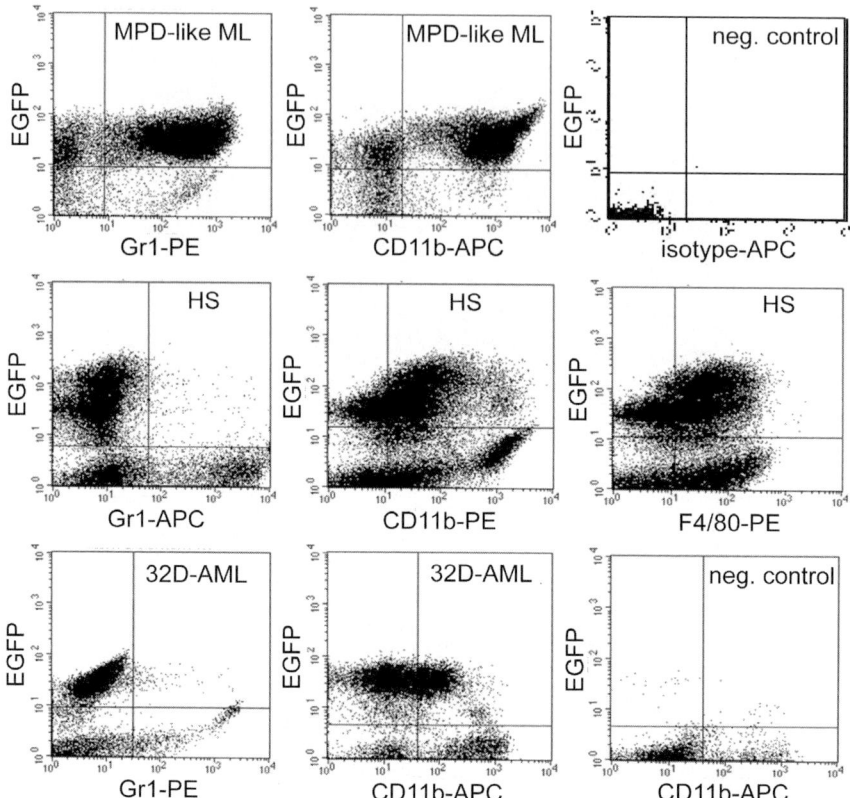

FIGURE 3. Flow cytometry of surface antigen expression in TAg-associated hematopoietic tumors. The MPD-like leukemia cells were homogenously positive for Gr1 and CD11b (case 420, upper panel). The middle panel shows expression of CD11b and F4/80, but not Gr1, on HS cells from case 402, a secondary recipient of 196. HS cells also expressed CD19 (B cell marker) and Sca1 (progenitor cell marker, data not shown). The lower panel shows expression of CD11b, but not Gr1, on 32D cell-derived AML cells. These were also positive for Ter119 (data not shown).

growth–factor-independent, in contrast to their ancestors (data not shown). Flow cytometry confirmed the myeloid origin of SF91TAg-associated malignancies (FIG. 3). All control animals (groups III, V, and VI) survived without signs of disease until the termination of the study.

Insertion Site Analysis Reveals Hits in Known Collaborative Pathways of TAg

Southern blot analysis of tumor DNA revealed evidence of 3 to 6 insertion sites in HS cases 194–197, and 2 insertions for AML cases 379–381 (FIG. 4A). All samples showed distinct insertion patterns except AML cases 379 and 381,

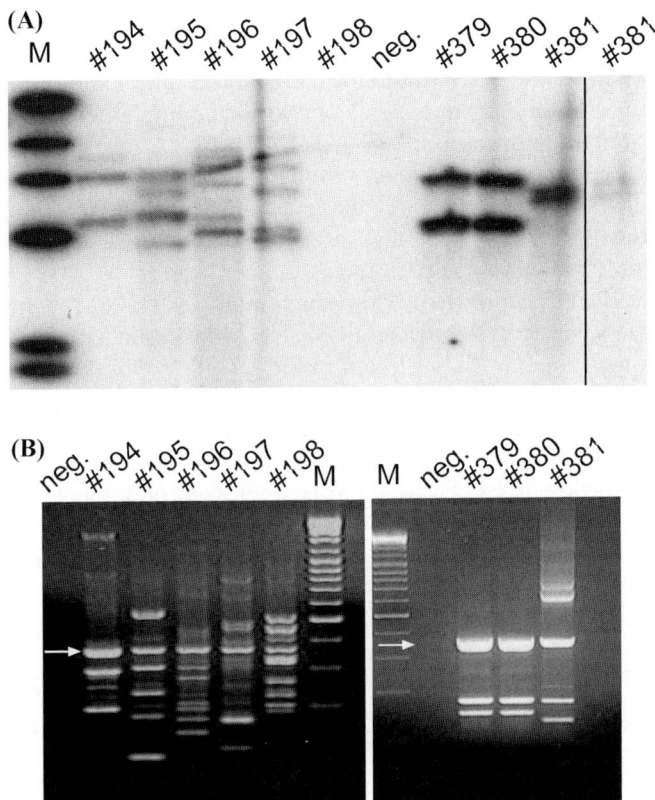

FIGURE 4. Southern blot (**A**) and LM-PCR (**B**) analysis of tumor samples. The Southern blot was probed with coding sequences of TAg. Bands of equal density are likely to be of clonal origin. Short exposure of the Southern blot reveals presence of two distinct bands in 380 (lane far right). The *white arrow* in (**B**) points to the internal control band of the LM-PCR (amplified vector sequence).

for which identical Southern blot results were observed, indicating that they arose from a clone that had self-renewed in the short culture period before injection (FIG. 4A). This demonstrated a selective advantage of this particular clone against the >100,000 competing 32D clones, all being distinguished by their retroviral insertion sites. We cannot rule out that the other tumors were of bi- or oligoclonal origin. Alternatively, they must have arisen from clones with multiple retroviral insertions, which might be triggered by TAg-mediated release of cell cycle checkpoint control. Cells containing more insertions would have a higher likelihood of random insertions in collaborating genes that contribute to clonal tumor evolution.

To identify the insertion sites, we used LM-PCR,[24] focusing on highly reproducible dominant bands, which were excised for direct sequencing (FIG. 4B). The results were consistent with the Southern blot data, in that between 2 and 6 insertions could be characterized per tumor (FIG. 4B). The loci that we found to be affected by retroviral insertions in TAg-associated tumors are listed in TABLE 2. Without exception, each tumor showed insertional events in genes belonging to at least one known complementation group of TAg-associated transformation.

Overall, we recovered 31 sequences from the 8 cases of malignancy available for these studies. As myeloid leukemias 379 and 380 were of identical clonal origin and thus shared two identical insertions (FIG. 4, TABLE 2), 29 different insertions could be annotated to unique genomic loci.

In tumors arising after retroviral transduction of HSC/HPC, potential collaboration partners were recovered as follows: HS 194 showed a hit in the tyrosine kinase receptor gene *Axl*, downstream of the growth factor gene *TGFβ1*, and in the growth-factor locus *Osm/Lif*. *Axl* signaling prevents apoptosis induced by E1A,[26] a TAg-related oncoprotein that also inhibits p53. Oncostatin M encoded by *Osm* and leukemia inhibitory factor (LIF) encoded by *Lif* as well as TGF-β1 have previously been found to be expressed or act as growth factors in cell lines immortalized by TAg,[27,28] suggestive of an autocrine loop. LIF is a potent survival factor for HSC/HPC.[29] This tumor had a third insertion in a poorly defined locus (RP24-108P23).

HS 195 had an insertion in the *Crk* proto-oncogene, encoding a cell cycle associated kinase, and another in *Sox4*, encoding a proto-oncogenic transcription factor involved in apoptosis prevention.[30] A third hit mapped to a gene encoding a purinergic receptor (*P2rx7*), not known to be a potential collaboration partner of TAg. This locus was recently listed as a retroviral insertion site in the retrovirus tagged cancer gene database (RTCGD, December 2006).[4] The RTCGD lists potential cancer genes as suggested from common insertion sites (CIS) in tumors induced by replication-competent retroviruses. However, if only one entry is listed for a given gene in this database (as is the case for *P2rx7*), this cannot be interpreted as a potential proto-oncogene.

BclX found in HS 196 is an antiapoptotic proto-oncogene, highly related to *Bcl2*.[18] The transcription factor allele *Fli1* was also found to be hit in this clone. *Fli1* is a typical insertion site in tumors elicited by replication-competent murine leukemia viruses,[4] and known to accelerate the development of HS in a transgenic mice expressing TAg under control of a myelomonocytic promoter.[17] Cells of this tumor had additional events in four signaling genes (*Ppil1, Ptger2, Srpk2, Suv39h*), the latter known to abrogate the function of AML1, a transcription factor that is required for normal myeloid differentiation.[31] Two of these four genes, *Ptger2 and Srpk2*, were recently introduced as CIS in the RTCGD, and could thus be considered as potential proto-oncogenes.

Two insertions from HS 197 happened in proto-oncogenes involved in the control of cellular survival and proliferation. One was in between the growth factor receptors *Pdgfrb* and *c-Fms* (*Csf1r*), a known collaborator of TAg.[15] The other was upstream of *BC031781/Lefty2*, the former representing a CIS, as defined by the RTCGD, with unknown function, the latter encoding a growth factor belonging to the TGF-β signaling pathway. Another potentially relevant hit occurred in a gene encoding a Src-like activity (*Sla*). The Src pathway is a well-known complementation group of TAg-mediated transformation.[16,32] A fourth insertion recovered from HS 197 mapped to a poorly defined locus (RP23-247J12).

HS 421 showed a hit in the *Pim2/Eras* proto-oncogene locus. The *Pim* family of serine/threonine kinases is involved in the antiapoptotic signaling of cytokines.[33] *Eras* encodes the only known constitutively active form of *Ras* present in the mammalian germ line.[34] Strikingly, another insertion into the same locus (only 10 kb apart) was found in one of the 32D-derived myeloid leukemias (case 381, below). Further insertions that might have promoted HS 421 occurred in the signaling genes *Dusp3* and *Plxna4*, the former recently listed as a single event RTCGD, the latter encoding a semaphorin-related receptor. Moreover, an insertion occurred into the locus *Syne1*, which encodes a nuclear envelope protein, also recently listed as a CIS in the RTCDG.

Sema4b, another gene of the semaphorin receptor family, was hit in MPD-ML 420. *Sema4b* is a CIS observed in hematopoietic malignancies induced by replicating retroviruses.[4] Tumor cells of 420 had two additional insertions in signaling genes (*Tceb3, Il1r1*). *Tceb3* encodes a transcriptional elongation factor potentially involved in cell cycle progression, and *Il1r1* the receptor of interleukin-1. Another insertion recovered from mouse 420 mapped to the *AB041803/2210408F21Rik* locus, which is poorly defined yet also present as a CIS in the RTCGD.

As described above, the 32D-based monocytic leukemias 379 and 380 were derived from the same clone, indicating that self-renewal had occurred already prior to transplantation and that its selection was not a random event. Southern blot and LM-PCR revealed two insertions in this clone (FIG. 4). One was in the locus of *Hspa4*, encoding a heat shock protein and thus a potential physical collaboration partner of TAg's Dnaj domain.[11] *Hspa4* is a CIS in the RTCGD. The second insertion was located in an ubiquitin-like protein potentially involved in cell proliferation control (*Uhrf1*).

The third AML derived from TAg-modified 32D cells showed the earliest disease manifestation (7 weeks) and contained two insertions. One was present in between *Eras* and *Pim2* (only 10kb apart from the insertion found in HS 421, see above). The other insertion had occurred in *Max*, and thus within the Myc pathway, another candidate cooperation partner of TAg.[35]

In contrast to animals suffering from TAg-vector induced leukemia/sarcoma, healthy animals did not show this strong overrepresentation of insertions in potential collaboration genes of TAg. In case 198 (healthy animal that only showed expression of dLNGFR but not TAg as suggested by the absence of the coexpressed EGFP), no known collaboration partner of TAg could be identified as an insertion site (TABLE 2). In three of the four control animals of group III (transplanted with hematopoietic cells marked with vectors that only

TABLE 2. Insertion sites recovered from tumors and healthy hematopoiesis

Case (diagnosis)	Insertions per Southern blot	Locus	Gene ID	Chromosome	Pos. to TSS (intron/exon)	Gene class
194 (HS)	3	Axl*/Tgfb1*	26362/21803	7A3-B1	5875(i4)/95535	POG
		Osm*/Lif*	18413/16878	11A1	1336 (i1)/−28296	POG
						POG
		RP24-108P23	NA	1		unknown
195 (HS)	4	Myo1c/Crk*	12928	11C-E1	−24174	POG
		Sox4	20677	13A3-A5	8988	POG
		P2rx7/P2rx4	18439	5F	39788(e14)	signaling
196 (HS)	6	BclX	12048	2H1	41359(i1)	POG
		Fli1	14247	9A4	37272 (i1)	POG
		Ppil 1*/BC004004	68816	17A3.3	−5575	signaling
		LOC625111/Ptger2*	19217	14B	105216	POG
		Srpk2*/LOC627225	20817	5A3	−21433	POG
		Suv39h	20937	X A1-A2	4858 (i3)	signaling
197 (HS)	5	Csf1r*/Pdgfrb*	12978/18596	18D	−8096/52370	POG
						POG
		Sla*/Tgn*	20491/21819	15D2	−10212 (i6)/140927 (i40)	signaling
						POG
		BC031781*/Lefty2*	208768/320202	1H4	−1225	POG
		RP23-247J12	NA	11	−44070	signaling
421 (HS&MPD-ML)	ND	Otud5/Pim2*/Eras*	18715/353283	X A1.1	−7328/57629	unknown
						POG
						POG
		Dusp3	72349	11D	3175 (e2)	signaling
		RP23-460A13		12D2		unknown
		Plxna4	243743	6 B1	76582 (i2)	signaling
		Syne1	432425	10A1	362 (i1)	POG
420 (MPD-ML)	ND	5430400N05Rik=Zfp710*/Idh2/Sema4b*	209225/20352	7D1	−123/−166040	POG
		Tceb3*/Rpl11	27224	4D3	−7161	signaling
		AB041803*/2210408F21Rik	232685	6A3.3	−587	POG
		Il1r1	16177	1 B	37457(i1)	signaling

Continued.

TABLE 2. Continued.

Case (diagnosis)	Insertions per Southern blot	Locus	Gene ID	Chromosome	Pos. to TSS (intron/exon)	Gene class
379=380 (AML)	2	Hspa4*/Zcchc10	15525	11 B1.3	−849	signaling
		Uhrf1 = NP95	18140	17 D-E1	−1086	signaling
381 (AML)	2	Otud5/Pim2*/Eras*	18715/353283	X A1.1	−17907/68208	POG
						POG
						POG
198 (healthy, dLNGFR)	ND	Max	17187	12D1-D3	−136018	signaling
		Ppia-ps1A_583.1	241074	1C2	39947	signaling
		Rfl1*/Rad51l3*	67338/19364	11C	1821 (i1)/21664	signaling
		Spc18*/Zfp592*	56529/233410	7D2	−385/−46744	signaling
		Supt4h1*/Bzrap1*	20922/207777	11C	16617/−22927	signaling
						POG
414 (healthy, EGFP)	ND	Lipc	15450	9D	95721(i1)	other
		1600020E01Rik*/ 2300003P22Rik	72012	6D1	62072	unknown
415 (healthy, EGFP)	ND	Pol dip2*/Tnfaip1*	67811	11B5	356 (i1)	other
			21927		23544	signaling
		Nt5c2*/Rnf134*	76952	19D2	1211(i1)	other
			71041		90335	signaling
		Adsl1/Siva/Akt1*	11651	12F1-F2	56662	POG
		Mgat1*/Zfp62*	17308/22720	11 B1.2	644 (i1)/40571	other
						signaling
416 (healthy, EGFP)	ND	Slc35f5	74150	1 E3	−1342	signaling
		LOC381528	381528	4B3	n.d.	unknown
		Pdlim2*/9930012K11Rik	213019	14D1	17175	signaling
		Arhgap4	171207	X A7.1	3145	POG
		Stc1	20855	14D1	−106936	signaling
		Dap	223453	15 B2	15448 (i2)	signaling

Insertion sites were mapped using LM-PCR and NCBI database research of sequence matches in the mouse genome (December 2004–December 2006). Gene ID = gene identification number; Pos. to TSS = position of insertion with respect to transcriptional start site; HS = histiocytic sarcoma; MPD-ML = myeloproliferative disorder-like myeloid leukemia; AML = acute myeloid leukemia (all of the monocytic subtype); ND = not determined; *located in gene dense region; selected for characterization; POG = proto-oncogene including common insertion sites (CIS) from RTCGD, December 2004–December 2006.

expressed EGFP), we recovered vector insertion sites by LM-PCR. Only one potential collaboration partner of TAg (*Akt1* gene in case 415, TABLE 2) was found among 12 loci.

In summary, SF91TAg-associated insertion sites in tumor samples showed a strong overrepresentation of insertions in proto-oncogenes and signaling genes that have previously been implicated as potential collaboration partners of TAg.

DISCUSSION

This study reveals that retroviral vector-mediated expression of TAg in HSC/HPC induces HS or MPD-ML, tumors formed by relatively mature hematopoietic cells. This suggests that the myeloid lineage is most susceptible to transformation by TAg, and that TAg does not block myelomonocytic or histiocytic differentiation. Only when TAg was introduced into pretransformed 32D cells, AML of the monocytic subtype was induced. The spectrum of hematopoietic tumors triggered by TAg is thus strongly dependent on the target cell type and their underlying differentiation potential. Finally, insertional mutagenesis is an important cofactor of clonal selection following tumor induction by retroviral expression of TAg.

Using another oncogene (*Sox4*), similar findings were recently reported in another study, and formal proof was obtained that insertionally deregulated cellular signaling genes collaborate with the retrovirally expressed proto-oncogene in tumor acceleration.[9] Together, these studies suggest that clonality and insertion sites should be investigated when using retroviral vectors to unravel the phenotypic consequences of (proto-)oncogene expression. Our study thus supports the hypothesis that the few cases of malignant transformation encountered to date with retroviral vectors encoding potentially therapeutic receptor molecules[5,6] involved the selection of insertional gene alterations that collaborated with ectopic signaling effects of the transgene. Therefore, the expression of signaling genes or proto-oncogenes in retroviral vector-mediated therapeutic interventions requires special caution.[7,8]

The genes recovered in this study not only match with existing literature regarding TAg's transformation partners; different tumor clones even revealed hits in identical loci (*Pim2/Eras*) or related complementation groups (growth factors, kinases, antiapoptotic genes, semaphorin receptors). Considering the rather small data set of this study, this is a remarkable finding. Further arguing for a stringent selection of a subset of appropriate cooperation partners is the phenotypic homogeneity of the tumors elicited by TAg-encoding retroviral vectors: All were of myeloid origin, and HS dominated after gene transfer into HSC/HPC. This was reminiscent of a transgenic mouse model where TAg was expressed from a myeloid promoter.[17] As the retroviral promoter used in our approach is of high activity in all hematopoietic lineages, rather mature myeloid cells must be most sensitive to TAg-mediated transformation. Accordingly, retroviral insertion sites must have been involved in the acceleration of clonal outgrowth, as we have previously observed in normal hematopoietic clones,[1] whereas the tumor phenotype was largely caused by the interplay of TAg with the developmental background of the target cell. However, we would not exclude that the insertion sites may sometimes even determine the malignant phenotype, as probably in the cases of a MPD-ML observed after retroviral transfer of TAg into HSC/HPC. In line with this hypothesis, both kinetics and phenotype of malignancies varied when we used dose-escalated retroviral vectors encoding non-oncogenic transgenes to mutagenize murine HSC/HPC,[2] reflecting that in these cases random insertional events were the major driving force of transformation.

That the retroviral insertion sites were indeed required for TAg-associated malignant transformation and not only accelerating tumor manifestation is supported by the observation that one animal (198) remained healthy and had eliminated TAg-modified cells, despite the transplantation of high numbers of transduced cells. Consequently, it might be of interest to clone the insertion sites in tumors containing integrated SV40 genomes.[10,12]

Interestingly, we found that tumors derived from primary HSC/HPC had more hits than those recovered from pretransformed 32D cells, suggesting that at least two insertional gene alterations are required for disease progression when targeting HSC/HPC with a retroviral vector encoding TAg. Alternatively, this might reflect bi- or oligoclonal origin of these tumors, suggesting that the insertion sites enhanced the competitive fitness of different clones to similar levels. Clonality was proven for two 32D cell-derived tumors that showed a retroviral insertion close to a heat shock protein, representing a cellular cofactor of TAg.[11]

Retroviral vectors might have several advantages over the use of replication-competent retroviruses for the discovery of tumorigenic networks, potentially alleviating the identification of cellular targets for future molecular therapies. First, the number of insertions can be limited, thus reducing the incidence of innocent bystander mutations recovered from tumor cells, and increasing the stringency of selection. Second, the minimal number of lesions required for induction of malignancy can be determined, because mutagenesis by retroviral vectors in contrast to the use of replicating retroviruses is self-limited. Third, for the same reason, the mutagenic impact can be restricted to a cell population of interest that can be enriched to high purity prior to shotgun mutagenesis. Thus, one might also be able to define the exact source of tumor stem cells, which in the hematopoietic system might arise from either HSC or committed HPC, probably depending on the underlying mutations.[36,37] Fourth, vectors can be more easily modified than replicating retroviruses to express (proto-)oncogenes at a certain level or in conjunction with additional genes of interest. Finally, mutagenesis using retroviral vectors is readily applicable in human cells, where the use of replicating retroviruses would raise safety concerns. In summary, we suggest to exploit the mutagenic potential of retroviral vectors for stringent analyses of transforming gene networks in defined cell types.

ACKNOWLEDGMENTS

This study was supported by the Deutsche Krebshilfe (grant: 10-2090-Li I) and DFG (KFO 110-A1; Ba1837-7/1 and Fe568/5-2). We are very grateful to Maimona Id and Nicole Brauer-Dewor for technical assistance, and Rolf Baumann, Hans Grundtke, and Bernd Polivka (Radiotherapy, Hannover Medical School) for irradiation of animals.

REFERENCES

1. KUSTIKOVA, O.S. et al. 2005. Clonal dominance of hematopoietic stem cells triggered by retroviral gene marking. Science **308:** 1171–1174.
2. MODLICH, U. et al. 2005. Leukemias following retroviral transfer of multidrug resistance 1 are driven by combinatorial insertional mutagenesis. Blood **105:** 4235–4246.
3. MIKKERS, H. & A. BERNS. 2003. Retroviral insertional mutagenesis: tagging cancer pathways. Adv. Cancer Res. **88:** 53–99.
4. AKAGI, K. et al. 2004. RTCGD: retroviral tagged cancer gene database. Nucleic Acids Res. **32:** D523–D527.
5. LI, Z. et al. 2002. Murine leukemia induced by retroviral gene marking. Science **296:** 497.
6. HACEIN-BEY-ABINA, S. et al. 2003. LMO2-associated clonal T cell proliferation in two patients after gene therapy for SCID-X1. Science **302:** 415–419.
7. BAUM, C. et al. 2003. Side effects of retroviral gene transfer into hematopoietic stem cells. Blood **101:** 2099–2114.
8. DAVÉ, U., N.A. JENKINS & N.G. COPELAND. 2004. Gene therapy insertional mutagenesis insights. Science **303:** 333.
9. DU, Y. et al. 2005. Cooperating cancer-gene identification through oncogenic-retrovirus-induced insertional mutagenesis. Blood **106:** 2498–2505.
10. SAENZ-ROBLES, M.T., C.S. SULLIVAN & J.M. PIPAS. 2001. Transforming functions of Simian Virus 40. Oncogene **20:** 7899–7907.
11. SULLIVAN, C.S. & J.M. PIPAS. 2002. T antigens of simian virus 40: molecular chaperones for viral replication and tumorigenesis. Microbiol. Mol. Biol. Rev. **66:** 179–202.
12. GAZDAR, A.F., J.S. BUTEL & M. CARBONE. 2002. SV40 and human tumours: myth, association or causality? Nat. Rev. Cancer **2:** 957–964.
13. DAVID, H. et al. 2001. Simian virus 40 is present in human lymphomas and normal blood. Cancer Lett. **162:** 57–64.
14. SHIVAPURKAR, N. et al. 2002. Presence of simian virus 40 DNA sequences in human lymphomas. Lancet **359:** 851–852.
15. RHIM, J.S. et al. 1990. Evidence for the multistep nature of *in vitro* human epithelial cell carcinogenesis. Cancer Res. **50:** S5653–S5657.
16. HAHN, W.C. & R.A. WEINBERG. 2002. Rules for making human tumor cells. N. Engl. J. Med. **347:** 1593–1603.

17. KONE, J. et al. 2002. F-MuLV acceleration of myelomonocytic tumorigenesis in SV40 large T antigen transgenic mice is accompanied by retroviral insertion at Fli1 and a novel locus, Fim4. Leukemia **16:** 1827–1834.
18. GILLET, R. et al. 2002. Effect of Bcl-2 expression on hepatic preneoplasia in mice. Cancer Lett. **177:** 189–195.
19. SCHWIEGER, M. et al. 2002. AML1-ETO inhibits maturation of multiple lympho-hematopoietic lineages and induces myeloblast transformation in synergy with ICSBP deficiency. J. Exp. Med. **196:** 1227–1240.
20. FEHSE, B. et al. 2000. CD34 splice variant: an attractive marker for selection of gene-modified cells. Mol. Ther. **1:** 448–456.
21. WILL, E. et al. 2006. HOXB4 inhibits cell growth in a dose-dependent manner and sensitizes cells towards extrinsic cues. Cell Cycle **5:** 14–22.
22. KRAUNUS, J. et al. 2004. Self-inactivating retroviral vectors with improved RNA processing. Gene Ther. **11:** 1568–1578.
23. LI, Z. et al. 2003. Predictable and efficient retroviral gene transfer into murine bone marrow repopulating cells using a defined vector dose. Exp. Hematol. **31:** 1206–1214.
24. SCHMIDT, M. et al. 2001. Detection and direct genomic sequencing of multiple rare unknown flanking DNA in highly complex samples. Hum. Gene Ther. **12:** 743–749.
25. MIZUKI, M. et al. 2000. Flt3 mutations from patients with acute myeloid leukemia induce transformation of 32D cells mediated by the Ras and STAT5 pathways. Blood **96:** 3907–3914.
26. LEE, W.P. et al. 2002. Akt is required for Axl-Gas6 signaling to protect cells from E1A-mediated apoptosis. Oncogene **21:** 329–336.
27. HOFMANN, M.C. et al. 2003. Establishment and characterization of neonatal mouse sertoli cell lines. J. Androl. **24:** 120–130.
28. LIU, Z.Y. et al. 1997. Characterization of signal transduction pathways in human bone marrow endothelial cells. Blood **90:** 2253–2259.
29. FLETCHER, F.A. et al. 1991. Leukemia inhibitory factor improves survival of retroviral vector-infected hematopoietic stem cells *in vitro*, allowing efficient long-term expression of vector-encoded human adenosine deaminase *in vivo*. J. Exp. Med. **174:** 837–845.
30. BUSSLINGER, M. 2004. Transcriptional control of early B cell development. Annu. Rev. Immunol. **22:** 55–79.
31. CAMERON, E.R. & J.C. NEIL. 2004. The Runx genes: lineage-specific oncogenes and tumor suppressors. Oncogene **23:** 4308–4314.
32. MORENO, C.S. et al. 2004. Signaling and transcriptional changes critical for transformation of human cells by simian virus 40 small tumor antigen or protein phosphatase 2A B56gamma knockdown. Cancer Res. **64:** 6978–6988.
33. WHITE, E. 2003. The pims and outs of survival signaling: role for the Pim-2 protein kinase in the suppression of apoptosis by cytokines. Genes Dev. **17:** 1813–1816.
34. TAKAHASHI, K., K. MITSUI & S. YAMANAKA. 2003. Role of ERas in promoting tumour-like properties in mouse embryonic stem cells. Nature **423:** 541–545.
35. CASILLAS, M.A. et al. 2003. Induction of endogenous telomerase (hTERT) by c-Myc in WI-38 fibroblasts transformed with specific genetic elements. Gene **316:** 57–65.
36. DICK, J.E. 2003. Stem cells: Self-renewal writ in blood. Nature **423:** 231–233.
37. JAMIESON, C.H. et al. 2004. Granulocyte-macrophage progenitors as candidate leukemic stem cells in blast-crisis CML. N. Engl. J. Med. **351:** 657–667.

NUP98 Dysregulation in Myeloid Leukemogenesis

M. A. S. MOORE,[a] K. Y. CHUNG,[b] M. PLASILOVA,[a] J. J. SCHURINGA,[c] J.-H. SHIEH,[a] P. ZHOU,[d] AND G. MORRONE[e]

[a]*Moore Laboratory, Cell Biology Program, Memorial Sloan-Kettering Cancer Center, New York, New York 10021, USA*

[b]*Department of Medicine, Memorial Sloan-Kettering Cancer Center, New York, New York 10021, USA*

[c]*Department of Hematology, University Medical Center Groningen, University of Groningen, the Netherlands*

[d]*Department of Pathology and Laboratory Medicine, Weill Medical College and Graduate School of Medical Sciences of Cornell University, New York, New York 10021, USA*

[e]*Department of Experimental and Clinical Medicine "Gaetano Salvatore," University of Catanzaro Magna Graecia, Catanzaro, Italy*

ABSTRACT: Nucleoporin 98 (NUP98) is a component of the nuclear pore complex that facilitates mRNA export from the nucleus. It is mapped to 11p15.5 and is fused to a number of distinct partners, including nine members of the homeobox family as a consequence of leukemia-associated chromosomal translocations. NUP98-HOXA9 is associated with the t(7;11)(p15;p15) translocation in acute myeloid leukemia (AML), myelodysplastic syndrome, and blastic crisis of chronic myeloid leukemia. Expression of NUP98-HOXA9 in murine bone marrow resulted in a myeloproliferative disease progressing to AML by 7–8 months. Transduction of NUP98 fusion genes into human CD34$^+$ cells confers a proliferative advantage in long-term cytokine-stimulated and stromal cocultures and in NOD-SCID engrafted mice, associated with a five- to eight-fold increase in hematopoietic stem cells. NUP98-HOXA9 expression inhibited erythroid and myeloid differentiation but enhanced serial progenitor replating. NUP98-HOXA9 upregulated a number of homeobox genes of the A and B cluster as well as MEIS1 and Pim-1, and downmodulated globin genes and C/EBPα. The HOXA9 component of the NUP98-HOXA9 fusion protein was protected from cullin-4A–mediated ubiquitination and subsequent proteasome-dependent degradation. In NUP98-HOX–transduced CD34$^+$ cells and cells from AML patients with t(7;11)(p15;p15) NUP98 was no longer associated with the

Address for correspondence: Dr. Malcolm A.S. Moore, Moore Laboratory, Cell Biology Program, Memorial Sloan-Kettering Cancer Center, New York, NY 10021. Voice: 212-639-7090; fax: 212-717-3618.

m-moore@ski.mskcc.org

nuclear pore complex but formed intranuclear aggregation bodies. Analysis of NUP98 allelic expression in AML and myelodysplastic syndrome showed loss of heterozygosity observed in 29% of the former and 8% of the latter. This was associated with poor prognosis.

KEYWORDS: homeobox genes; nucleoporin 98; stem cells; leukemia; ubiquitination; loss of heterozygosity

LEUKEMOGENIC POTENTIAL OF TRANSLOCATIONS INVOLVING NUCLEOPORIN 98 (NUP98)

One of the most promiscuous fusion partner genes detected in acute myeloid leukemia (AML) is the NUP98 gene, located on chromosome 11p15.5, which to date has been observed fused to 21 different fusion partners (TABLE 1).[1] The association of Nup98 gene arrangements with therapy-associated AML/myelodysplastic syndrome (MDS) suggests that the NUP98 locus may be susceptible to gross chromosomal rearrangements induced by genotoxic agents. NUP98 encodes a 98-kD protein that is an important component of the nuclear pore complex (NPC). The nuclear pore is involved in the transport of proteins larger than 40 kD, RNAs, and a subset of DNA viruses.[2,3] The NPC is composed of a central eight-fold, radially symmetrical ring-and-spoke assembly, a filamentous nuclear basket with cytoplasmic fibers and NUP family proteins present on the nuclear, cytoplasmic, or both sides of the NPC.[4,5] The NUP98 protein is synthesized directly from its specific transcripts, or an alternative transcript generates a precursor NUP186 protein that is then proteolytically cleaved into NUP98 and NUP96.[6,7]

Recent studies show that NUP98 is localized on both sides of the NPC through interaction with NUP214/NUP88/NUP62 on the cytoplasmic side and NUP160/NUP133/NUP107/NUP96 on the nuclear side.[8] NUP98 is the only NUP protein to have GLFG repeats as its docking site and these sites exhibit high affinity for exportins, such as CRM1, a member of the karyopherin-3 family.[9] NUP98 is involved in transcriptional regulation and GLFG repeats have transcriptional activation potency mediated in part by interaction with CBP/p300.[10] When proteins with a nuclear export signal are exported from the nucleus, they form a complex with CRM1 and combine with RanGTP and then interact with the NUP98 docking site. Following hydrolysis of RanGTP to RanGDP by RanGAP the transported protein is then released into the cytoplasm. NUP98 is a mobile NUP that is reported to translocate from the NPC to the nuclear body during active transcription or to the nucleoli when overexpressed or coexpressed with Rev, suggesting that it may have other, yet to be determined, functions.[3,7] NUP98 is normally moved between the novel nuclear body (GLFG body), and the NPC and the GLFG domain is itself responsible for targeting NUP98 to the GLFG body.[11] NUP98 is implicated in the modulation of nuclear RNA, and export and injection of NUP98-specific

TABLE 1. Translocations in leukemia associated with NUP98 fusion

Cytogenetics	Fusion partner genes	Product	Disease
t(7;11)(p15;p15)	HOXA9, 11, 13,	Homeodomain	MDS, AML, CML
t(11;12)(p15;q13)	HOXC11, 13	Homeodomain	AML
t(2;11)(q31;p15)	HOXD11, 13	Homeodomain	MDS, AML
t(1;11)q23;p15)	PMX1,	Homeodomain	AML, MDS, CML
T(9;11)(q34;p15)	PMX2	Homeodomain	AML
Inv11(p15;q22)	DDX10	DEAD-box RNA helicase	MDS, AML, CML
t(11;20)(p15;q11)	TOP1	Topoisomerase	MDS, AML
t(3;11)(p24;p15)	TOP2	Topoisomerase	AML
t(6;11)(q24;p15	C6ORF80	Uncharacterized	AML, T-ALL
t(9;11)(p22;p15)	LEDGF	Transcriptional coactivator	AML, CML
t(5;11)(q35;p15)	NSD1	Transcription cofactor	AML
t(8;11)(p11;p15)	NSD3	Transcription cofactor	AML
t(11;17)(p15;p13)	PHF23	Plant homeodomain protein	AML
t(4;11)(q21;p15)	RAP1GDS1	Guanine exchange factor	T-ALL
t(10;11)(q25;p15)	ADD3	Adductin3. Skeletal protein for cell membrane	AML, T-ALL
t(11;21)(p12p13)	JARIDA1A	Retinoblastoma binding protein	AML
t(11;18)(p15;q12)	SETBP1	SET binding protein	T-ALL

antibodies into Xenopus oocyte nuclei selectively inhibits nuclear export of multiple RNA species, including snRNAs, 5S RNA, large ribosomal RNAs, and mRNA, while leaving protein import intact.[12] The N-terminal M9-, GLEB, and FG domains of NUP98 bind to RNA export factors, including RaeI, Kapβ, and TAP, which are essential components of multiple RNA export pathways.[13]

Homozygous NUP98 knockout mice die at 6.5–7.5 days postconception but cell lines established at earlier stages from knockout mice proliferate extensively and exhibit no gross structural abnormalities of the NPC; however, the capacity of their NPC to bind karyopherins was significantly reduced.[14]

The fusion partners of NUP98 observed in leukemia form two distinct groups: homeobox genes (homeodomain [HD] proteins HOXA9, 11, 13, HOXC11, 13, and HOXD11, 13, and class-2 homeobox genes PMX1 and 2) and nonhomeobox genes (TABLE 1).[1] All NUP98 fusions join the N-terminal GLFG repeats of NUP98 to the C-terminal portion of the partner gene, which, in the case of the homeobox gene partners, includes the HD (FIG. 1). The GLFG repeat is required to target NUP98 to nuclear structures termed *GLFG bodies*, whose function is obscure. GLFG also binds TAP and may facilitate interaction with the transcriptional coactivators CREB-binding protein (CBP) and p300.[15] This latter interaction may provide a link between NUP98 protein mobility and active transcription. The GLFG repeats of NUP98 in NUP98-PMX1 and probably of other NUP98-containing fusion proteins are capable of recruiting histone deacetylase 1 (HDAC1), as well as CBP, suggesting that these fusion proteins can act as both transcriptional coactivators and corepressors.[16]

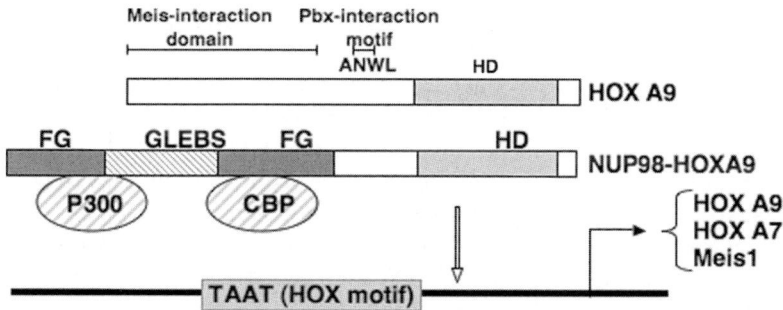

FIGURE 1. Comparison of the structure of HOXA9 and the NUP98-HOXA9 fusion protein. The transcriptional regulatory region of HOXA9 is replaced by the C-terminal FG region of NUP98. Note the retention of the N-terminal DNA binding homeodomain (HD) and the PBX-interaction domains of HOXA9 in the chimeric protein.

The NUP98-HOX fusion also preserves the HOXA sequences that encode both a DNA-binding HD and a short ANWL motif N-terminal to the HD that binds the PBX family of transcriptional cofactors (FIG. 1). NUP98-HOXA9 could also heterodimerize with endogenous PBX, which forms both heterodimers and triple complexes with HOXA9 and MEIS1 proteins. NUP98-PMX1 appears to have a dominant-negative effect on PMX1/RF-mediated c-FOS activation.[16] c-FOS is dramatically increased when leukemic cells are induced to differentiate and blocking c-FOS expression using antisense oligomers impairs normal myeloid differentiation.

There is no homology among nonhomeobox partners for NUP98; however, these unrelated molecules possess common coiled-coil domains.[17] Such domains have been shown to be critical for the leukemogenic function of PML-RARα and AML1-ETO, and both form high MW complexes within the cell that may promote excessive recruitment of CBP/p300. Thus the mechanism involved in the oncogenicity of NUP98-HOX fusions is complex, involving both gene activation and repression.

MURINE MODEL OF NUP98 FUSION WITH HOMEOBOX GENES

Calvo et al.[18,19] showed that murine myeloid progenitors immortalized in vitro by HOXA9 exhibit specific differentiation characteristics: they become neutrophils in response to G-CSF, macrophages in response to M-CSF, and they die in the presence of c-Kit ligand (KL) despite the fact that they express cell surface c-Kit (SCF-R, CD117). MEIS1, a TALE HD gene, alters the differentiation of HOXA9-immortalized cells by suppressing their G-CSF-induced granulocytic differentiation and by allowing them to proliferate in response to KL. These novel MEIS1 functions are proposed to cooperate with HOXA9 to promote leukemic potential.[18] Mouse myeloid progenitors immortalized with NUP98-HOXA9 exhibited cytokine-specific differentiation responses distinct from those of progenitors immortalized by HOXA9, namely, differentiation arrest in response to G-CSF or GM-CSF as well as proliferation in response to KL alone.[19] Using $HOX9^{-/-}$ marrow, Calvo et al.[19] demonstrated that expression of HOXA9 is not required for myeloid immortalization by NUP98-HOXA9. Rapid leukemogenesis by NUP98-HOXA9 may therefore result from both the intrinsic functions of NUP98-HOXA9 as well as functions of coexpressed HOX and MEIS1 genes. MEIS1 binds N-terminal HOX9 sequences that are eliminated by fusion with NUP98 (TABLE 1), leading to speculation that the stronger transforming function of NUP98-HOXA9 relative to HOXA9 is that it consolidates, within a single oncoprotein, the collective biochemical functions of HOXA9 and MEIS1. NUP98-HOXA9 also enforced strong transcription of the cellular HOXA9, HOXA7, and MEIS1 genes at levels similar to those found in mouse leukemias generated by proviral activation of HOX9 and MEIS1. NUP98-HOXA9 increases the level of PBX2, most likely through a mechanism of MEIS1-mediated protein stabilization. NUP98-HOXA9 mutants were generated that were unable to physically bind or cooperate with PBX proteins and unable to transactivate. Despite their inability to physically bind or functionally cooperate with PBX proteins, they retained ability to immortalize IL-3-dependent progenitors with the same efficiency and kinetics as did wild-type NUP98-HOXA9.[19] These mutants retained the ability to upregulate endogenous HOXA9 and MEIS1.[19] The genetic targets through which NUP98-HOXA9 mediates both differentiation arrest and enforcement of HOX9 and MEIS1 transcription are unlikely to require PBX proteins as heterodimer partners. NUP98-HOXA9 activates or maintains transcription of a select subset of HOX genes and, significantly, this subset contains both of the HOX genes (A7 and A9) selected by proviral integration to cooperate with MEIS1 in BXH-2 mouse leukemia development. In this context, HOXA9, HOXA7, and MEIS1 are also expressed persistently in the vast majority of human AML.

Expression of NUP98-HOXA9 in murine bone marrow resulted in a myeloproliferative disease (MPD) in transplanted mice, with neutrophil leukocytosis and extramedullary hematopoiesis progressing to AML by 7–8 months. Ten percent of NUP98-HOXA9 transgenic mice developed AML after a long latency, and NUP98-HOXA9 leukemia exhibits a more differentiated phenotype than wild-type HOXA9 AML.[20] When these NUP98-HOXA9 transgenic mice were crossed with the BXH2 strain (which shows a high incidence of leukemia by 1 year of age due to a horizontally transmitted B-ecotropic murine leukemia virus), AML onset was earlier (4–8 months).[21] In this model, retroviral insertional mutagenesis identified a number of cofactors that interacted with NUP98-HOXA9 in leukemia progression. The most frequent was MEIS1 followed by Fcr2B/Fcr1 (immunoglobulin Fc receptor) and DNAL4 (dynein light chain) and CON1 (SCR domain protein). Collaboration of MEIS1 with

NUP98-HOXA9 reduced the median latency of AML development to 4–5 months.[21] NUP98-HOXD13 transgenic mice develop a uniformly fatal MDS that recapitulates all of the key features of clinical MDS, including peripheral blood cytopenias, bone marrow dysplasia and apoptosis, and transformation at approximately 12 months to leukemia (erythroleukemia, chronic megakaryocytic leukemia with myelofibrosis progressing to acute megakaryocytic leukemia).[22] Progression to AML was accelerated by the combination of NUP98-HOXD13 with MEIS1.[23] NUP98-PMX1 transgenic mice showed late development of myeloid leukemia, predominantly chronic myeloid leukemia (CML).[24] Artificial fusions of NUP98 to HOXA10 and HOXB3 (HOX genes not previously identified as fusion proteins) showed transforming activity but fusion with HOXB4 did not.[25] This indicates the intrinsic leukemogenic potential of many HOX genes that correlates with ability to block hematopoietic differentiation and enhance self-renewal. The strong transforming potential of NUP98-HOXD13 or NUP98-HOXA10 allowed development of preleukemic myeloid lines from bone marrow that faithfully replicated the first step of HOX-induced leukemogenesis.[26] These lines contain early granulo-monocytic progenitors with extensive *in vitro* self-renewal capacity, short-term myeloid repopulating activity, and low propensity for spontaneous leukemic conversion. MEIS1 efficiently induces leukemic progression of these lines independent of its direct binding to DNA and likely reflects its ability to increase the repopulating capacity of the preleukemic cells by increasing their self-renewal/proliferative potential.

NUP98-HOXA9 has been detected in blastic crisis of Philadelphia chromosome positive CML indicating that cooperation between NUP98-HOXA9 and the BCR-ABL signaling pathway via receptor tyrosine kinase may be important in the progression of CML.[27,28] Two independent groups have reported oncogenic interaction between NUP98-HOXA9 and BCR-ABL or Tel-PDGFRβR chimeras in retroviral-mediated gene transfer experiments.[29,30] In these studies, coexpression of both chimeras significantly accelerated acute disease. Overexpression of the wild-type Flt3 receptor collaborates with NUP98-HOXA10 and NUP98-HOXD13 to introduce aggressive AML in a mouse transplant model, and MEIS1 and HOXA9 overexpression both lead to marked elevation of Flt3.[31]

NUP98 fusion with the nonhomeobox gene TOP1 also leads to an MPD with high WBC associated with neutrophil maturation and development of AML by a median of 7.5 months.[32] The transcriptional coactivator property of TOP1 provides a possible functional similarity to HOX genes.

NUP98 chimeras may have acquired oncogenicity via (*a*) transcriptional activation of downstream genes achieved by interaction between GLFG and coactivators such as CBP/p300; (*b*) disruption of the N-terminal GLFG domain resulting in nuclear retention and/or intranuclear translocation of certain proteins, such as HOX cofactor MEIS1, to enhance cooperation between HOX and MEIS1; and (*c*) impaired mRNA export due to mislocalization of the GLFG motif, resulting in a decrease in protein expression of potential tumor suppressor genes. These are not mutually exclusive. Also NUP98-HOXA9 chimeras may act in a dominant-negative fashion for wild-type NUP98.

INHIBITION OF MYELOID AND ERYTHROID DIFFERENTIATION IN NUP98-HOXA9–EXPRESSING HUMAN CD34$^+$ CELLS

A significant reduction in erythroid progenitors was observed in BFU-E clonogenic assays of CD34$^+$ cells retrovirally transduced with an NUP98-HOXA9 vector and stimulated by erythropoietin alone or by Epo+ KL or a combination of cytokines (TABLE 2).[33] Morphologically, there was a predominance of immature blast promyelocytes relative to the mature erythroid elements predominating in control erythroid colonies.[34] Myeloid colony formation was also reduced in NUP98-HOXA9 cultures stimulated by G-CSF or GM-CSF alone but not in cultures with multiple cytokines including IL-3 (TABLE 2).[33] In cytokine-stimulated suspension cultures of sorted NUP98-HOXA9–transduced CD34$^+$ cells, there was maturation arrest in the neutrophil lineage with absence of terminally differentiated neutrophils (TABLE 2).[33] *In vivo*, following engraftment into NOD-SCID IL2Rγ-/- mice, NUP98-HOXA9 cells differentiated into lymphoid CD8$^+$ cells as well as myeloid lineage although the absolute T cell engraftment was significantly less than that seen with control cells.[33] This observation reinforces the view that NUP-98-HOXA9 targets a stem cell with lymphomyeloid potential.

ENHANCEMENT OF STEM CELL PROPERTIES IN NUP98-HOX–TRANSDUCED HUMAN CD34$^+$ CELLS

Replating of primary colonies from cultures stimulated with a combination of cytokines yielded secondary myeloid colonies but no BFU-E, with NUP98-HOXA9–expressing primary colonies generating significantly more secondary colonies than MIGR1 control colonies ($P < 0.01$).[33] Tertiary passage was obtained only with NUP98-HOXA9 colonies. This suggests that the expression of NUP98-HOXA9 resulted in acquisition of some measure of self-renewal by immature myeloid progenitors.

The selective advantage for human CD34$^+$ cells transduced with NUP98 fused to either HOXA9, HOXA11, or HOXA13 was evaluated *in vitro* in long-term competition cultures on stroma or with cytokines alone (FIG. 2A). Initially, the transduced GFP+ cells and control cells showed comparable proliferation but by 3 weeks the cells with NUP-HOX fusion progressively outgrew the control cells, eventually predominating in cultures from 5 to 8 weeks (FIG. 2A). These kinetics are consistent with enhanced proliferation

TABLE 2. Impairment of myeloid and erythroid differentiation in clonogenic and suspension culture assay of NUP98-HOXA9–transduced cord blood CD34+ cells[33]

	Clonogenic assay[a]			
		Colonies/500 cells		
Progenitor	Stimulus	Control	NUP98-HOXA9	Percentage of control
CFU-GM	GM-CSF (20 ng/mL)	20.0 ± 2.0	4.0 ± 2.0	20%
CFU-GM	G-CSF (2 ng/mL)	17.5 ± 2.0	4.0 ± 2.0	23%
CFU-GM	G-CSF (20 ng/mL)	35.0 ± 2.0	10.0 ± 2.0	29%
CFU-GM	IL-3, G-CSF, KL, IL-6, Epo	27.0 ± 3.0	30.0 ± 2.0	111%
BFU-E	Epo (0.6 U/mL)	6.0 ± 1.0	2.0 ± 1.0	33%
BFU-E	Epo (6.0 U/mL)	40.0 ± 2.2	20.0 ± 3.0	50%
BFU-E	Epo (0.6 U/mL) KL (20 ng/mL)	37.0 ± 2.0	22.0 ± 2.5	55%
BFU-E	Epo (0.6 U/mL) KL (20 ng/mL)	49.0 ± 3.0	21.0 ± 2.5	43%
BFU-E	IL-3, G-CSF, KL, IL-6, Epo	54.0 ± 4.0	30.0 ± 4.0	55%

	Week 3 suspension culture[b]		
		Percentage of total cells	
Morphology	Stimulus	Control	NUP98/HOXA9
Blast/promyelocytes	Flt-3L, Tpo, KL (100 ng/mL)	32.0 ± 3.0%	76.0 ± 6.0%
Myelocyte/metamyelocyte	Flt-3L, Tpo, KL (100 ng/mL)	35.0 ± 4.0%	23.0 ± 4.0%
PMN (bands, segmented)	Flt-3L, Tpo, KL (100 ng/mL)	33.0 ± 2.0%	1.0 ± 0.5%

[a]500 NUP98-HOXA9-GFP+ transduced and FACS sorted cord blood CD34+ cells plate in 1 mL methylcellulose cultures in triplicate with indicated cytokine. Colonies scored at day 14.

[b]Week 3 suspension culture of 10,000 sorted NUP98-HOXA9-GFP FACS sorted CD34+ cells in T12.5 flasks with indicated cytokines. Cytokines added weekly.

at the stem cell level.[33] In a more rigorous evaluation of hematopoietic stem cell (HSC) function, CD34+ cells transduced with NUP98-HOXA9, NUP98-HOXA11, or NUP98-HOXA13 were evaluated for HSC function using the limiting dilution week 5–6 CAFC assay on MS-5 stromal cells (FIG. 2B). NUP98-HOXA9–expressing cells began to form cobblestone areas (CAs) by week 1, and these progressively increased in number until approaching confluence by weeks 3–4. By weeks 5–6, greater numbers (four- to eightfold more) of these candidate HSC were detected in NUP98-HOXA9 cultures compared to control with a four- to fivefold expansion seen with NUP98-HOXA11– and NUP98-HOXA13–transduced cells, indicating an expansion of the primitive stem/progenitor cell pool[33] (M.A. Moore & M. Plasilova, unpublished data). Experiments where sorted HSC-enriched CD34+/CD38− and progenitor-enriched CD34+/CD38+ NUP98-HOXA9–transduced populations were assayed in MS-5 stromal cocultures revealed that only the former subset was able to generate week 5–6 CAFCs, suggesting that the proliferative advantage and enhanced self-renewal of NUP98-HOXA9–expressing cells reflects an effect at the level of primitive HSCs.[33] Replating of dissociated week 5 CAs

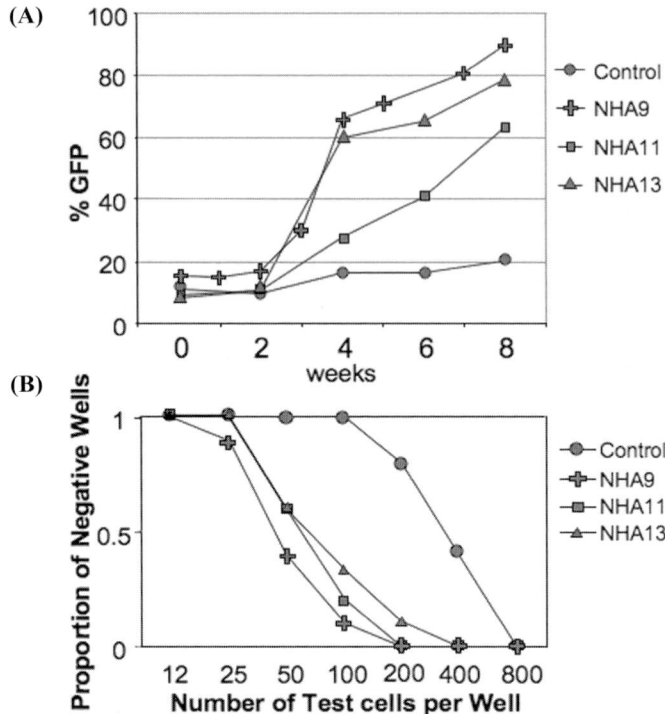

FIGURE 2. NUP98-HOXA9 impairs myeloid and erythroid differentiation and enhances stem cell expansion in CD34$^+$ cells. (**A**) Long-term suspension culture of cord blood CD34$^+$ cells transduced with NUP98-HOXA9, NUP98-HOXA11, NUP98-HOXA13, or control empty vectors. Eight percent to 18% of input cells were EGFP$^+$ and the percentage of EGFP$^+$ cells increased rapidly and progressively by 3 weeks of suspension culture in the presence of Flt3L, KitL, and Tpo (100 ng/mL). (**B**) CAFC assay for week 5 CA formation. Cord blood CD34$^+$ cells were transduced with NUP98-HOX fusion vectors or control empty vector, and EGFP$^+$ cells sorted and plated at limiting dilution (0–800 cells per well) on MS-5 stroma.

onto fresh MS-5stroma showed that approximately 30% NUP98-HOXA9 CAs, but not control areas, were able to generate secondary and tertiary CAs. Colony assays with sorted EGFP$^+$ week 5 cobblestone-derived cells highlighted (*a*) the presence of a much higher number of progenitors in the NUP98-HOXA9 CAs than in those derived from control transduced cells, and (*b*) a significant proportion of CFCs was BFU-Es and CFU-GEMMs in the NUP98-HOXA9 CAs but not in the control. This implies that enforced expression of the fusion protein, particularly in the context of the hematopoietic microenvironment, maintains the transduced cells in a more immature pluripotent state.

The *in vitro* proliferative advantage of enforced NUP98-HOXA9 expression in CD34$^+$ cells persisted *in vivo*, equivalent numbers of nonsorted NUP98-HOXA9– and control vector–transduced CD34$^+$ cells were transplanted into

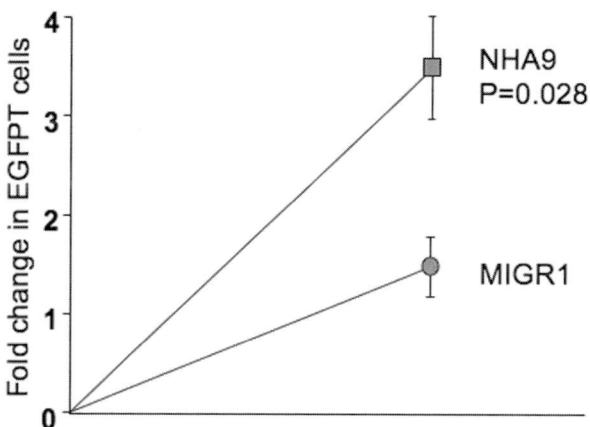

FIGURE 3. NUP98-HOXA9 enhances human hematopietic engraftment in an immunodeficient mouse model. Comparison of the expansion of human CD45$^+$, EGFP$^+$ cells in bone marrow following intravenous or intrafemoral injection of 100,000 unsorted NUP98-HOXA9-EGFP$^-$ transduced or control transduced cord blood CD34$^+$ cells into NOD/SCID, NOD/SCID IL2Rγ^{null} and NOD/SCIDβ2null mice. Transduction efficiency of the input CD34$^+$ population was 9%–20%. Data expressed as fold change in percentage of EGFP$^+$ cells engrafted in the bone marrow after 5–7 weeks. 12 mice in experimental and control groups. The NUP98-HOXA9 mean increase was 3.31-fold, significantly greater ($P = 0.02$) than the control.

irradiated NOD/SCID, NOD/SCID $\beta_2 M^{null}$, and IL2Rγ^{null} mice and engraftment measured 5–7 weeks posttransplantation by fluorescence-activated cell sorting (FACS) analysis of human CD45 and EGFP expression in bone marrow mononuclear cells.[33] In all three models, human hematopoietic engraftment was obtained in bone marrow (2%–42%) and a selective *in vivo* proliferative advantage of NUP98-HOXA9–expressing cells relative to the nontransduced cells coinjected was revealed by a mean 3.3-fold increase in the percentage of human CD45 cells recovered at 5–6 weeks that expressed EGFP relative to the EGFP percentage in the input population (9%–15%) (FIG. 3). This was significantly greater ($P < 0.028$) than the average expansion of EGFP cells seen in control mice (1.6-fold). The proliferative advantage of NUP98-HOXA9–expressing cells was also accompanied by extramedullary engraftment in the spleen and liver. These mononuclear EGFP$^+$ cells were predominantly myelomonocytic blasts. Additional colony assays from engrafted CD34$^+$-immunopurified mononuclear cells from bone marrow of the engrafted mice revealed a predominance of CFU-GEMM and CFU-GM colonies. The greatest degree of human engraftment (31%–34%) was observed in NOD/SCID IL2Rγ^{null} mouse marrow with 20%–28% of the human cells expressing the myelomonocytic marker CD14, 3%–5% expressing the erythroid marker glycophorin A, and 6%–9% expressing lymphoid markers CD7, CD8, or CD4. The

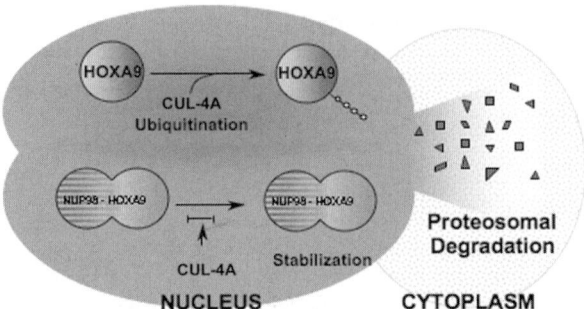

FIGURE 4. NUP98-HOXA9 protein is resistant to ubiquitination. HOXA9 protein is subject to rapid ubiquitination and proteosomal degradation mediated by the CUL-4 ubiquitin ligase complex. HOXA fused to NUP98 is resistant to CUL-4A ubiquitination and the protein has a three- to fivefold longer half-life.

expression of EGFP in CD8 T cells (43% in control, 19–29% in the NUP98-HOXA9–engrafted mice) indicated that expression of the fusion protein was not incompatible with T lymphocyte differentiation although this pathway of differentiation may be quantitatively impaired.[33]

ENHANCED STABILITY OF THE NUP98-HOXA9 FUSION PROTEIN COMPARED TO WILD-TYPE HOXA9 IS MEDIATED BY DECREASED SENSITIVITY TO CULLIN-4A–DEPENDENT UBIQUITINATION

In murine studies Calvo et al.[19] monitored the stability of NUP98-HOXA9 and HOXA9 proteins by quantitating anti-HOXA9 immunostaining of total lysates from cells treated with cyclohexamide over a 6-h time course. HOXA9 exhibited a half-life of 35 min, whereas that of NUP98-HOXA9 was 105 min. Zhang et al.[35] showed that cullin-4A (CUL-4A) ubiquitin ligase–dependent HOX protein degradation plays a central role in determining HOX protein levels, which in turn regulate the balance between HSC self-renewal and differentiation. To examine whether the stability of the NUP98-HOXA9 fusion protein was also subjected to regulation by CUL-4A, HeLa cells were transiently transfected with both *HA-tagged HOXA9* and *HA-NUP98-HOXA9,* together with increasing doses of *CUL-4A*. HA-HOXA9 was readily degraded in a CUL-4A dose-dependent manner.[33] In striking contrast, the steady-state levels of HA-NUP98-HOXA9 were not affected significantly by the increased expression of CUL-4A. Pulse chase analysis confirmed the difference in protein half-lives. These results indicate that the NUP98-HOXA9 fusion protein is resistant to CUL-4A–mediated proteolysis (FIG. 4). The transforming potential of NUP98-HOXA9 may thus, in part, be related to its enhanced stability, with a three- to fivefold longer half-life of the fusion HOXA9 protein relative to wild-type HOXA9. Enhanced NUP98-HOXA9 stability would also allow for more robust expression of endogenous homeobox genes. The ubiquitination site on the first helix of HOXA9 is present in the fusion protein, but it is possible that some form of steric hindrance mediated by the NUP98 component, dimerization of NUP-HOXA9, or its binding to wild-type NUP98, block interaction with CUL-4A. An alternative mechanism may involve a CBP/p300 bound to FG-repeats on NUP98 mediating acetylation of lysine residues required for ubiquitination. p53 and its structural and functional homolog p73 are protected from ubiquitination by such an acetylation process.[36,37] Competition between ubiquitination and acetylation of overlapping lysine residues constitutes a novel mechanism regulating protein stability.[38]

DISCOVERY OF POTENTIAL DOWNSTREAM TARGETS OF NUP98-HOXA9 BY MICROARRAY ANALYSIS

Abberant HD-containing proteins, such as NUP98-HOXA9, may cause leukemia by interfering with the transcriptional programs of hematopoietic proliferation and differentiation. In this context, over 102 cytoplasmic mRNAs were significantly altered in K562 myeloid leukemic cells transduced with NUP98-HOXA9, 92 being increased and only 10 decreased.[39] A similar analysis of wild-type HOXA9 revealed 13 target genes, 12 of which were upregulated and 1 downregulated. Two studies have been reported on comparative microarray analysis of RNA isolated from freshly sorted CD34$^+$ NUP98-HOXA9–expressing CD34$^+$ cells and empty vector control cells, one using cord blood CD34$^+$ cells[33] (TABLE 3) and one using G-CSF mobilized adult CD34$^+$ cells from pretreated patients with multiple myeloma. In the former study, at 3 days posttransduction 50 genes were upregulated and 15 downregulated, and in the latter study, 60 genes were upregulated and 4 downregulated (TABLE 3).[33,34] Interestingly, there was no overlap of the gene-expression profile in NUP98-HOXA9–transduced CD34$^+$ cells with the transcriptosome of NUP98-HOXA9–transduced human myeloid leukemic cells that had already undergone leukemic transformation with BCR-ABL.[39] Consistent with murine studies,[19] expression of NUP98-HOXA9 upregulated the expression of many homeobox genes, including endogenous HOXA5, HOXA6, HOXA7, HOXA9, and HOXB5, and TALE HD genes MEIS1 and PBX3 in the cord blood study[33] and HOXA9, HOXB2, HOXC6, PBX3, and MEIS1 in the adult CD34 study.[34] Over 200 genes are reported regulated by HOXA9 in leukemic cell lines or CD34$^+$ cells,[43,44] but the genes do not include the HOX/MEIS1 genes upregulated by NUP98-HOXA9. HOXA9 plays an important role in normal hematopoiesis and HOXA9, HOXA7, and MEIS1 are expressed in primary myeloid leukemic cells and early self-renewing CD34$^+$ cells and downregulate with differentiation.[45,46] Thrombopoietin, a key regulator of HSC proliferation, enhanced HOXA9 nuclear import and interaction with MEIS1 in HSC

TABLE 3. Differential gene expression in cord blood CD34+ cells expressing NUP98-HOXA9[a]

Gene symbol	Gene	Fold change	Ref. 34[b]	Other CD34/CD133 arrays		
				Ref. 40[c]	Ref. 41[d]	Ref. 42[e]
FGF18	Fibroblast growth factor 18	21.11	+			
UTS2	Urotensin 2	13.00	+			
FOSB	FBJ murine osteosarcoma viral oncogene homolog B	12.13				+
HOXA9	Homeobox A9	6.50			+	
HOXA5	Homeobox A5	4.92		+		
CD44	CD44-Human CD44 antigen precursor	4.92				
CD69	CD69 (p60, early T cell activation antigen)	4.59	+			
H2BFA	H2B histone family, member A	4.00				
PCLO	Piccolo (presynaptic cytomatrix protein)	4.00				
PLA2G4A	Phospholipase A2, group IVA	3.73				
CDC2L5	Cell division cycle 2–like 5	3.73				
DDX17	DEAD/H box polypeptide 17 (72 kD)	3.73				
HOXB5	Homeobox B5	3.48				
KIAA1110	KIAA1110 protein	3.48				
SRRM1	Serine/arginine repetitive matrix 1	3.25				
RC3	Rabconnectin-3	3.25				
SPUVE	Protease, serine 23	3.03				
PBX3	Pre-B-cell leukemia transcription factor	3.03	+			
PIM1	Pim-1 oncogene	3.03				
RBPMS	RNA-binding protein gene with multiple splicing	3.03		+	+	+
TPX1	Testis specific protein 1	3.03				

Continued.

TABLE 3. Continued.

Gene symbol	Gene	Fold change	Ref. 34[b]	Other CD34/CD133 arrays			
				Ref. 40[c]	Ref. 41[d]		Ref. 42[e]
HNRPA1	Heterogeneous nuclear ribonucleoprotein A1	3.03					
IGFBP4	Insulin-like growth factor binding protein 4	2.83					
SCA7	Spinocerebellar ataxia 7	2.83					
PRKWNK1	Protein kinase, lysine deficient 1	2.83					
BPAG1	Bullous pemphigoid antigen 1	2.83					
SOX4	Sex-determining region Y (SRY)-box 4	2.83	+				
POLR2B	Polymerase II polypeptide B	2.83					
H2AFO	H2A histone family member O	2.83					
NR1P1	Nuclear receptor interacting protein 1	2.64		+			
HLF	Hepatic leukemia factor	2.64		+	+		
IL1A	Interleukin 1, alpha	2.64					+
HOXA6	Homeobox A6	2.64	+				
STAT1	Signal transducer and activator of transcription 1	2.64	+				
PTPN13	Protein tyrosine phosphatase, nonreceptor type 13	2.46					
HPIP	Hematopoietic PBX-interacting protein	2.46					
EIF4G3	Eukaryotic translation initiation factor 4 gamma, 3	2.30					
GAPCENA	Rab6 GTPase activating protein	2.30					
MLLT3	Myeloid/lymphoid or mixed-lineage leukemia; t(3)/AF9	2.30	+	+			
LEPR	Leptin receptor	2.30					
ISGF3G	Interferon-stimulated transcription factor 3, gamma	2.14	+				
HMGCS1	3-Hydroxy-3-methylglutaryl-coenzyme A synthase 1	2.14					
ADD3	Adducin 3 (gamma)	2.14					
HOXA7	Homeobox A7	2.14	+				
RUNX1	Runt-related transcription factor 1	2.14		+	+		+
ALDH1A1	Aldehyde dehydrogenase 1 family, member A1	2.14		+	+		
MEIS1	Meis 1	2.00	+	+	+		+

Continued.

TABLE 3. Continued.

Gene symbol	Gene	Fold change	Ref. 34[b]	Other CD34/CD133 arrays		
				Ref. 40[c]	Ref. 41[d]	Ref. 42[e]
IL7	Interleukin 7	2.00				
DUSP1	Dual-specificity phosphatase 1	1.87				
BAX	BCL2-associated X protein	1.87				
CEBPA	CCAAT/enhancer binding protein (C/EBP), alpha	−1.87	+			
CD9	CD9 (p24)	−2.00				
PECAM1	Platelet/endothelial cell adhesion molecule (CD31 antigen)	−2.00				
MT1G	Metallothionein 1G	−2.14				
MT1X	Metallothionein 1X	−2.14				
MT2A	Metallothionein 2A	−2.14				
S100A8	S100 calcium binding protein A8	−2.64				
WNT5B	Wingless-type MMTV integration site family, member 5B	−2.64				
HBA1	Hemoglobin, Alpha 1	−2.83				
HBA2	Hemoglobin, Alpha 2	−3.03				
HBE1	Hemoglobin, Epsilon 1	−3.25				
HBG1	Hemoglobin, Gamma A	−4.00				
HBD	Hemoglobin, Delta	−4.59				
HBB	Hemoglobin, Beta	−4.92				
PF4	Platelet factor 4	−8.00				

[a]Chung et al.[33] Cord blood CD34+ cells 3 days posttransduction with NUP98-HOXA9. CB CD34+ cells were prestimulated for 48 h in QBSF60 medium with KL, FL, and TPO (100 ng/mL of each) followed by three transduction rounds in the next 48 h on retronectin. GFP+ cells were sorted and total RNA was isolated and used to hybridize gene arrays. Data shown is the comparison of GFP+ NUP98-HOXA9 vs. GFP+ MIGR1 cells. A change in gene expression was only considered significant when the fold change was >1.87 with a statistical P-value of <0.05 and a signal value of >200.
[b]Takeda et al.[34] (human G-CSF mobilized CD34+ from treated myeloma patients 3 days posttransduction with NUP98-HOXA9).
[c]Jaatinen et al.[40] (cord blood CD133+ cells).
[d]He et al.[41] (cord blood CD34+ and CD133+ cells).
[e]Hemmoranta et al.[42] (cord blood CD34+ and CD133+ cells).

in a MAPKinase-dependent fashion.[47] HOXA9 has been shown to bind DNA cooperatively with either PBX1 or MEIS1, two other members of the HD family.[48] MEIS1 was originally described as a HOX cofactor that alters HOX-DNA–binding specificity and affinity, and increases HOX-transcriptional activity.[48] Targeted disruption of HOXA9 in mice leads to reduced numbers of progenitor cells and a profound defect in HSC.[49] Conversely, enforced expression of HOXA9 promoted proliferative expansion of HSC and progenitor cells and subsequently inhibited their differentiation. These data highlight the importance of precise control of HOXA9 protein levels during hematopoiesis. HOXA9 is one of the top 20 genes distinguishing AML from acute lymphocytic leukemia (ALL) and correlates with poor prognosis.[47] HOXA9, HOXA7, and MEIS1 genes are coexpressed strongly in all but the acute promyelocytic subset of AML.[50] HOXA9 behaves as an oncogene in leukemia following mutations that induce its persistent expression or that convert it into a persistent transcriptional activator. Upregulation of HOXA5 may also play a role in the observed suppression of erythropoiesis as reflected in the downregulation of a number of globin genes (Hb alpha, beta, delta gamma, epsilon) in cord blood $CD34^+$ cells[33] (TABLE 3). Enforced expression of HOXA5 in human $CD34^+$ cells has been demonstrated to preferentially support myeloid differentiation, with a reduced frequency of erythroid progenitors (BFU-E).[51,52] NUP98-HOXA9 upregulated HOXB5, and upregulation of this gene has also been implicated in leukemia[53] with high expression levels associated with poor outcome.[54]

Transformation of myeloid progenitors by the mixed lineage leukemia (MLL) oncoprotein is dependent on HOXA7 and HOXA9[55] and leukemias associated with MLL translocations show uniform activation of HOXA9, which may be the common pathway that unifies diverse initiating events in many myeloid leukemias.[56] Although no individual HOX gene is essential, Kumar et al.[57] proposed that the "HOX code," minimally defined by the HOXA5-A9 cluster, is central to MLL leukemogenesis. Chimeric MLL-AF9 is reported in AML with t(9;11)(p22;q23),[56] and we observed AF9 upregulation in the NUP98-HOXA9 transcriptosome.[33] MLL-AF9 knock-in mice develop leukemia phenotypically similar to naturally occurring leukemia in humans.[58,59] In the fusion protein, it is the C-terminal 91–amino acid region of AF9 that renders the fusion leukemogenic. Both AF9 and MLL share sequence similarity and function as transcription activators. The C terminus of AF9 physically interacts with the BCL-6 corepressor molecule (Bcor), originally identified as a protein that interacts with the BCL-6 oncoprotein, and augments its activity as a transcriptional repressor.[60] MLL fusion partners AF4 and AF9 interact at subnuclear foci and may participate in a web of protein interactions with a common functional goal, likely involving chromatin remodeling.[61] Disruption of this web, for example, by fusion with MLL or upregulation of AF9 by NUP98-HOXA9 may be important in leukemogenesis.

The serine-threonine kinase Pim-1 is a proto-oncogene and is the most overexpressed gene in the transcriptosome of HOXA9-transduced human hematopoietic cells.[62] It is also highly expressed in NUP98-HOXA9–transduced cord blood $CD34^+$ cells (TABLE 3). It has been noted that the hematological phenotypes of $HOXA9^{-/-}$ and $Pim-1^{-/-}$ deficient mice are strikingly similar and that Pim-1 is an important downstream component of the HOXA9 signaling pathway.[63] The HOXA9 protein binds to the Pim-1 promoter and induces Pim-1 mRNA and protein in hematopoietic cells. Induction of Pim-1 increases phosphorylation and inactivation of proapoptotic BAD protein and because unphosphorylated BAD normally binds to and inactivates antiapoptotic proteins, such as $Bcl-X_L$ and Bcl-2, this would be an antiapoptotic event.[63] In add-back experiments, Pim-1 restored the proliferative potential of $HOXA9^{-/-}$ cells. Pim-1 also physically interacts with the Runx family of transcription factors, phosphorylating and enhancing the transcriptional activity of Runx1.[64] Runx1 is also upregulated in the NUP98-HOXA9 transcriptosome and is a critical regulator of hematopoietic development and a frequent target for chromosomal translocation in leukemia.[65] It enhances gene transcription by interacting with transcriptional coactivators, such as p300 and CBP. RUNX1 expression peaks in early hematopoietic progenitors and HSCs, and decreased levels of RUNX1 have been shown to reduce the number of HSCs (loss of a single AML1 allele resulted in a 50% reduction in long-term repopulating stem cells), suggesting that RUNX1 plays a role in HSC homeostasis.[66,67] Both Pim-1 and Runx genes cooperate with Myc genes in tumor formation. Overexpression of Pim-1 in human epithelial cancer has been reported to dysregulate cyclin B1 protein expression and induce genomic instability by subverting the mitotic spindle checkpoint.[68] Overexpressing cells have abnormal spindles, centrosome amplification, and chromosome mis-segregation, leading to aneuploidy and polyploidy. Pim-1 is also upregulated at the translational level by eukaryotic translation initiation factor 4E (eIF-4E).[69] The eIF4E/eIF4G complex has a central role in the regulation of gene expression at the level of translation initiation.[70] EIF4G was upregulated in the NUP98-HOXA9 transcriptosome and eIF4E is upregulated in blastic CML, in M4/5 AML and in a murine model of blastic crisis of CML produced by cotransduction of hematopoietic progenitors with BCR-ABL and NUP98-HOXA9.[71,72] HOXA9 and eIF4E colocalize in abnormally large eIF4E/eIF4G nuclear bodies characteristic of FAB M4/M5 leukemias. Inhibition of the interaction between eIF4E and eIF4G by small molecule inhibitors has been shown to induce apoptosis in cancer cell lines.[70] eIF4E promotes the selective transport of specific mRNAs such as cyclin D1 and ornithine decarboxylase (ODC). It also alters gene expression at multiple levels: in cytoplasm it acts as the rate limiting step of translation initiation and in the nucleus it facilitates export of a subset of mRNAs, and both functions contribute to eIF4E's ability to oncogenically transform cells.[71,72] The mTOR inhibitor rapamycin also inhibits this pathway as does the antiviral drug ribavarin that directly binds eIF4E and impedes colony formation by subsets of AML with upregulated eIF4E.[70]

PBX3, a member of the PBX family of TALE homeobox genes, is upregulated in both NUP98-HOXA9–transduced adult and cord blood CD34$^+$ cells (TABLE 3).[33,34] A number of isoforms of PBX3 have been identified and PBX3D is favored in normal cells but PBX3C expression is favored in leukemic cells.[73] Functional studies showed that PBXC and PBX3D are unable to interact with the PBX-interacting factor PREP1 and weakly interact with MEIS1 proteins, suggesting that PBX3C and D may affect PBX3-mediated transcriptional activity by acting in opposition to the known PBX proteins through alternative PBX3 complex formation. A novel hematopoietic PBX-interacting protein (HPIP), identified by its ability to bind to and regulate all members of the PBX family, was also upregulated by NUP98-HOXA9.[33] It is present in hematopoietic cells localized to cytoskeletal fibers, but it can shuttle between nucleus and cytoplasm and can inhibit the transcriptional activity of the oncogene E2A-PBX1.[74]

NUP98-HOXA9 upregulation of hepatic leukemia factor (HLF) is consistent with upregulation of HLF reported in normal CD34$^+$ and CD133$^+$ cells[40,45] and in human myeloid leukemic stem cells[75,76] HLF is a basic leucine zipper protein defined by a PAR domain that plays a critical role in hematopoietic-specific expression of the LMO2 gene.[77] A role for HLF in HSC self-renewal is supported by studies showing that ectopic expression of HLF-enhanced HSC engraftment and inhibited apoptosis.[78]

The gene for the RNA-binding protein with multiple splicing (RBPMS) was upregulated in NUP98-HOXA9–transduced cord blood CD34$^+$ cells and also in the transcriptosomes of normal CD34$^+$ and CD133$^+$ cells.[33,40,41,45] RBPMS interacts with the TGF-β receptor type 1, and the presence of TGF-β increases phosphorylation of the C-terminal SSXS region in Smad 2 and 3, promoting nuclear accumulation of Smad2, Smad3, and Smad 4.[79] Overexpression of RBPMS enhances Smad-dependent transcriptional activity in a TGF-dependent manner, whereas knockdown of RBPMS decreases this activity. TGF-β has pleotropic effects on diverse cell types but has been shown to mediate a cytostatic effect on human hematopoietic stem and progenitor cells via upregulation of p57, a cyclin-dependent kinase inhibitor.[80] p57 was reported upregulated in adult CD34$^+$ cells transduced with NUP98-HOXA9.[34] Leukemic cells are generally more resistant to TGF-β inhibition possibly due to the fact that p57 is silenced by promoter methylation in 30%–50% of all patients with acute leukemia.[80] The Smad signaling pathway is shared by numerous ligands and the system is inherently redundant, which may explain why TGF-β type I receptor conditional knockout mice exhibit normal hematopoiesis under steady-state and stressed conditions.[81] Blocking the entire pathway by overexpression of the inhibitory Smad7 (normally involved in a negative feedback loop in response to TGF-β) in murine, HSC is reported to cause increased

HSC self-renewal.[82] Smad4 has also been shown to be critical for HSC self-renewal in an inducible Smad4-deletion model.[83] Smad4 was necessary for maintenance of self-renewal and reconstituting capacity, leaving homing potential, viability, and differentiation intact. Smad4 deletion was associated with downregulation of Notch1 and c-myc, placing Smad4 within the network of genes involved in the regulation of HSC renewal. We have recently reported a previously unknown alternative pathway for TGF-β signaling in which TIF1γ competes with Smad4 to form complexes with Smad2 and 3.[84] This establishes an equilibrium in which TGF-β-induced R-Smad-Smad4 complexes keep hematopoietic progenitors quiescent, whereas R-Smad-TIF1γ complexes direct human hematopoietic cells toward erythroid differentiation.[84] Knockdown of TIF1γ-enhanced HSC self-renewal and inhibited erythroid differentiation[84] (D. Dorn, W. He, J. Massague & M.A. Moore, unpublished data) and it is possible that upregulated RPBMS tilts the balance in TGF-β signaling to the Smad4 rather than the TIF1γ pathway.

A striking feature of the gene profile of NUP98-HOXA9 in adult CD34$^+$ cells was the upregulation of large numbers (36) of interferon (IFN)-inducible genes.[33] Activation of the IFN signaling cascade is normally associated with suppression of hematopoietic cell proliferation. In contrast only two IFN inducible genes (STAT1 and ISGF3G) were upregulated following NUP98-HOXA9 transduction of cord blood CD34$^+$ cells.[33] IFN-stimulated gene factor 3 (ISGF3G) is involved in activation of IFN-α stimulated genes, including STAT1. In a CML model of P210 BCR/ABL transduction, a third of upregulated genes were IFN responsive, including ISGF3G and STAT1.[85] STAT1 has been shown to act as a tumor promoter for leukemia development by maintaining high MHC class 1 expression that protects against immunosurveillance by NK cells.[86]

Aldehyde dehydrogenase is upregulated by NUP98-HOXA9, HOXA9, and HOXA10 in cord blood CD34$^+$ cells (TABLE 3).[34,40,41] Multilineage repopulating HSC are exclusively aldehyde dehydrogenase positive.[87] The protein confers drug resistance to cyclophosphamide and participates in oxidation of retinol to all-*trans* retinoic acid, a molecule that has complex and profound effects on proliferation of and differentiation of HSC.

A number of growth factors are upregulated by NUP98-HOXA9, including FGF18, IL-1, IGFB4,[33,34] renin, thymic stromal lymphopoietin, and angiopoietin 1 and 2.[34] Fibroblast growth factor 18 is among the most upregulated genes in the NUP98-HOXA9 transcriptosome but has not previously been associated with hematopoiesis or leukemia. The gene is upregulated in colorectal cancer and is a direct downstream target of the canonical WNT signaling pathway in progenitor cells via the beta-catenin/Tcf4 complex.[88] Its elevated expression promoted the growth of colon cancer cells *in vitro* and its downregulation suppressed growth or survival.

CD44 is also upregulated by NUP98-HOXA9.[33] This protein mediates adhesive cell–cell and cell–extracellular matrix interactions through binding to its main ligand hyaluronan (HA), a glycosaminoglycan highly concentrated in the endosteal region. CD44 and HA together with the chemokine SDF1 are involved in the trafficking of human CD34$^+$ stem/progenitor cells to bone marrow.[89] CD44 is also required for homing and engraftment of BCR-ABL-expressing leukemic stem cells.[90] Elevated CD44v expression on AML cells including leukemic stem cells has been documented and the expression of certain variants has been associated with poor prognosis.[91] Ligation of CD44 by activating antibodies can inhibit proliferation and reverse the blockage in differentiation of AML blasts of subtypes M1–M5 and inhibited leukemic engraftment.[91]

The transcription factor CCAAT enhancer binding protein-α (C/EBPα), a tumor suppressor gene and a crucial regulator of granulopoiesis through inhibition of c-JUN, was downmodulated by NUP98-HOXA9 (TABLE 3).[33,34] Disruption of C/EBPα, including dominant negative mutations of CEBPα, are found in AML.[92] In t(8;21) AML1-ETO positive AML, blast cells have eight-fold lower level of C/EBPα mRNA and undetectable protein levels.[93] C/EBPα deficiency in mice results in progenitor hyperproliferation and macrophage and neutrophil differentiation defects marked by absence of receptors for M-CSF and G-CSF.[94] Conditional C/EBPα knockout in mice blocked the differentiation of the common myeloid progenitor with myeloblast accumulation in marrow, absence of neutrophils, and enhancement of HSC competitive repopulating capacity and self-renewal.[95] We have recently presented data suggesting that C/EBPα might be a critical downstream gene whose downregulation in CD34$^+$ cells following Flt3-ITD- or STAT5A-constitutive activation accounts for the proliferative and differentiative events observed.[96] We have also coexpressed an estrogen-inducible C/EBPα-ER protein together with NUP98-HOXA9 in CB CD34$^+$ cells and showed that reexpression of C/EBPα counteracted the stem cell proliferation and cell expansion associated with C/EBPα downregulation.[33]

The platelet endothelial cell adhesion molecule PECAM1 is also downregulated by NUP98-HOXA9.[34] This cell adhesion molecule is expressed on hematopoietic and endothelial cells and is implicated in trans-endothelial migration of leukocytes, integrin activation and cell survival. In some AML and ALL cases leukemic cells have low levels of PECAM expression, which may contribute to decreased adhesion of leukemic cells, and loss of PECAM1 expression may confer a growth advantage on hematopoietic cells.[97]

DISRUPTION OF NUP98 LOCALIZATION FOLLOWING TRANDUCTION WITH NUP98-HOX VECTORS OR IN PRIMARY LEUKEMIC CD34$^+$ CELLS EXPRESSING NUP98-HOXA9

Nakamura[98] observed that NUP98-HOXA9 and NUP98-PMX1 localization differs from their normal counterparts, forming nuclear aggregates that

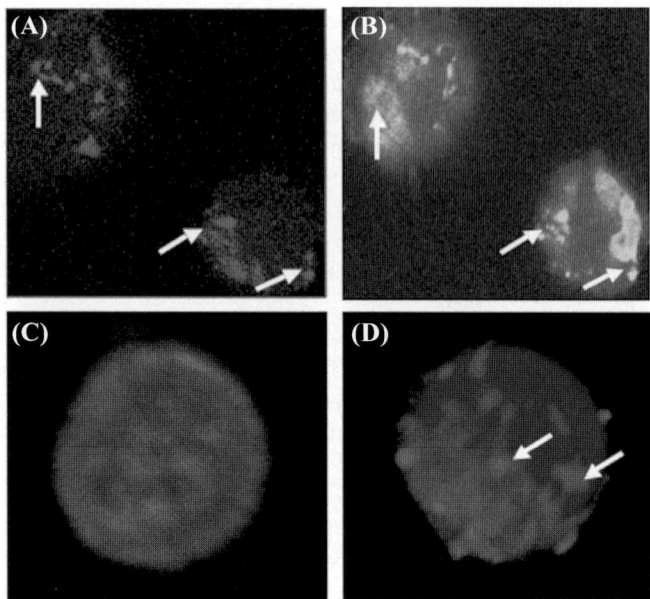

FIGURE 5. Dysregulation of NUP98 protein localization in cells expressing NUP98-HOXA9. Immunostaining of K562 cells transduced with NUP98-HOXA **(A-B)**. Nuclear aggregates (*arrowed*) stained for the HE-tag on NUP98 **(A)** or NUP98 **(B)**. Note absence of NUP98 localization in the NPC of the nuclear membrane. **(C)** NUP98 immunostaining of normal CD34$^+$ cells. Note localization of the protein in the nuclear membrane and absence of nuclear aggregates. **(D)** NUP98 immunostaining of a leukemic cell from the bone marrow of a patient with the NUP98-HOXA9 translocation. Note the presence of numerous NUP98$^+$ nuclear aggregates and impairment in NUP98 association with the NPC.

might perturb the DNA binding of HOX and PMX1 proteins as well as their interaction with specific cofactors. Immunostaining for NUP98 or HE-tagged NUP98-HOXA9 or NUP98-HOXA11 revealed that in transduced CD34$^+$ hematopoietic progenitors or cell lines, NUP98 was no longer associated with the NPC but was found in large aggregates within the nucleus (FIG. 5, Plasilova M and Moore MA, unpublished observation).[99] A similar disruption with nuclear aggregate formation characterized CD34$^+$ cells from AML patients with the t(7;11)(p15;p15) NUP98-HOXA9 translocation (FIG. 5).[99]

NUP98 LOSS OF HETEROZYGOSITY IS A FREQUENT EVENT IN AML AND IS ASSOCIATED WITH ADVERSE PROGNOSIS

LOH in the 11p15 region has been observed in hematological malignancies and in a variety of solid tumors, suggestive of a tumor suppressor function

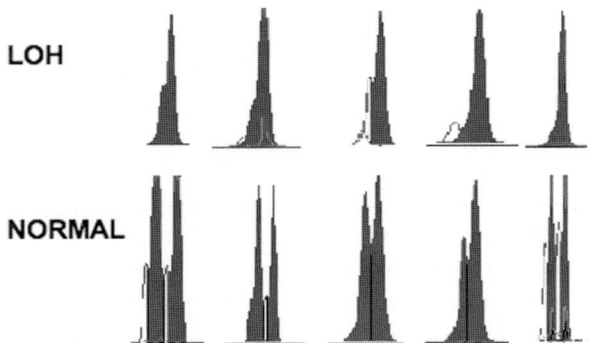

FIGURE 6. Analysis of LOH at the NUP98 locus of chromosome 11.5p in a representative group of AML patients. LOH analysis was performed by PCR for NUP98 allelic markers and is shown as a graphic display of the amount of DNA run through the column (Intensity peak) of the *y*-axis and retention time of the DNA on the column along the *x*-axis. Note samples displaying LOH and those showing expression of both alleles.

located in this region.[100,101] We analyzed loss of heterozygosity (LOH) at five markers in and around the NUP98 on the 11p region in AML and MDS patients and leukemic cell lines using highly polymorphic microsatellite repeat markers.[99] We found a high prevalence of 11p15 LOH of the NUP98 region in 41/140 patients with overt AML and in 4/20 leukemic cell lines in comparison to low frequency in MDS (4/49) patients and normal control $CD34^+$ cells derived from cord blood (0/12) (FIG. 6). The NUP98 markers were located in a portion of the NUP98 gene that is preserved in fusion genes and thus does not interfere with identification of the frequency of LOH. We did not observe the silencing of the NUP98 gene by methylation of its promoter in either control or leukemic samples, suggesting the importance of this region in gene function regulation.[99] Real-time quantitative PCR analysis of genomic DNA for the NUP98 gene using primers spanning the C terminus of the gene confirmed the LOH analysis showing expression of only one copy of the gene with no evidence of uniparental disomy. NUP98 immunostaining of two AML lines, KG1, with both NUP98 alleles expressed, and KG1a, with NUP98 LOH (the latter a more undifferentiated factor-independent derivative of the factor-dependent KG1 line) revealed a quantitative impairment in NUP98 expression in the nuclear membrane associated with LOH.[99] Nup98 LOH was associated with the copresence of an NUP98-HOXA9 translocation in one patient with AML whereas in a second patient NUP98-HOXA9 was present with normal expression of NUP98 by the other allele.[99] Since NUP98 appeared not to associate with the nuclear membrane in the latter patient despite one intact NUP98 allele (FIG. 5), we postulate that the NUP98-HOXA fusion protein may be dimerizing with the remaining wild-type NUP98 protein, sequestering it in the nuclear aggregates.

The presence of NUP98 LOH significantly correlated with decreased disease-free interval ($P < 0.002$) and overall patient survival ($P < 0.004$) compared to a group of AML patients without the LOH.[99] The presence of LOH was not associated with age, gender, cytogenetic findings, presence of Flt3 mutations, FAB subtype, prior history of MDS, and WBC or platelet counts. Overall, these data support the hypothesis that NUP98 allelic loss may be an important event in leukemogenesis. The lower frequency of LOH in MDS suggests that allelic loss may be a secondary event associated with leukemia progression. Two agents with documented antileukemic activity, IFN-γ[102] and deguelin, a naturally occurring rotenoid,[103] are reported to upregulate the expression of NUP98 and may have potential for treatment of AML patients with NUP98 depletion or disruption.

CONCLUSION

The NUP98-HOXA9 fusion transforms normal HSCs, enhancing their self-renewal and impairing their myeloid and erythroid differentiation. This appears to be achieved by multiple mechanisms. In the transcriptional activation model there is an aberrant transcription upregulating a variety of genes that favor self-renewal and potentially impair differentiation, as well as a downregulation of genes that also results in block differentiation and enhance proliferation. The NUP98 component of the chimeric protein further enhances aberrant transcription by the binding p300/CBP to its FG repeats. This aberrant transcription mediated via the HOXA9 homeodomain is mediated and is promoted by the increased stability of the chimeric HOX protein that is resistant to ubiquitin-mediated degradation. The NUP98-HOXA9 fusion protein appears to act in a dominant negative manner, potentially binding and sequestering remaining wild-type NUP98 protein in nuclear aggregates, leaving the NPC devoid of NUP98. The resulting impairment in mRNA export may likely disrupt pathways that are involved in the normal process of HSC differentiation. The observation that the NUP98 locus has allelic loss in nearly a third of AML patients implies that the gene normally plays a role as a tumor suppressor. The association of NUP98 allelic loss with poor prognosis suggests that NUP98 dysregulation plays a significant role in leukemia development and progression. It has been proposed that the development of acute leukemia requires the cooperation of at least two leukemogenic mutations, the first conferring a proliferation of survival advantage while a second primarily interferes with differentiation and apoptosis. Although additional mutations would seem to be needed to produce an AML phenotype, the NUP98-HOXA chimeric protein has multiple proleukemic actions that impact critical pathways involved not only in HSC self-renewal but also differentiation.

ACKNOWLEDGMENTS

M.A.S.M. was supported by a Leukemia and Lymphoma Society of America Specialized Center of Research grant and The Gar Reichman Fund of the Cancer Research Institute. J.J.S. was supported by a grant EMBO ALTF-412-2001. P.Z. is a scholar of the Leukemia Lymphoma Society and was supported by NCI grants CA118085 and CA 92792. G.M. was supported by the Italian Ministry for University and Research Interlink and the Italian Association for Cancer Research.

REFERENCES

1. ROMANA, S.P. et al. 2006. NUP98 rearrangements in hematopoietic malignancies: a study of the Groupe Francophone de Cytogenetique Hematologique. Leukemia **20:** 696–706.
2. NAKIELNY, S. & G. DREYFUSS. 1999. Transport of proteins and RNAs in and out of the nucleus. Cell **99:** 677–690.
3. ZOLOTUKHIN, A.S. & B.K. FELBER. 1999. Nucleoporins nup98 and nup214 participate in nuclear export of human immunodeficiency virus type 1 Rev. J. Virol. **73:** 120–127.
4. ROUT, M.P. et al. 2000. The yeast nuclear pore complex: composition, architecture, and transport mechanism. J. Cell. Biol. **148:** 635–651.
5. CRONSHAW, J.M. et al. 2002. Proteomic analysis of the mammalian nuclear pore complex. J. Cell. Biol. **158:** 915–927.
6. ROSENBLUM, J.S. & G. BLOBEL. 1999. Autoproteolysis in nucleoporin biogenesis. Proc. Natl. Acad. Sci. USA **96:** 11370–11375.
7. FONTOURA, B.M., G. BLOBEL & M.J. MATUNIS. 1999. A conserved biogenesis pathway for nucleoporins: proteolytic processing of a 186-kilodalton precursor generates Nup98 and the novel nucleoporin, Nup96. J. Cell. Biol. **144:** 1097–1112.
8. GRIFFIS, E.R., S. XU & M.A. POWERS. 2003. Nup98 localizes to both nuclear and cytoplasmic sides of the nuclear pore and binds to two distinct nucleoporin subcomplexes. Mol. Biol. Cell. **14:** 600–610.
9. PRITCHARD, C.E. et al. 1999. RAE1 is a shuttling mRNA export factor that binds to a GLEBS-like NUP98 motif at the nuclear pore complex through multiple domains. J. Cell. Biol. **145:** 237–254.
10. KASPER, L.H. et al. 1999. CREB binding protein interacts with nucleoporin-specific FG repeats that activate transcription and mediate NUP98-HOXA9 oncogenicity. Mol. Cell. Biol. **19:** 764–776.
11. GRIFFIS, E.R. et al. 2002. Nup98 is a mobile nucleoporin with transcription-dependent dynamics. Mol. Biol. Cell. **13:** 1282–1297.
12. POWERS, M.A. et al. 1997. The vertebrate GLFG nucleoporin, Nup98, is an essential component of multiple RNA export pathways. J. Cell. Biol. **136:** 241–250.
13. BLEVINS, M.B. et al. 2003. Complex formation among the RNA export proteins Nup98, Rae1/Gle2, and TAP. J. Biol. Chem. **278:** 20979–20988.

14. Wu, X. et al. 2001. Disruption of the FG nucleoporin NUP98 causes selective changes in nuclear pore complex stoichiometry and function. Proc. Natl. Acad. Sci. USA **98:** 3191–3196.
15. NAKAMURA, T. et al. 1999. NUP98 is fused to PMX1 homeobox gene in human acute myelogenous leukemia with chromosome translocation t(1;11)(q23;p15). Blood **94:** 741–747.
16. BAI, X.T. et al. 2006. Trans-repressive effect of NUP98-PMX1 on PMX1-regulated c-FOS gene through recruitment of histone deacetylase 1 by FG repeats. Cancer Res. **66:** 4584–4590.
17. HUSSEY, D.J. & A. DOBROVIC. 2002. Recurrent coiled-coil motifs in NUP98 fusion partners provide a clue to leukemogenesis. Blood **99:** 1097–1098.
18. CALVO, K.R. et al. 2001. Meis1a suppresses differentiation by G-CSF and promotes proliferation by SCF: potential mechanisms of cooperativity with Hoxa9 in myeloid leukemia. Proc. Natl. Acad. Sci. USA **98:** 13120–13125.
19. CALVO, K.R. et al. 2002. Nup98-HoxA9 immortalizes myeloid progenitors, enforces expression of Hoxa9, Hoxa7 and Meis1, and alters cytokine-specific responses in a manner similar to that induced by retroviral co-expression of Hoxa9 and Meis1. Oncogene **21:** 4247–4256.
20. KROON, E. et al. 2001. NUP98-HOXA9 expression in hemopoietic stem cells induces chronic and acute myeloid leukemias in mice. Embo. J. **20:** 350–361.
21. IWASAKI, M. et al. 2005. Identification of cooperative genes for NUP98-HOXA9 in myeloid leukemogenesis using a mouse model. Blood **105:** 784–793.
22. LIN, Y.W. et al. 2005. NUP98-HOXD13 transgenic mice develop a highly penetrant, severe myelodysplastic syndrome that progresses to acute leukemia. Blood **106:** 287–295.
23. PINEAULT, N. et al. 2003. Induction of acute myeloid leukemia in mice by the human leukemia-specific fusion gene NUP98-HOXD13 in concert with Meis1. Blood **101:** 4529–4538.
24. WANG, Y. et al. 2004. [Development of human myeloid leukemia-like phenotype in NUP98-PMX1 transgenic mice]. Zhonghua Xue Ye Xue Za Zhi **25:** 262–265.
25. PINEAULT, N. et al. 2004. Differential and common leukemogenic potentials of multiple NUP98-Hox fusion proteins alone or with Meis1. Mol. Cell. Biol. **24:** 1907–1917.
26. PINEAULT, N., C. ABRAMOVICH & R.K. HUMPHRIES. 2005. Transplantable cell lines generated with NUP98-Hox fusion genes undergo leukemic progression by Meis1 independent of its binding to DNA. Leukemia **19:** 636–643.
27. YAMAMOTO, K. et al. 2000. Expression of the NUP98/HOXA9 fusion transcript in the blast crisis of Philadelphia chromosome-positive chronic myelogenous leukaemia with t(7;11)(p15;p15). Br. J. Haematol. **109:** 423–426.
28. AHUJA, H.G. et al. 2001. NUP98 gene rearrangements and the clonal evolution of chronic myelogenous leukemia. Genes Chromosomes Cancer **30:** 410–415.
29. DASH, A.B. et al. 2002. A murine model of CML blast crisis induced by cooperation between BCR/ABL and NUP98/HOXA9. Proc. Natl. Acad. Sci. USA **99:** 7622–7627.
30. MAYOTTE, N. et al. 2002. Oncogenic interaction between BCR-ABL and NUP98-HOXA9 demonstrated by the use of an in vitro purging culture system. Blood **100:** 4177–4184.
31. PALMQVIST, L. et al. 2006. The Flt3 receptor tyrosine kinase collaborates with NUP98-HOX fusions in acute myeloid leukemia. Blood **108:** 1030–1036.

32. GUREVICH, R.M., P.D. APLAN & R.K. HUMPHRIES. 2004. NUP98-topoisomerase I acute myeloid leukemia-associated fusion gene has potent leukemogenic activities independent of an engineered catalytic site mutation. Blood **104:** 1127–1136.
33. CHUNG, K.Y. *et al.* 2006. Enforced expression of NUP98-HOXA9 in human CD34(+) cells enhances stem cell proliferation. Cancer Res. **66:** 11781–11791.
34. TAKEDA, A., C. GOOLSBY & N.R. YASEEN. 2006. NUP98-HOXA9 induces long-term proliferation and blocks differentiation of primary human CD34+ hematopoietic cells. Cancer Res. **66:** 6628–6637.
35. ZHANG, Y. *et al.* 2003. CUL-4A stimulates ubiquitylation and degradation of the HOXA9 homeodomain protein. Embo. J. **22:** 6057–6067.
36. PEARSON, M. *et al.* 2000. PML regulates p53 acetylation and premature senescence induced by oncogenic Ras. Nature **406:** 207–210.
37. BERNASSOLA, F. *et al.* 2004. Ubiquitin-dependent degradation of p73 is inhibited by PML. J. Exp. Med. **199:** 1545–1557.
38. GRONROOS, E. *et al.* 2002. Control of Smad7 stability by competition between acetylation and ubiquitination. Mol. Cell. **10:** 483–493.
39. GHANNAM, G. *et al.* 2004. The oncogene Nup98-HOXA9 induces gene transcription in myeloid cells. J. Biol. Chem. **279:** 866–875.
40. JAATINEN, T. *et al.* 2006. Global gene expression profile of human cord blood-derived CD133+ cells. Stem Cells **24:** 631–641.
41. HE, X. *et al.* 2005. Differential gene expression profiling of CD34+ CD133+ umbilical cord blood hematopoietic stem progenitor cells. Stem Cells Dev. **14:** 188–198.
42. HEMMORANTA, H. *et al.* 2006. Transcriptional profiling reflects shared and unique characters for CD34+ and CD133+ cells. Stem Cells Dev. **15:** 839–851.
43. FERRELL, C.M. *et al.* 2005. Activation of stem-cell specific genes by HOXA9 and HOXA10 homeodomain proteins in CD34+ human cord blood cells. Stem Cells **23:** 644–655.
44. DORSAM, S.T. *et al.* 2004. The transcriptome of the leukemogenic homeoprotein HOXA9 in human hematopoietic cells. Blood **103:** 1676–1684.
45. LAWRENCE, H.J. *et al.* 1999. Frequent co-expression of the HOXA9 and MEIS1 homeobox genes in human myeloid leukemias. Leukemia **13:** 1993–1999.
46. KAWAGOE, H. *et al.* 1999. Expression of HOX genes, HOX cofactors, and MLL in phenotypically and functionally defined subpopulations of leukemic and normal human hematopoietic cells. Leukemia **13:** 687–698.
47. KIRITO, K., N. FOX & K. KAUSHANSKY. 2004. Thrombopoietin induces HOXA9 nuclear transport in immature hematopoietic cells: potential mechanism by which the hormone favorably affects hematopoietic stem cells. Mol. Cell. Biol. **24:** 6751–6762.
48. SHEN, W.F. *et al.* 1999. HOXA9 forms triple complexes with PBX2 and MEIS1 in myeloid cells. Mol. Cell. Biol. **19:** 3051–3061.
49. LAWRENCE, H.J. *et al.* 1997. Mice bearing a targeted interruption of the homeobox gene HOXA9 have defects in myeloid, erythroid, and lymphoid hematopoiesis. Blood **89:** 1922–1930.
50. GOLUB, T.R. *et al.* 1999. Molecular classification of cancer: class discovery and class prediction by gene expression monitoring. Science **286:** 531–537.
51. CROOKS, G.M., J. FULLER, D. PETERSEN, *et al.* 1999. Constitutive HOXA5 expression inhibits erythropoiesis and increases myelopoiesis from human hematopoietic progenitors. Blood **94:** 519–528.

52. FULLER J.F. et al. 1999. Characterization of HOX gene expression during myelopoiesis: role of HOX A5 in lineage commitment and maturation. Blood **93:** 3391–3400.
53. NEBEN, K., S. SCHNITTGER, B. BRORS, et al. 2005. Distinct gene expression patterns associated with FLT3- and NRAS-activating mutations in acute myeloid leukemia with normal karyotype. Oncogene **24:** 1580–1588.
54. BULLINGER, L., K. DOHNER, E. BAIR, et al. 2004. Use of gene-expression profiling to identify prognostic subclasses in adult acute myeloid leukemia. N. Engl. J. Med. **350:** 1605–1616.
55. AYTON, P.M. & M.L. CLEARY. 2003. Transformation of myeloid progenitors by MLL oncoproteins is dependent on Hoxa7 and Hoxa9. Genes Dev. **17:** 2298–2307.
56. MEYER, C. et al. 2006. The MLL recombinome of acute leukemias. Leukemia **20:** 777–784.
57. KUMAR, A.R. et al. 2004. Hoxa9 influences the phenotype but not the incidence of Mll-AF9 fusion gene leukemia. Blood **103:** 1823–1828.
58. DOBSON, C.L. et al. 1999. The mll-AF9 gene fusion in mice controls myeloproliferation and specifies acute myeloid leukaemogenesis. Embo. J. **18:** 3564–3574.
59. SOMERVAILLE, T.C. & M.L. CLEARY. 2006. Identification and characterization of leukemia stem cells in murine MLL-AF9 acute myeloid leukemia. Cancer Cell. **10:** 257–268.
60. SRINIVASAN, R.S., A.C. DE ERKENEZ & C.S. HEMENWAY. 2003. The mixed lineage leukemia fusion partner AF9 binds specific isoforms of the BCL-6 corepressor. Oncogene **22:** 3395–3406.
61. ERFURTH, F. et al. 2004. MLL fusion partners AF4 and AF9 interact at subnuclear foci. Leukemia **18:** 92–102.
62. DORSAM S.T. et al. 2004. The transcriptome of the leukemogenic homeoprotein HOXA9 in human hematopoietic cells. Blood **103:** 1676–1684.
63. HU, Y.L. et al. 2007. Evidence that the Pim1 kinase gene is a direct target of HOXA9 and a mediator of HOXA9's effects on hematopoiesis. Blood [Epub ahead of print].
64. AHO, T.L. et al. 2006. Pim-1 kinase phosphorylates RUNX family transcription factors and enhances their activity. BMC Cell. Biol. **7:** 21.
65. NIMER, S.D. & M.A. MOORE. 2004. Effects of the leukemia-associated AML1-ETO protein on hematopoietic stem and progenitor cells. Oncogene **23:** 4249–4254.
66. LORSBACH, R.B. et al. 2004. Role of RUNX1 in adult hematopoiesis: analysis of RUNX1-IRES-GFP knock-in mice reveals differential lineage expression. Blood **103:** 2522–2529.
67. SUN, W. & J.R. DOWNING. 2004. Haploinsufficiency of AML1 results in a decrease in the number of LTR-HSCs while simultaneously inducing an increase in more mature progenitors. Blood **104:** 3565–3572.
68. ROH, M. et al. 2005. Chromosomal instability induced by Pim-1 is passage-dependent and associated with dysregulation of cyclin B1. J. Biol. Chem. **280:** 40568–40577.
69. HOOVER, D.S. et al. 1997. Pim-1 protein expression is regulated by its 5′-untranslated region and translation initiation factor eIF-4E. Cell Growth Differ. **8:** 1371–1380.
70. MOERKE, N.J. et al. 2007. Small-molecule inhibition of the interaction between the translation initiation factors eIF4E and eIF4G. Cell **128:** 257–267.

71. CULJKOVIC, B., I. TOPISIROVIC & K.L. BORDEN. 2007. Controlling gene expression through RNA regulons: the role of the eukaryotic translation initiation factor eIF4E. Cell Cycle **6**: 65–69.
72. TOPISIROVIC, I. *et al*. 2003. Aberrant eukaryotic translation initiation factor 4E-dependent mRNA transport impedes hematopoietic differentiation and contributes to leukemogenesis. Mol. Cell. Biol. **23**: 8992–9002.
73. MILECH, N., U.R. KEES & P.M. WATT. 2001. Novel alternative PBX3 isoforms in leukemia cells with distinct interaction specificities. Genes Chromosomes Cancer **32**: 275–280.
74. ABRAMOVICH, C *et al*. 2002. Functional characterization of multiple domains involved in the subcellular localization of the hematopoietic Pbx interacting protein (HPIP). Oncogene **21**: 6766–6771.
75. MOORE, M.A. 2005. Converging pathways in leukemogenesis and stem cell self-renewal. Exp. Hematol. **33**: 719–737.
76. MOORE, M.A. *et al*. 2007. Constitutive activation of Flt3 and STAT5A enhances self-renewal and alters differentiation of hematopoietic stem cells. Exp. Hematol. **35**(Suppl 1): 105–116.
77. CRABLE, S.C. & K.P. ANDERSON. 2003. A PAR domain transcription factor is involved in the expression from a hematopoietic-specific promoter for the human LMO2 gene. Blood **101**: 4757–4764.
78. SHOJAEI, F. *et al*. 2005. Hierarchical and ontogenic positions serve to define the molecular basis of human hematopoietic stem cell behavior. Dev. Cell. **8**: 651–663.
79. SUN, Y. *et al*. 2006. Potentiation of Smad-mediated transcriptional activation by the RNA-binding protein RBPMS. Nucleic Acids Res. **34**: 6314–6326.
80. SCANDURA, J.M. *et al*. 2004. Transforming growth factor β-induces cell cycle arrest of human hematopoietic cells requires p57KIP2 up-regulation. Proc. Natl. Acad. Sci. USA **42**: 15231–15236.
81. LARSSON, J. *et al*. 2003. TGF-beta signaling-deficient hematopoietic stem cells have normal self-renewal and regenerative ability *in vivo* despite increased proliferative capacity *in vitro*. Blood **102**: 3129–3135.
82. BLANK, U. *et al*. 2006. Smad7 promotes self-renewal of hematopoietic stem cells *in vivo*. Blood **108**: 4246–4255.
83. KARLSSON, G. *et al*. 2007. Smad4 is critical for self-renewal of hematopoietic stem cells. J. Exp. Med. **204**: 467–474.
84. HE, W. *et al*. 2006. Hematopoiesis controlled by distinct TIF1g and Smad4 branches of the TGFβ pathway. Cell **125**: 929–941.
85. HAKANSSON, P. *et al*. 2004. Identification of genes differentially regulated by the P210 BCR̃/ABL1 fusion oncogene using cDNA microarrays. Exp. Hematol. **32**: 476–482.
86. KOVACIC, B. *et al*. 2006. STAT1 acts as a tumor promoter for leukemia development. Cancer Cell **10**: 77–87.
87. HESS, D.A., T.E. MEYERROSE, L. WIRTHLIN, *et al*. 2004. Functional characterization of highly purified human hematopoietic repopulating cells isolated according to aldehyde dehydrogenase activity. Blood **104**: 1648–1655.
88. SHIMOKAWA, T, Y. FURUKAWA, M. SAKAI, *et al*. 2003. Involvement of the FGF18 gene in colorectal carcinogenesis, as a novel downstream target of the beta-catenin/T-cell factor complex. Cancer Res. **63**: 6116–6120.

89. AVIGDOR, A, P. GOICHBERG, S. SHIVTIEL, et al. 2004. CD44 and hyaluronic acid cooperate with SDF-1 in the trafficking of human CD34 +stem/progenitor cells to bone marrow. Blood **103**: 2981–2989.
90. KRAUSE, D.S., K. LAZARIDES, U.H. VON ANDRIAN & R.A. VAN ETTEN. 2006. Requirement for CD44 in homing and engraftment of BCR-ABL-expressing leukemic stem cells. Nat. Med. **12**: 1175–1180.
91. JIN, L., K.J. HOPE, Q. ZHAI, et al. 2006. Targeting of CD44 eradicates human acute myeloid leukemic stem cells. Nat. Med. **12**: 1167–1174.
92. FROHLING, S. & H. DOHNER. 2004. Disruption of C/EBPa function in acute myeloid leukemia. N. Engl. J. Med. **351**: 2370–2372.
93. PABST, T., B.U. MUELLER, N. HARAKAWA, et al. 2001. AML1-ETO downregulates the granulocytic differentiation factor C/EBPalpha in t(8;21) myeloid leukemia. Nat. Med. **7**: 444–451.
94. HEATH, V., H.C. SUH, M. HOLMAN, et al. 2004. C/EBPα deficiency results in hyperproliferation of hematopoietic progenitor cells and disrupts macrophage development *in vitro* and *in vivo*. Blood **104**: 1639–1647.
95. ZHANG, P., J. IWASAKI-ARAI, H. IWASAKI, et al. 2004. Enhancement of hematopoietic stem cell repopulating capacity and self-renewal in the absence of the transcription factor C/EBPα. Immunity **21**: 853–863.
96. WIERENGA, A.T., H. SCHEPERS, M.A. MOORE, et al. 2006. STAT5-induced self-renewal and impaired myelopoiesis of human hematopoietic stem/progenitor cells involves down-modulation of C/EBP{alpha}. Blood **107**: 4326–4333.
97. DE WAELE, M., W. RENMANS, K. JOCHMANS, et al. 1999. Different expression of adhesion molecules on CD34+ cells in AML and B-lineage ALL and their normal bone marrow counterparts. Eur. J. Haematol. **63**: 192–201.
98. NAKAMURA, T. 2005. NUP98 fusion in human leukemia: dysregulation of the nuclear pore and homeodomain proteins. Int. J. Hematol. **82**: 21–27.
99. PLASILOVA, M., H.B. OMMEN, J. JELINEK, et al. 2006. Loss of heterozygosity (LOH) of the NUP98 gene is an adverse prognostic factor in acute myeloid leukemia (AML). Blood **108**: 668abs.
100. KRSKOVA-HONZATKOVA, L., J. CERMAK, J. SAJDOVA, et al. 2001. Loss of heterozygosity and heterogeneity of its appearance and persistingin the course of acute myeloid leukemia and myelodysplastic syndrome. Leuk. Res. **25**: 45–53.
101. KARNIK, P., M. PARIS, B.R. WILLIAMS, et al. 1998. Two distinct tumor suppressor loci within chromosome 11p15 implicated in breast cancer progression and metastasis. Hum. Mol. Genet. **7**: 895–903.
102. ENNINGA, J., D.E. LEVY, G. BLOBEL & B.M. FONTOURA. 2002. Role of nucleoporin induction in releasing an mRNA nuclear export block. Science **295**: 1523–1525.
103. LIU, H.L., Y. CHEN, G.H. CUI, et al. 2005. Deguelin regulates nuclear pore complex proteins Nup98 and Nup88 in U937 cells *in vitro*. Acta Pharmacol. Sin. **26**: 1265–1273.

LEF-1 Is a Decisive Transcription Factor in Neutrophil Granulopoiesis

JULIA SKOKOWA AND KARL WELTE

Department of Pediatric Hematology and Oncology, Medical School Hannover, 30625 Hannover, Germany

ABSTRACT: We found that lymphoid enhancer-binding factor 1 (LEF-1) is a decisive transcription factor in granulopoiesis controlling proliferation, proper lineage commitment, and granulocytic differentiation via regulation of its target genes C/EBP-α, cyclin D1, c-myc, and survivin. Myeloid progenitor cells of patients with severe congenital neutropenia (CN) showed a severe downregulation of LEF-1 and its target genes expression. Expression of neutrophil elastase (NE) is also severely reduced in CN myeloid progenitors. Intriguingly, ELA2 gene promoter is positively regulated by direct binding of LEF-1 or LEF-1 target gene C/EBP-α. In summary we demonstrated that LEF-1 is not only crucial in lymphopoiesis, but also in myelopoiesis, documenting new functions of LEF-1.

KEYWORDS: LEF-1; severe congenital neutropenia; ELA2; HAX1; C/EBP-α

INTRODUCTION

Granulopoiesis is a lifelong multistage process with continuous generation of a large number of mature neutrophils ($>10^6$ cells/min/kg body weight) from a small number of hematopoietic stem cells. To maintain continuous constitutive granulopoiesis, all these events must be closely regulated by varieties of intrinsic transcription factors (such as RUNX, PU.1, C/EBP-α) and distinct cytokines (e.g., G-CSF, GM-CSF, IL-3). Recently, we found that lymphoid enhancer-binding factor 1 (LEF-1) plays a key role in the regulation of proliferation and granulocytic differentiation of hematopoietic cells.[1]

LEF-1 belongs to the LEF-1/TCFs (T cell factors) family of high mobility group domain containing transcription factors.[2] Although the gene structure of all these family members is remarkably similar, characterization of the full-length human LEF-1 gene locus and its complete set of mRNA products

Address for correspondence: Dr. Karl Welte, Department of Pediatric Hematology and Oncology, Hannover Medical School, Carl-Neuberg-Str. 1, 30625 Hannover, Germany. Voice: +49-511-532-6710; fax: +49-511-532-9120.

Welte.Karl.H@mh-hannover.de

Ann. N.Y. Acad. Sci. 1106: 143–151 (2007). © 2007 New York Academy of Sciences.
doi: 10.1196/annals.1392.012

showed that LEF-1 exists as a unique set of alternatively spliced isoforms, and is functionally different from other TCFs.[3] LEF-1/TCFs transcription factors are considered to be members of the canonical Wnt signaling pathway, performing their functions in complexes with β-catenin.[4] LEF-1 may also acquire diverse regulatory functions by association with Wnt-/β-catenin-independent stimuli (e.g., TGF-β and Notch pathways).[5,6] Moreover, one of the unique features of LEF-1 is to regulate target gene expression by DNA bending and helical phasing of transcription-binding sites, acting as an "architectural" transcription factor.[7,8] To date, analysis of hematopoietic functions of LEF-1 was mostly restricted to the lymphoid compartment. It has been shown that LEF-1 is highly expressed in pro- and pre-B cells and in thymocytes and is downregulated in mature lymphocytes in a mouse model. LEF-1 has a context-dependent activation domain and in complex with its co-activator, ALY, contributes to maximal function of the T cell receptor-α (TCR-α) enhancer in T cell precursors independent of β-catenin.[9] In pro- and pre-B lymphocytes LEF-1 binds and activates the recombination-activating gene-2 (RAG-2) promoter together with c-Myb and Pax-5.[10,11] Less is known about the role of LEF-1 in human hematopoiesis. In contrast to mouse system, LEF-1 is expressed not only in thymocytes, but also in mature T cells and is downregulated during antigen encountering and engagement of the TCR or IL-15 receptor in naïve T cells.[12] In B lymphopoiesis LEF-1 is highly upregulated as the cells commit to the B lineage, and the expression is kept continuously high until the cells become immature B cells, where LEF-1 level is reduced to the same as in uncommitted progenitors and disappear in B lymphocytes from peripheral blood.[13]

By comparative analysis of transcription factors in myeloid precursor cells of healthy individuals and patients suffering from severe congenital neutropenia (CN), we identified LEF-1 as a key regulator of myelopoiesis and as pathognomonic factor for defective granulopoiesis in CN.[1] In healthy controls, we found varying LEF-1 expression levels at all stages of myelopoiesis, with peak expression in promyelocytes. LEF-1 mRNA expression in promyelocytes of individuals with CN was significantly lower than in healthy controls and in some cases LEF-1 was even absent (FIG. 1), resulting in defective expression of the LEF-1 target genes cyclin D1, c-myc, and survivin. Moreover, C/EBP-α, a key transcription factor in granulopoiesis, was severely affected in CN promyelocytes and was directly regulated by LEF-1. Reconstitution of LEF-1 in early hematopoietic progenitors of two individuals with CN corrected the defective myelopoiesis and resulted in the differentiation of these progenitors into mature granulocytes. These observations indicate that LEF-1 is a decisive factor regulating neutrophilic granulopoiesis. Its absence plays a critical role in the defective maturation program of myeloid progenitors in individuals with CN.

CN is a hematological disorder, which is characterized by "maturation arrest" of granulopoiesis in bone marrow, absence of mature granulocytes in peripheral blood, and severe recurrent bacterial infections. The bone marrow

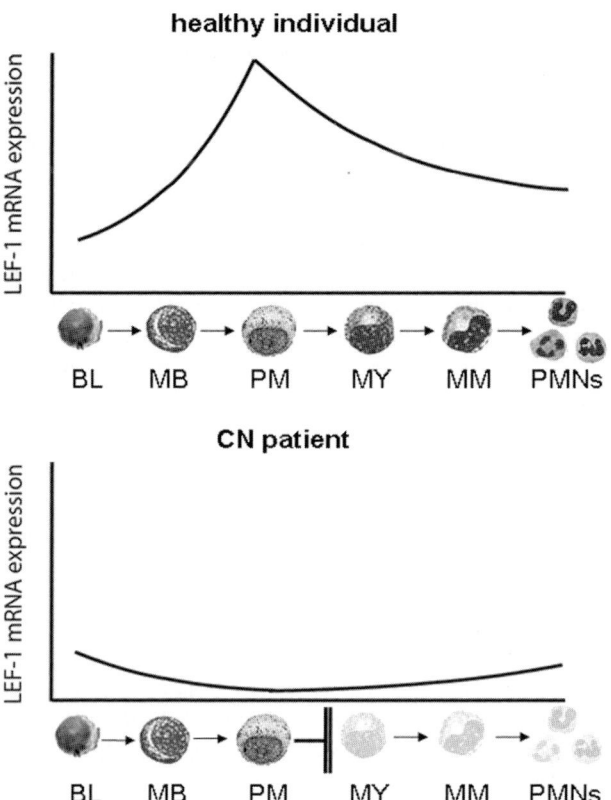

FIGURE 1. LEF-1 mRNA variably expressed in different stages of granulopoiesis of healthy individuals with a peak at the promyelocitic stage and is abrogated in CN. BL = blast; MB = myeloblast; PM = promyelocyte; MY = myelocyte; MM = metamyelocyte; PMNs = polymorphonuclear granulocytes.

of CN patients shows a "maturation arrest" of neutrophil precursors at the promyelocyte–myelocyte stage independent of the inheritance subtypes.[14,15] Although CN is a rare hematopoietic disorder (with an estimated frequency of one per 200,000 population), the characteristic phenotype that is common for all CN patients represents an excellent model to study the regulation of granulocytic differentiation in human system.

CN is a heterogeneous syndrome as judged by inherited mutations in several genes in subgroups of patients.[14] Autosomal dominant mutations in elastase 2 (ELA2) gene encoding neutrophil elastase (NE) protein have been found in approximately 60% of CN patients.[16] Recently, our group identified novel autosomal recessive mutations in the HAX1 gene encoding mitochondrial protein HAX1 in approximately 30% of patients, including individuals from the original Kostmann pedigree.[17] There are also few CN cases not associated with any known mutations.

The heterogeneous genetic origin and concomitantly identical bone marrow phenotype in CN patients allow us to hypothesize that there is a common transcription factor/signaling mechanism downstream of different genetic alterations leading to neutropenia. The candidate downstream factor is supposed to be: (1) essential in granulopoiesis (with the highest activity in promyelocytes); (2) specifically affected only in CN, but not in other types of neutropenia; and (3) its restoration should overcome the maturation arrest of promyelocytes of CN patients. LEF-1 fulfils all these criteria.

To study the upstream regulators of LEF-1, CN is a unique clinical model ideal to investigate myeloid-specific signaling pathways. LEF-1 expression was downregulated in all studied CN patients irrespective of mutation status. Therefore, the downstream mechanisms associated with mutations in ELA2 or HAX1 genes which cause downregulation of LEF-1 in CN have to be investigated. What is known about ELA2, HAX-1, and their pathogenic involvement in CN? NE is a protease stored in granules of neutrophilic granulocytes, which is formed during the promyelocytic phase and cleaves a variety of different proteins, such as collagen, elastin, and cytokines, including granulocyte colony-stimulating factor (G-CSF). ELA2 mutations in CN have been described initially in 1999, but a causal connection between these mutations and defective granulopoiesis in CN is still unresolved. There are several disputed observations: (1) the same ELA2 mutations have been observed in two different syndromes: CN and cyclic neutropenia[16,18]; (2) there is no clear genotype–phenotype correlation and more than 20 different ELA2 mutations confer widely disparate effects on NE enzymatic activity[19]; (3) analysis of mutated recombinant NE proteins has shown that there is no evident changes in the stability, substrate specificity, no consistent effects on glycosylation and proteolytic activity retained wild-type levels in some mutants; (4) gene targeting of ELA2 has failed to reproduce neutropenic phenotype in mice.[20] Nevertheless, some studies proposed that mutant ELA2 triggers accelerated apoptosis of granulocyte precursors.[21,22] In addition, it has been proposed that ELA2 mutations lead to cytoplasmatic accumulation of the altered protein, disturbance of intracellular trafficking, and activation of the unfolded protein response and induction of apoptosis.[23] Less is known about HAX1 functions in myelopoiesis. HAX1 is the mitochondrial protein, which participates in the signal transduction, in the cytoskeletal control, and has antiapoptotic functions. Recently, Klein et al. showed the importance of HAX1 in the maintenance the inner mitochondrial membrane potential and in the protection of myeloid cells against apoptosis.[17] Since the arrested CN promyelocytes show impaired proliferation and differentiation in response to G-CSF, as well as accelerated apoptosis[24,25] and a majority of patients reveal mutations in genes associated with granulocytic differentiation and apoptosis, namely ELA2 and HAX-1, it is obvious to hypothesize that ELA2 and HAX-1 are upstream regulators of LEF-1.

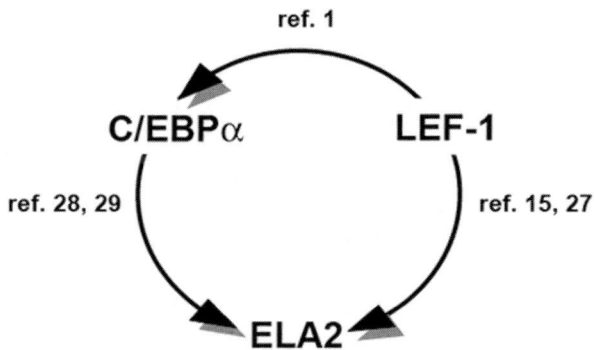

FIGURE 2. LEF-1 regulates ELA2 transcription (see references).

We and others[15,26] observed severely reduced expression of ELA2 mRNA and NE protein in myeloid progenitors of CN patients, in comparison to healthy individuals. Similar to LEF-1, ELA2 was downregulated in both groups of CN patients carrying either ELA2 or HAX1 mutations. At the same time, patients with cyclic neutropenia, in which ELA2 is mutated in most cases, ELA2 levels were comparable with that of healthy individuals (Ref. 26 and Skokowa, data not shown). Therefore, we assume that ELA2 mutations may not be involved in the downregulation of ELA2 transcription. It has been shown that LEF-1 binds and activates promoter of ELA2 gene.[27] Additionally, ELA2 promoter is regulated by a direct binding of LEF-1 target gene C/EBP-α.[28,29] Overexpression of full-length LEF-1 or its dominant negative form that lacks β-catenin-binding domain (dnLEF-1) into U937 cells resulted in a dramatic upregulation of ELA2 mRNA/protein (Skokowa, data not shown). Since both, C/EBP-α and LEF-1 bind to the ELA2 promoter, the reduced expression of elastase-RNA and -protein in CN myeloid cells is explainable by a direct LEF-1-dependent and/or indirect LEF-1:C/EBP-α-dependent regulation of ELA2 (FIG. 2).

Recently, Wang et al. showed that C/EBP-α and G-CSF receptor (G-CSFR) signals synergistically induce ELA2 expression.[30] There is clear evidence that G-CSF-dependent granulocytic differentiation is severely affected in CN patients: (1) serum of these patients contains normal or increased levels of biologically active G-CSF and G-CSFRs that are expressed on CN myeloid cells at normal levels[31,32]; (2) CN patients responded only to the regular injections of pharmacological doses of recombinant human G-CSF (100–1000 times greater than physiological levels)[15]; and (3) *in vitro* growth of granulocyte colonies in CFU assays is defective with few colonies formed in spite of maximal stimulation with G-CSF.[33] In our study the reconstitution of LEF-1 in CN progenitor cells led to the upregulation of C/EBP-α and correction of G-CSF-triggered defective granulopoiesis. This finding proposes that LEF-1 seems to be involved in G-CSF:G-CSFR signaling. At the same time, G-CSF treatment of $CD34^+$ CN cells led to the slight upregulation of C/EBP-α and G-CSFR expression in the absence of LEF-1. It explains residual LEF-1-independent granulopoiesis under hyperactivation of G-CSF:G-CSFR signaling in CN patients *in vivo*: more than 90% of CN patients respond to G-CSF with absolute neutrophil counts above 1000/μL, independent from the CN subtype. Involvement of LEF-1 in G-CSFR signaling is a subject of our recent study.

We also found that LEF-1 mediates proliferation and survival of immature $CD34^+$ hematopoietic progenitor cells. Repression of endogenous LEF-1 by specific short hairpin RNA inhibited expression of LEF-1 target genes (c-myc, cyclin D1, survivin), reduced proliferation, and induced apoptosis of $CD34^+$ progenitors from healthy individuals. According to our findings in myelopoiesis, others have demonstrated that LEF-1 regulates proliferation and survival of lymphoid progenitors, namely pro-B cells, pre-B cells, and early thymic progenitors.[8–13] From a broader perspective, LEF-1 is important at a particular precursor stage of both lymphopoiesis and granulopoiesis.

Interestingly, overexpression of LEF-1 in $CD34^+$ progenitors from healthy individuals with subsequent culture in stem cell expansion medium without G-CSF or GM-CSF resulted in elevated proliferation. At the same time, LEF-1 mRNA levels reached a maximum at the promyelocytic stage of differentiation and declined during last steps of granulocyte maturation, indicating LEF-1-dependent signaling to be important at the definitive time-restricted, or/and stage-specific window. Therefore, based on these data we assume that there are two diverse biphasic functions of LEF-1 on proliferation and differentiation with two peaks of activity at different stages of cell life span. We hypothesize that first peak of LEF-1 activity corresponds to the stage of undifferentiated $CD34^+$ hematopoietic progenitor cells, which triggers proliferation and may be β-catenin dependent. The findings that activated β-catenin induces self-renewal and proliferation of hematopoietic stem cells[34,35] support this hypothesis. However, it remains unclear which of LEF-1/TCFs family members is involved in this process as β-catenin interaction partner. To achieve neutrophil differentiation in granulocyte precursor cells, LEF-1 expression must be elevated at the promyelocytic stage and LEF-1 function is β-catenin-independent, but is under G-CSF control. Similar stage-specific biphasic function of LEF-1 has been described in the context of hair follicle development. It has been shown that the dynamic expression of LEF-1 and TCF-3 (with expression patterns reciprocally to LEF-1) is necessary for proper hair follicle formation.[36] Similar regulatory principles were described for homeobox genes (HOXs, SOXs families) in mammalian and drosophila development including blood formation and have to be analyzed more precisely for the LEF-1 signaling system.

In summary, we found that LEF-1 is an important transcription factor regulating proliferation and differentiation of myeloid progenitors to mature neutrophils. We also identified a common pathologic mechanism of "maturation

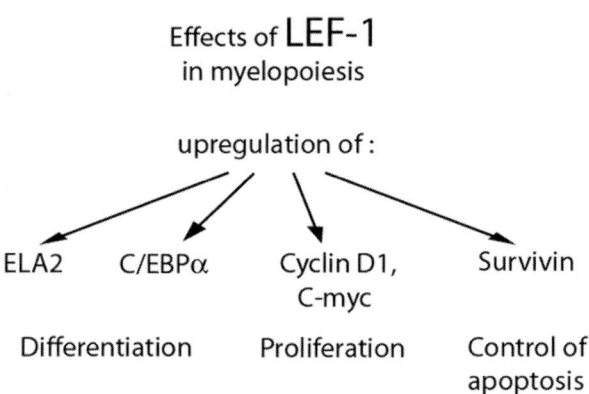

FIGURE 3. Effects of LEF-1 in myelopoiesis.

arrest" of promyelocytes in CN, namely loss of LEF-1 and its target genes expression. There is a dual role of LEF-1 in myelopoiesis by activation of: (1) granulocyte-specific proteins C/EBP-α and ELA2; and (2) proliferation- and apoptosis-regulated molecules cyclin D1, c-Myc, and survivin (FIG. 3).

ACKNOWLEDGMENTS

This work was supported by the German Network "Congenital Bone Marrow Failure Syndromes" of the Federal Ministry of Education and Research (BMBF), Deutsche Forschungsgemeinschaft (SFB 566), Madeleine-Schickedanz-Kinderkrebsstiftung, and Jose Carreras Leukaemie-Stiftung e.V.

REFERENCES

1. SKOKOWA, J., G. CARIO, M. UENALAN, et al. 2006. LEF-1 is crucial for neutrophil granulocytopoiesis and its expression is severely reduced in congenital neutropenia. Nat. Med. **12:** 1191–1197.
2. CLEVERS, H. & M. VAN DE WETERING. 1997. TCF/LEF factor earn their wings. Trends Genet. **13:** 485–489.
3. ARCE, L., N.N. YOKOYAMA & M.L. WATERMAN. 2006. Diversity of LEF/TCF action in development and disease. Oncogene **25:** 7492–7504.
4. NOVAK, A. & S. DEDHAR.1999. Signaling through beta-catenin and Lef/Tcf. Cell. Mol. Life Sci. **56:** 523–537.
5. NAWSHAD, A. & E.D. HAY. 2003. TGFbeta3 signaling activates transcription of the LEF1 gene to induce epithelial mesenchymal transformation during mouse palate development. J. Cell. Biol. **163:** 1291–1301.
6. ROSS, D.A. & T. KADESCH. 2001. The notch intracellular domain can function as a coactivator for LEF-1. Mol. Cell. Biol. **21:** 7537–7544.

7. LOVE, J.J., X. LI, D.A. CASE, et al. 1995. Structural basis for DNA bending by the architectural transcription factor LEF-1. Nature **376:** 791–795.
8. GIESE, K., C. KINGSLEY, J.R. KIRSHNER, et al. 1995. Assembly and function of a TCR alpha enhancer complex is dependent on LEF-1-induced DNA bending and multiple protein-protein interactions. Genes Dev. **9:** 995–1008.
9. BRUHN, L., A. MUNNERLYN & R. GROSSCHEDL. 1997. ALY, a context-dependent coactivator of LEF-1 and AML-1, is required for TCRalpha enhancer function. Genes Dev. **11:** 640–653.
10. JIN, Z.X., H. KISHI, X.C. WEI, et al. 2002. Lymphoid enhancer-binding factor-1 binds and activates the recombination-activating gene-2 promoter together with c-Myb and Pax-5 in immature B cells. J. Immunol. **169:** 3783–3792.
11. REYA, T., M. O'RIORDAN, R. OKAMURA, et al. 2000. Wnt signaling regulates B lymphocyte proliferation through a LEF-1 dependent mechanism. Immunity **13:** 15–24.
12. WILLINGER, T., T. FREEMAN, M. HERBERT, et al. 2006. Human naive CD8 T cells down-regulate expression of the WNT pathway transcription factors lymphoid enhancer binding factor 1 and transcription factor 7 (T cell factor-1) following antigen encounter *in vitro* and *in vivo*. J. Immunol. **176:** 1439–1446.
13. DOSEN, G., E. TENSTAD, M.K. NYGREN, et al. 2006. Wnt expression and canonical Wnt signaling in human bone marrow B lymphopoiesis. BMC Immunol. **7:** 13.
14. WELTE, K., C. ZEIDLER, D.C. DALE. 2006. Severe congenital neutropenia. Semin. Hematol. **43:** 189–195.
15. SKOKOWA, J., M. GERMESHAUSEN, C. ZEIDLER, et al. 2007. Severe congenital neutropenia: inheritance and pathophysiology. Curr. Opin. Hematol. **14:** 22–28.
16. HORWITZ, M., K.F. BENSON, R.E. PERSON, et al. 1999. Mutations in ELA2, encoding neutrophil elastase, define a 21-day biological clock in cyclic haematopoesis. Nat. Genet. **23:** 433–436.
17. KLEIN, C., M. GRUDZIEN, G. APPASWAMY, et al. 2007. HAX1 deficiency causes autosomal recessive severe congenital neutropenia (Kostmann disease). Nat. Genet. **39:** 86–92.
18. DALE, D.C., R.E. PERSON, A.A. BOLYARD, et al. 2000. Mutations in the gene encoding neutrophil elastase in congenital and cyclic neutropenia. Blood **96:** 2317–2322.
19. THUSBERG, J., & M. VIHINEN. 2006. Bioinformatic analysis of protein structure-function relationships: case study of leukocyte elastase (ELA2) missense mutations. Hum. Mutat. **27:** 1230–1243.
20. GRENDA, D.S., S.E. JOHNSON, J.R. MAYER, et al. 2002. Mice expressing a neutrophil elastase mutation derived from patients with severe congenital neutropenia have normal granulopoiesis. Blood **100:** 3221–3228.
21. MASSULLO, P., L.J. DRUHAN, B.A. BUNNELL, et al. 2005. Aberrant subcellular targeting of the G185R neutrophil elastase mutant associated with severe congenital neutropenia induces premature apoptosis of differentiating promyelocytes. Blood **105:** 3397–3404.
22. APRIKYAN, A.A., T. KUTYAVIN, S. STEIN, et al. 2003. Cellular and molecular abnormalities in severe congenital neutropenia predisposing to leukemia. Exp. Hematol. **31:** 372–381.
23. KÖLLNER, I., B. SODEIK, S. SCHREEK, et al. 2006. Mutations in neutrophil elastase causing congenital neutropenia lead to cytoplasmic protein accumulation and induction of the unfolded protein response. Blood **108:** 493–500.

24. KONISHI, N., M. KOBAYASHI, S. MIYAGAWA, *et al.* 1999. Defective proliferation of primitive myeloid progenitor cells in patients with severe congenital neutropenia. Blood **94:** 4077–4083.
25. CARIO, G., J. SKOKOWA, Z. WANG, *et al.* 2005. Heterogeneous expression pattern of pro- and anti-apoptotic factors in myeloid progenitor cells of patients with severe congenital neutropenia treated with granulocyte colony-stimulating factor. Br. J. Haematol. **129:** 275–278.
26. SERA, Y., H. KAWAGUCHI, K. NAKAMURA, *et al.* 2005. A comparison of the defective granulopoiesis in childhood cyclic neutropenia and in severe congenital neutropenia. Haematologica **90:** 1032–1041.
27. LI, F.Q., R.E. PERSON, K. TAKEMARU, *et al.* 2004. Lymphoid enhancer factor-1 links two hereditary leukemia syndromes through core-binding factor alpha regulation of ELA2. J. Biol. Chem. **279:** 2873–2884.
28. OELGESCHLAGER, M., I. NUCHPRAYOON, B. LUSCHER, *et al.* 1996. C/EBP, c-Myb, and PU.1 cooperate to regulate the neutrophil elastase promoter. Mol. Cell. Biol. **16:** 4717–4725.
29. LAUSEN, J., S. LIU, M. FLIEGAUF, *et al.* 2006. ELA2 is regulated by hematopoietic transcription factors, but not repressed by AML1-ETO. Oncogene **25:** 1349–1357.
30. WANG, W., X. WANG, A.C. WARD, *et al.* 2001. C/EBPalpha and G-CSF receptor signals cooperate to induce the myeloperoxidase and neutrophil elastase genes. Leukemia. **15:** 779–786.
31. MEMPEL, K., T. PIETSCH, T. MENZEL, *et al.* 1991. Increased serum levels of granulocyte colony-stimulating factor in patients with severe congenital neutropenia. Blood. **77:** 1919–1922.
32. KYAS, U., T. PIETSCH, K. WELTE. 1992. Expression of receptors for granulocyte colony-stimulating factor on neutrophils from patients with severe congenital neutropenia and cyclic neutropenia. Blood. **79:** 1144–1147.
33. NAKAMURA, K., M. KOBAYASHI, N. KONISHI, *et al.* 2002. Defects of granulopoiesis in patients with severe congenital neutropenia. Hiroshima J. Med. Sci. **51:** 63–74.
34. JAMIESON, C.H., L.E. AILLES, S.J. DYLLA, *et al.* 2004. Granulocyte-macrophage progenitors as candidate leukemic stem cells in blast-crisis CML. N. Engl. J. Med. **351:** 657–667.
35. SCHELLER, M., J. HUELSKEN, F. ROSENBAUER, *et al.* 2006. Hematopoietic stem cell and multilineage defects generated by constitutive beta-catenin activation. Nat. Immunol. **7:** 1037–1047.
36. MERRILL, B.J., U. GAT, R. DASGUPTA, *et al.* 2001. Tcf3 and Lef1 regulate lineage differentiation of multipotent stem cells in skin. Genes Dev. **15:** 1688–1705.

Role of Thrombopoietin in Mast Cell Differentiation

ANNA RITA MIGLIACCIO,[a,b] ROSA ALBA RANA,[c]
ALESSANDRO M. VANNUCCHI,[d] AND FRANCESCO A. MANZOLI[e]

[a]*Department of Hematology, Oncology and Molecular Medicine, Istituto Superiore Sanità, Roma, Italy*

[b]*Department of Medicine and MPD-RC, University of Illinois at Chicago, Chicago, Illinois, USA*

[c]*Department of Biomorphology, University G. D'Annunzio, Chieti, Italy*

[d]*Department of Hematology, University of Florence, Florence, Italy*

[e]*Department of Normal Human Anatomy, University of Bologna, Bologna, Italy*

> ABSTRACT: Mast cells are important elements of the body response to foreign antigens, being those represented either by small molecules (allergic response) or harbored by foreign microorganisms (response to parasite infection). These cells derive from hematopoietic stem/progenitor cells present in the marrow. However, in contrast with most of the other hematopoietic lineages, mast cells do not differentiate in the marrow but in highly vascularized extramedullary sites, such as the skin or the gut. Mast cell differentiation in the marrow is activated as part of the body response to parasites. We will review here the mast cell differentiation pathway and what is known of its major intrinsic and extrinsic control mechanisms. It will also be described that thrombopoietin, the ligand for the Mpl receptor, in addition to its pivotal rule in the control of thrombocytopoiesis and of hematopoietic stem/progenitor cell proliferation, exerts a regulatory function in mast cell differentiation. Some of the possible implications of this newly described biological activity of thrombopoietin will be discussed.
>
> KEYWORDS: thrombopoietin; Mpl; *Gata1*; mast cells; apoptosis

INTRODUCTION

Mast cells are hematopoietic cells localized mainly in highly vascularized extramedullary sites, such as the connective region of the skin and the mucosa of

Address for correspondence: Anna Rita Migliaccio, Ph.D., Department of Medicine and MPD-RC, University of Illinois at Chicago, Chicago, IL, and Hematology, Oncology and Molecular Medicine, Istituto Superiore Sanità, Viale Regina Elena 299, 00161 Rome, Italy. Voice: 0039-0649902690; fax: 0039-0649902530.

migliar@iss.it

the gut, where they engage themselves in the process of allergic response and in the immune reaction against parasites.[1,2] As all of the other hematopoietic cells, murine mast cells derive from multipotent c-$Kit^{low}CD34^{low}Sca$-1^{pos} progenitor cells present in the marrow and, to a lesser extent, in the spleen.[3,4] During fetal development, multilineage c-$Kit^{low}CD34^{low}Sca$-1^{pos} progenitor cells give rise to mast cell precursors, characterized by the phenotype c$Kit^{high}CD34^{low}Sca$-$1^{pos/neg}$, that circulate in the blood to colonize extramedullary sites.[5] These precursors possess extensive proliferation capacity, small cytoplasmic granules, and absent, or low, surface expression of the receptor that binds the Fc portion of the IgE antibody with high affinity (FcϵRI^{neg}).[4,5] Once in the extramedullary sites, c-$Kit^{high}CD34^{low}Sca1^{pos/neg}$ cells express their capacity to differentiate into mast cells with tissue-restricted mast cell-specific protease (MMCP) profiling (connective, mucosal, and serosal mast cells in the skin, gut, and peritoneum, respectively) throughout the life of the animals.[1,2,6] It is not clear at the moment whether each tissue-restricted mast cell type represents a different cell population in its own right or whether these lineages can *trans*-differentiate into one another once transferred into the right environment. The electron microscopy of a representative connective mast cell is presented in FIGURE 1.

Recently mast cell-restricted progenitor (MCP) cells, characterized by the phenotype c-$Kit^{pos}Sca$-$1^{neg}T1/ST2^{pos}$, have been purified from the marrow of adult mice.[7] As the multilineage cells described earlier, MCP generate both connective and serosal mast cells when transplanted into mast cell–deficient mice. In animals engrafted with MCP, formation of mast cells is detected as soon as 7 days after transplantation, whereas longer times (usually 14– 21 days) are required to observe mast cell reconstitution in animals transplanted with multilineage progenitor cells.[7] This observation, and the fact that c-$Kit^{low}CD34^{low}Sca$-1^{pos} cells generate *in vitro* cells similar in phenotype and function to MCP isolated from the marrow, suggests that MCP are generated, as are all other lineage-restricted progenitor cells, during the commitment process of multilineage progenitor cells.[8] Although the frequency of MCP in the marrow is comparable to that of the other lineage-restricted progenitor cells (2– 3%), mastocytopoiesis remains limited in marrow throughout adult life. In fact, under steady-state conditions, the number of mast cells (defined by the phenotype c-$Kit^{high}CD34^{pos}Fc\epsilon RI^{pos}$) detectable in the marrow is low (0.02% of the total population) compared to that of other hematopoietic precursors, such as megakaryocytes (~6%) and erythroblasts (20–25%).[9,10] Furthermore, these marrow precursors do not contain cytoplasmic granules, and, although expressing mRNA for both the α and β chain of FcϵRI, express low levels of FcϵRI on their surface.[9] The molecular mechanism that, in normal mice, restricts the mastocytopoietic potential of stem/progenitor cells to extramedullary sites is overcome as part of the reaction against parasites, as demonstrated by the fact that experimentally induced infection with the intestinal worm *Trichinella spiralis* restores mast cell differentiation in the mouse marrow; at day 5 after

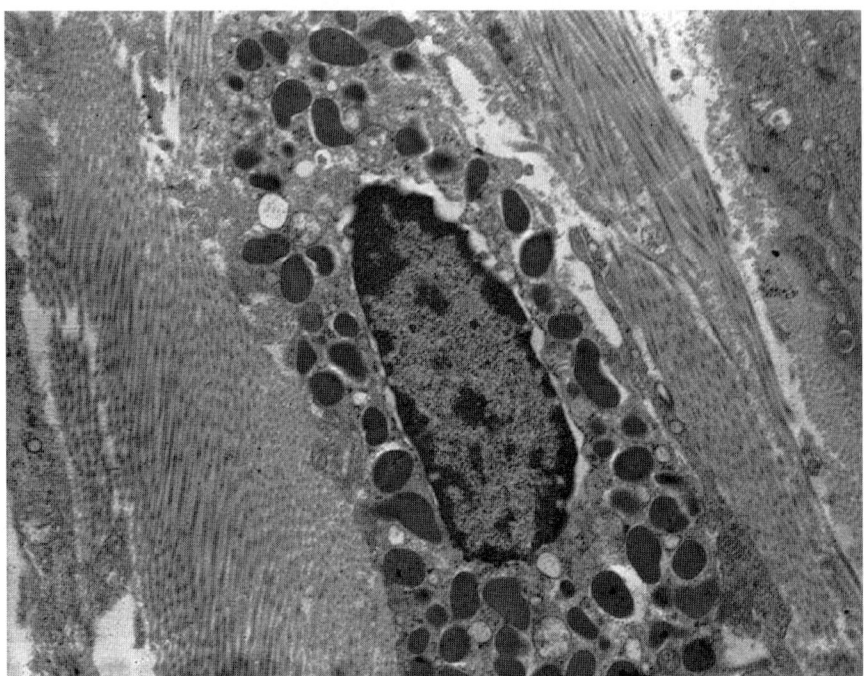

FIGURE 1. Transmission electron microscopy of a section from the skin at the base of the ear of a wild-type mouse showing the morphology of a typical connective mast cell. To be noted the presence, in the cell cytoplasm, of numerous granules whose heterogeneous electron density is characteristic of mature mast cells. Magnification X4400.

infection, there is an increase of c-Kit^{high} mast cell precursors in the marrow.[11] These cells begin their differentiation (as indicated by the complexity of their site/forward cytofluorimetric chart) in the marrow and migrate in the blood to colonize the gut. Increased number of mast cells became detectable in the gut at days 8–10 from parasite infection.[11,12]

EXTRINSIC AND INTRINSIC CONTROL OF MAST CELL DIFFERENTIATION

As for all the other hematopoietic lineages, the differentiation of stem/progenitor cells into mast cells is controlled by both extrinsic (growth factors produced by the microenvironment) and intrinsic (transcription factors that bind to DNA of stem/progenitor cells) pathways. Instrumental to the clarification of these pathways has been the analyses of the phenotype of mice harboring specific gene mutations. A further contribution to the identification of the extrinsic pathway has been provided by the definition of culture conditions necessary to observe mast cell differentiation *in vitro*.[13–16]

The extrinsic regulatory pathway includes the growth factors stem cell factor (SCF) and interleukin-3 (IL-3).[15,16] In fact, marrow (and spleen) hematopoietic progenitors generate, within 14 days of culture stimulated with optimal concentrations of SCF and IL-3, cells with the phenotypical (c-Kit^{high} CD34^{low} FcϵRI^{pos} cells with small alcian blue^{neg} cytoplasmic granules),[17] and functional (reconstitute connective and mucosal mast cells when transplanted into mast cell–deficient animals)[18] profile of mast cell precursors (see also FIG. 2). By day 21, these cells have completed their maturation into berberine sulfate^{pos} mature mast cells[14] that biochemically resemble connective mast cells[19] capable both of taking up and releasing serotonin after IgE/αIgE stimulation.[15,18] In these cultures, named bone marrow–derived mast cell cultures (BMMC), SCF sustains cell proliferation,[20,21] whereas either SCF[21,22] or IL-3[23] prevent apoptosis. The expression of neither SCF nor IL-3 is restricted to the marrow. In fact, SCF is produced by marrow and skin fibroblasts alike.[24] Furthermore, both the membrane and, at higher concentrations, the soluble form of SCF are capable of sustaining mast cell differentiation in BMMC cultures.[15] On the other hand, IL-3 is mainly produced by immune cells in response to specific stimuli. As such, IL-3 is not detectable *in vivo* under steady-state conditions.

MCP cultured under BMMC conditions also gives rise to mature mast cells.[7,25] However, in contrast with the culture of multilineage progenitors described earlier, cKit^{pos}/FcϵRI^{pos} cells are generated by MCP in only 7 days (FIG. 3). The mature cells obtained in these culture express a robust mast cell phenotype that includes high expression of all the MMCP analyzed. The MMCP that is expressed the most by these cells is MC-CPA, the MMCP that, *in vivo*, is primarily expressed by dermal mast cells.

Purification of MCP Growth of MCP Expression Profile of c-Kit^{pos} FcϵRI^{pos} Cells from MCP

The essential role played by SCF in mastocytopoiesis is further confirmed by the phenotype expressed by mice carrying alterations in the genes encoding SCF,[26] the SCF receptor c-Kit[27] and PI3K,[28] an important element of the SCF/c-Kit signaling.[29] Because of the pivotal role played by SCF/c-Kit interactions throughout development, the majority of the mutations at the *Scf* (steel) or *c-Kit* (also called white, W) locus are embryonically lethal. As such, these mutants do not provide information of the effects of SCF on a lineage that is primarily expressed in adult life. However, two *c-Kit* mutants, the double W/W^v and the Kit^{W-sh}/Kit^{W-sh} mutants, are viable and have played an important role in the characterization of SCF functions in mastocytopoiesis. Heterozygous W/W^v mice (i.e., expressing the W, truncated, and W^v, kinase defective, form of *c-Kit*[26]) are born with thrombocytopenia, macrocytic anemia, and mast cell deficiency (see FIG. 4). The hypomorphic Kit^{W-sh} mutation is a large

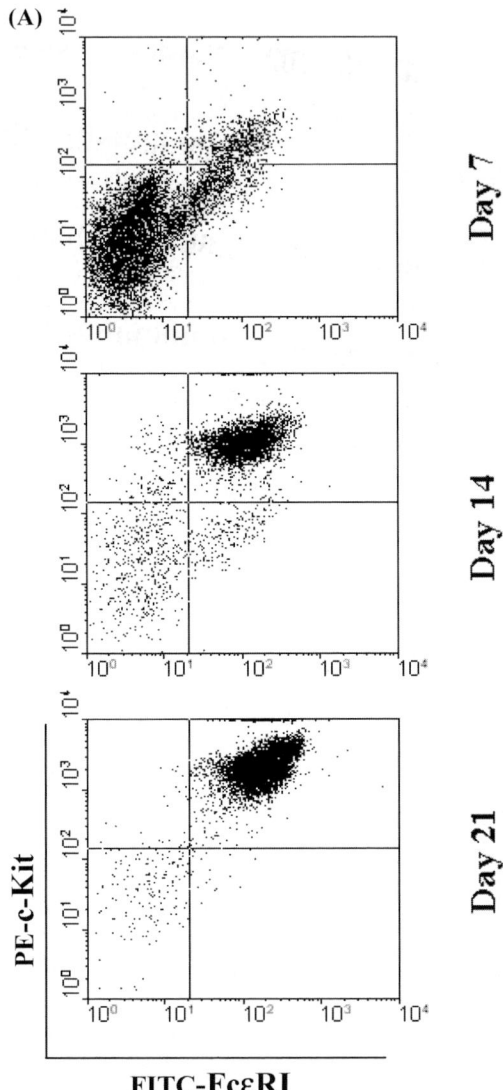

FIGURE 2. Representative flow cytometric analysis for c-Kit and FcεRI expression **(A)** and GATA1 immunostaining **(B)** of cells obtained in BMMC seeded with multilineage progenitor cells from wild-type mice. Cells were analyzed either at day 7, 21, or 26 of culture, as indicated. Negative controls for flow cytometry analysis were represented by cells labeled with irrelevant antibodies, and are not presented for convenience. In panel B, the green fluorescence in the left panels corresponds to GATA1-immunostaining, whereas the blue fluorescence in the right ones corresponds to DAPI staining of the nucleus. The GATA staining is localized predominantly in the cell cytoplasm until day 7–14 of culture to became predominantly localized in the nucleus of mature cells obtained by day 21. Magnification X100. (Panel A was modified by permission from Ghinassi et al.[25])

(B)

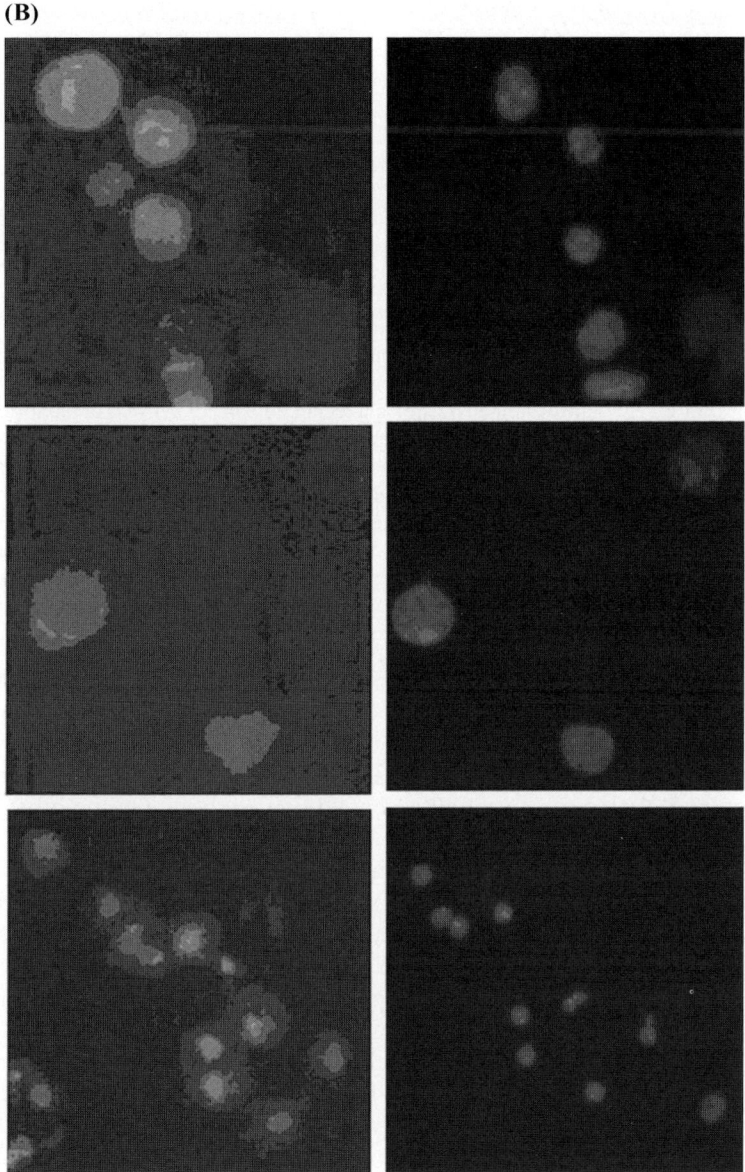

FIGURE 2. Continued.

deletion of the region of DNA from –23 to –154 Kb upstream of the *c-Kit* gene that contains the regulatory sequences specifically required for gene expression in mast cells.[30] Therefore, homozygous Kit^{W-sh}/Kit^{W-sh} mutants have normal blood cell counts but do not express mast cells. Because of the

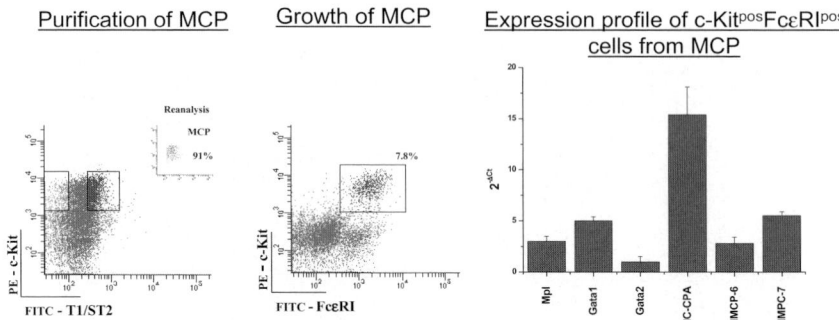

FIGURE 3. Mast cell differentiation occurs only in 7 days when BMMC are seeded with purified MCP. Left and middle panels describe the purification of marrow MCP according to the cKit^{pos}T1/ST2^{pos} gate described by the right rectangle. The re-analyses of the sorted cells for purity is presented in the insert. The frequency of mature mast cells (c-Kit^{pos}FcεRI^{pos} cells) obtained after 7 days in BMMC seeded with these purified MCP is presented in the middle panels. To be noted the different staining profile of progeny obtained from MEP versus those obtained from multilineage progenitors presented in FIGURE 2B. Approximately 8% of the MCP progeny express already the mature c-Kit^{pos}FcεRI^{pos} mast cell phenotype by day 7. By expression profiling, mast cells obtained in these culture express robust levels of all MMCPs, as indicated by the quantitative RT-PCR analyses presented in the right panel. To our surprise, these cells also express detectable levels of Mpl, the thrombopoietin receptor. (Partially modified by permission from Ghinassi et al.[25]) (In color in Annals online.)

exquisite mast cell deficiency induced by the mutation, these mutants are the recipients of choice for mast cell reconstitution studies.[8] On the other hand, under steady-state conditions, IL-3 is not produced in mice. Therefore, it is not surprising that mice lacking the *Il-3* gene do not have mast cell deficiency.[31] IL-3-deficient mice, however, respond poorly when challenged with parasites.[31] As said earlier, IL-3 expression is activated *in vivo* by several immune reactions. In particular, mast cells are induced to produce IL-3 in an autocrine/paracrine fashion on interaction of FcεRI receptor with IgE produced by activated B cells.[32] The phenotype of the *Il-3* knockout mice confirms the importance for appropriate immune response of B cell recruitment to the infected sites that will, in turn, increase the numbers of mast cells locally.[33] Further positive regulators of mast cell differentiation are represented by IL-4[34] and IL-9.[35] The only negative regulators for mast cell differentiation reported so far are represented by glucocorticoids.[36,37]

Genetic mouse mutants have also contributed to the identification of elements of the intrinsic regulatory pathway for mast cell differentiation. The mast cell-specific transcription factors identified up to know are represented by *Mitf*,[38] encoded by the microphtalmia locus,[39] *Gata2*,[40,41] and *Gata1*.[42,43] *Mitf* encodes a helix–loop–helix protein involved in the activation of *c-Kit* expression[44] that can have either positive (MMCP-6)[45] or negative[46] (MMCP-7, the protease most abundantly expressed by serosal mast cells) (FIG. 4) effects on the expression of MMCPs. *Gata2* and *Gata1* are both members of the GATA family of transcription factors.[47] In mast cell differentiation, as in other hematopoietic lineages, expression of *Gata2* precede that of *Gata1*[41] and is associated with the rate of cell proliferation.[40,48] Cells obtained from purified stem/progenitor cells under BMMC culture conditions express GATA1 throughout their differentiation,[43] although the protein becomes prevalently localized in the nucleus at later stages of maturation (FIG. 2). Functional *Gata1* binding sites are present in the regulatory region of Carboxypeptidase A[49] (MC-CPA, the protease-specific for dermal mast cells and primarily produced by mast cells in culture) (FIG. 2) and of the α[50] and β[51] chain of the FcϵRI and mast cells from genetically deficient *Gata1* mice remain FcϵRI^{neg}.[52]

The importance of *Gata1* for mast cell differentiation was first identified on the basis of the phenotype of mice harboring the hypomorphic *Gata1*^{low} mutation, a targeted deletion that removes upstream enhancer and promoter elements of the *Gata1* gene.[43] This mutation affects all stages of mast cell differentiation. More specifically, it alters the commitment process by inducing the formation of high numbers of a unique class of progenitor cells that committed toward the erythroid, megakaryocytic, and mast cell lineage.[43] Surprisingly, these progenitor cells do not have the antigenic profile of MCP, that are not detectable in tissues from these mutants, but rather that of progenitor cells that would normally be restricted for the megakaryocytic and erythroid lineage, the MEP.[25] In addition to their abnormal differentiation program, *Gata1*^{low} MEP have profound proliferation abnormalities and $\sim$10% of them generate a progeny that give rise with 95% efficiency to growth factor–dependent clonal cell lines. Therefore, in *Gata1*^{low} mice, the mast cell generating activity is abnormally acquired by MEP, that, are antigenically, but not functionally, equivalent to wild-type MEP. At the precursor level, the *Gata1*^{low} mutation induces an increase in proliferation that is coupled by increased cell death, but on balance generating more live cells than dead.[43] Finally, the *Gata1*^{low} mutation impairs expression of FcϵRI receptor and of MC-CPA in mature mast cells (Refs. 25, 43 and FIG. 4).

IN VIVO TREATMENT WITH THROMBOPOIETIN REDUCES GATA1 EXPRESSION AND INDUCES SIMILAR ALTERATIONS BOTH IN MEGAKARYOCYTES AND IN MAST CELLS

There are striking similarities between the extrinsic and intrinsic control pathway of megakaryocyte and mast cell differentiation. In fact, appropriate differentiation of both lineages is dependent on the extrinsic growth factors SCF (in synergy with thrombopoietin,[53] or IL-3,[14] respectively), and on the

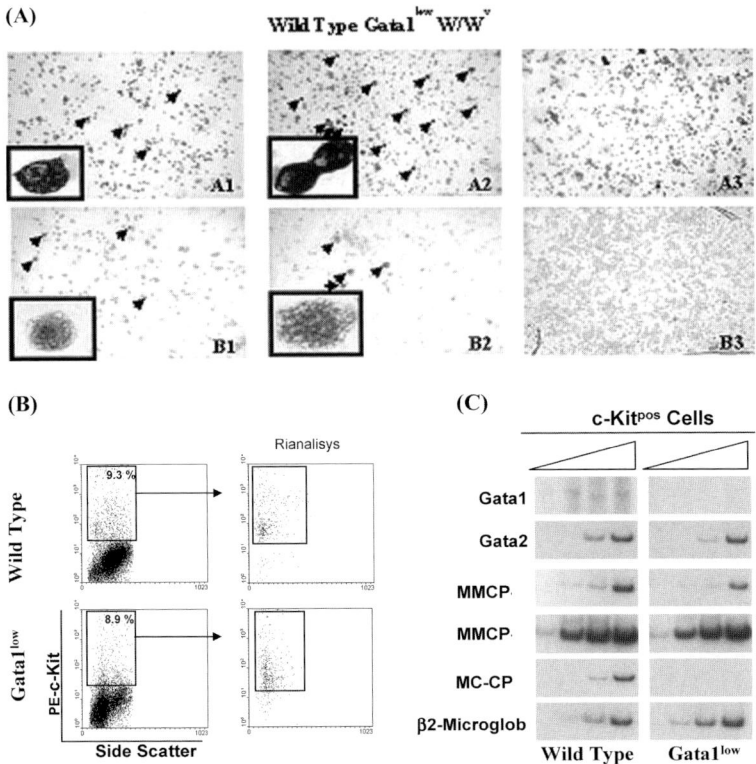

FIGURE 4. Frequency and expression profiling of serosal mast cells in peritoneal lavages from wild-type mice and from mice carrying either the hypomorphic $Gata1^{low}$ or W/W^v mutation, as indicated. **(A)** Toluidine blue (*top panels*) and alcian blue (bottom panels) staining of cytocentrifuged smears of peritoneal lavages. **(B)** Purification by cell sorting of wild-type and $Gata1^{low}$ serosal mast cells (c-Kit^{pos}). **(C)** Expression profiling analyzed by semiquantitative RT-PCR of cells purified in panel B. Semi-quantitative estimations of gene expression levels were based on the amount of product amplified at sequential time points during the PCR reaction, as indicated by the triangles. These results indicate that the frequency of mast cell precursors (*arrowheads* in the panels stained with toluidine blue) is higher in the peritoneum of $Gata1^{low}$ mice than in that of wild-type littermates. On the other hand, although the frequency of mature mast cells (*arrowheads* in the panels stained with alcian blue) is normal, $Gata1^{low}$ mast cells express barely detectable levels of MC-CPA (see panel C). In contrast, both mast cells and their precursors are absent in the peritoneum of W/W^v mice. Original magnification X40 in all of the panels and X200 in the inserts. (In color in Annals online.)

intrinsic transcription factor $Gata1$.[54,43] As shown earlier, not only the same regulatory sequences (the first enhancer of $Gata1$, the DNase hypersensitive site I (HSI), deleted by the $Gata1^{low}$ mutation) lead appropriate expression of the gene both in megakaryocytes and mast cells, but the abnormalities induced by experimental deletion of HSI (i.e., by reduced *Gata1* expression) in both lineages are very similar.[43,54–56] As mentioned earlier, reduced levels of *Gata1* impairs the differentiation potential of progenitor cells common for the megakaryocyte and mast cell lineage and increases proliferation while blocking maturation of their precursor cells. Some similarities are also observed between the abnormalities induced by the mutation in erythroid and mast cells. In fact, the abnormal progenitor cell population generated by the mutation is also restricted toward the erythroid lineage[25] and reduced expression of *Gata1* renders both erythroid and mast cell precursors more susceptible to apoptosis.[43,57]

Further similarities between megakaryocytic and mast cell lineages were discovered during the analysis of the effects of acute treatment with thrombopoietin on the hematopoiesis expressed by $Gata1^{low}$ mice. $Gata1^{low}$ mice compensate the defective hematopoiesis induced by the mutation by establishing a complex homeostatic mechanism that includes increased growth factor gene expression (such as TGF-β and PDGF) in the marrow microenvironment[58] and extensive extramedullary hematopoiesis in the spleen.[57] With age, $Gata1^{low}$ mice develop a phenotype very similar to the syndrome expressed by patient with idiopathic myelofibrosis.[58] For this reason, these animals have been proposed as a model for the human disease.

Also experimental manipulations that increase the concentration of thrombopoietin in animals (TPO^{high} mice) induce development of a rapid and fatal myelofibrosis syndrome. In fact, transgenic mice expressing high levels of the human thrombopoietin gene in their liver,[59] animals transplanted with hematopoietic stem cells transduced with a retrovirus carrying the human gene[60,61] or simply mice[62] and rats[63] injected over time with high doses of recombinant growth factor develop a myelofibrosis-like syndrome. Megakaryocytes from TPO^{high} mice express the same abnormalities presented by cells developed in hematopoietic tissues from $Gata1^{low}$ mice.[56,62] These abnormalities include accumulation of P-selectin on the demarcation membrane region, increased neutrophil emperipolesis within the megakaryocytes and death of the megakaryocytes by para-apoptosis. Because these same abnormalities are found in megakaryocytes from myelofibrotic patients,[64] they are thought to play an important role in the pathogenesis of the disease. These considerations suggested that may be the TPO^{high} and $Gata1^{low}$ model might represent elements of a common pathogenetic pathway for the disease.[65,66]

To clarify the relationship, if any, between thrombopoietin and GATA1 in the pathobiological mechanism leading to myelofibrosis in mice, we first compared thrombopoietin metabolism and efficacy in wild-type, as well as in $Gata1^{low}$ littermates. No difference was found between the two animals' groups in terms of thrombopoietin expression in liver, the major site of thrombopoietin production,[67] and of thrombopoietin concentration in blood (∼0.75 ng/mL).[62] Next we compared the response of wild-type and $Gata1^{low}$ mice to a thrombopoietin-administration schedule that transiently raised the thrombopoietin plasma levels up to ~5.7–6.0 ng/mL by day 7, a value similar to those (~3–100 ng/mL) observed in TPO^{high} animals because of transgenic expression of the human thrombopoietin gene in the liver[59] or of transplantation with stem cells infected with a retrovirus containing human thrombopoietin.[60,61] To our surprise, the two animals have clearly different response to treatment. Unexpectedly, in wild-type mice, thrombopoietin treatment reduced GATA1 content in megakaryocyte and induced abnormalities very similar to those expressed by megakaryocytes in the marrow from *Gata1*^{low} mice and from idiopathic myelofibrosis patients (Ref. 62 and FIG. 5A). Interestingly, several TdT-mediated dUTP nick end labeling (TUNEL^{pos}) cells, possibly erythroblasts, were detected in the spleen from thrombopoietin-treated wild-type mice, an indication of low *Gata1* expression also in these cells (FIG. 5A). In contrast, in *Gata1*^{low} mice, thrombopoietin treatment increased GATA1 content in megakaryocytes and, consequently, restored thrombocytopoiesis and reduced marrow fibrosis (Ref. 62 and FIG. 5A). On the other hand, the number of TUNEL^{pos} erythroblasts in the spleen of these mice was greatly reduced by the treatment, indicating that GATA1 content had been restored also in these cells (FIG. 5A). Such strict association between occurrence of the disease and low GATA1 content in megakaryocytes suggests that thrombopoietin is upstream of *Gata1* in a common pathobiological pathway leading to the development of myelofibrosis in mice and, possibly, of idiopathic myelofibrosis in men. It also suggests that HIS might function as a linker between thrombopoietin and GATA1 in the control of megakaryocytopoiesis.

The GATA1- and thrombopoietin-dependent control pathways of megakaryocytopoiesis had been indirectly linked by the observation that megakaryocytes from *Gata1*^{low} mice express reduced levels of Mpl,[55] suggesting that GATA1 is required for proper expression of the *mpl* gene in these cells. The observation that thrombopoietin treatment alters the *Gata1* mRNA expressed by megakaryocytes from normal mice suggests that the two pathways might also be linked by the presence of thrombopoietin-responsive elements among the regulatory regions of the gene. The fact that thrombopoietin had opposite effects on *Gata1* mRNA levels in megakaryocytes from wild-type and *Gata1*^{low} mice suggests that these putative thrombopoietin-response elements should include both enhancer and suppressor sequences, the suppressor ones being specifically located within HSI, the region deleted by the *Gata1*^{low} mutation.

The notion that HSI, the putative target of the thrombopoietin signaling cascade,[62] also plays a crucial role in the control of *Gata1* expression in mast cells[43] prompted us to analyze the number and differentiation state of mast cells in extramedullay site of mice being treated with thrombopoietin. As shown in FIGURE 5B, thrombopoietin treatment–induced changes in the dermal mast cell compartment similar to those induced in the megakaryocyte and erythroid compartment. It increased by twofold the number of mast cells in the ear of wild-type mice (from 266 ± 18 to 533 ± 35/mm^2), but reduced the number of mast cells expressing *Gata1*, and almost all the cells from thrombopoietin-treated animals became TUNEL^{pos} (from 16 ± 1 to $600 \pm 43/mm^2$) (FIG. 5B). As described earlier for megakaryocytes and erythroid cells, paradoxical effects were induced by thrombopoietin treatment in mast cells from *Gata1*^{low} mice. Thrombopoietin treatment did not significantly alter the number of mast cells detectable in the ear of these mice (450 ± 30 versus $360 \pm 25/mm^2$, respectively), but restored GATA1-immunostaining of the cells and greatly reduced the number of TUNEL^{pos} cells (from 600 ± 45 versus $60 \pm 2/mm^2$) (FIG. 5B). All of these similarities suggested to us that the putative thrombopoietin-responsive regulatory regions of *Gata1* might be functional also in mast cells. Mast cells would, then, represent a new cell target regulated for the biological activity of this growth factor.

Mpl Is Expressed by Serosal Mast Cells

A corollary of the hypothesis that thrombopoietin represents a new regulatory factor for mastocytopoiesis is that mast cells, and their precursors, must express Mpl, the thrombopoietin receptor. Indeed, the expression profiling presented in FIGURE 3 already indicates that mast cells obtained *in vitro* under BMMC conditions express levels of Mpl detectable by quantitative RT-PCR. Here, we pursue this analyses demonstrating that serosal wild-type mast cells, and their precursors, express detectable levels of Mpl mRNA and surface protein.

By quantitative RT-PCR, wild-type MCP purified from the marrow express level of *Mpl* within the range expressed by hematopoietic progenitor cells ($2^{-\Delta Ct} = 1.3 \pm 0.4 \times 10^{-2}$). On the other hand, serosal mast cells express level of *Mpl* that, although 2-log lower than those expressed by megakaryocytes ($2^{-\Delta Ct} = 5.1 \pm 0.2 \times 10^{-2}$), are easily detectable by this technique ($2^{\Delta Ct} \sim 7.2 \times 10^{-4}$). As control, *Mpl* mRNA is not detectable by serosal mast cells purified from mice genetically deprived of the *Mpl* gene (Mpl^{null} mice).

By flow cytometry analysis, the AMM2 Mpl-specific antibody detected a signal clearly above that of isotype controls on the surface of serosal mast cells (as defined by the phenotype c-Kit^{pos}/FcεRI^{pos}) from the peritoneum of wild-type mice (FIG. 6). Again, as further negative control, the antibody did not react with serosal mast cells (or CD41^{pos} megakaryocytes, not shown) from Mpl^{null} mice (FIG. 6).

Targeted Deletion of the Mpl *Gene Increases Mast Cell Differentiation in Mice*

To complete our characterization of the effects of thrombopoietin on mast cell differentiation, we analyzed the mast cells present in tissues from Mpl^{null} mice.[68] The results are summarized in FIGURE 7.

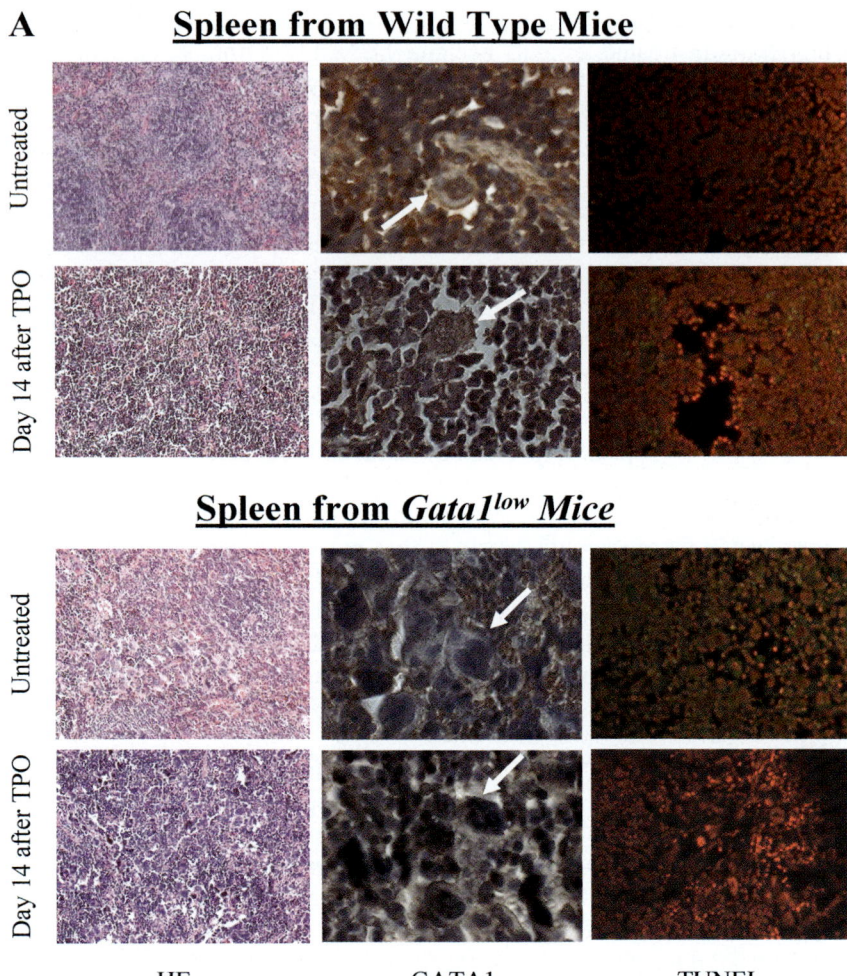

FIGURE 5A. Hematoxylin-eosin (HE), GATA1, and TUNEL staining of spleen sections from untreated wild-type and $Gata1^{low}$ littermates and from animals that had been treated 14 days earlier with thrombopoietin (100 μg/kg/day per 5 days), as indicated. HE staining (*left panels*) reveals massive megakaryocyte hyperplasia both in the spleen section from wild-type and $Gata1^{low}$ mice 14 days after thrombopoietin treatment. By GATA1-immunostaining (*middle panels*), megakaryocytes (indicated by arrows) from both untreated wild-type animals and from thrombopoietin-treated $Gata1^{low}$ littermates contain GATA1. In stead, megakaryocytes from untreated $Gata1^{low}$ and from thrombopoietin-treated wild-type animals do not react with the antibody. This result indicates that thrombopoietin has a paradoxical effect on GATA1 expression in megakaryocytes from the two animals: it inhibits $Gata1$ expression from the wild-type gene but restores expression from the mutated gene. Similar paradoxical effects were observed in the induction of erythroblast apoptosis (*right panels*). TUNEL[pos] cells are virtually non-detectable in spleen from wild-type mice while, as expected (Ref. 57), numerous TUNEL[pos] erythroblasts are detectable in spleen

It has been already reported that the Mpl^{null} mutation decreases the frequency of hematopoietic progenitors in the marrow.[68] The MPC progenitors are no exception. Their frequency in the marrow (and spleen) from Mpl^{null} mice is 1-log lower than in controls (FIG. 7). In contrast, although only 0.2% of the total cell population in the marrow of wild-type animals has a c-$Kit^{pos}Fc\epsilon RI^{pos}$ mast cell precursor phenotype, the frequency of c-$Kit^{pos}Fc\epsilon RI^{pos}$ cells in the marrow of Mpl^{null} mice is greater than 10-fold higher than in controls.

Increased numbers of mast cells are also present in the connective tissues of the ear and in the peritoneum from Mpl^{null} mice (FIG. 7A). Electron microscopy examination reveals that both dermal and serosal mast cells from Mpl^{null} mice are twice as big as the corresponding wild-type cells and contain more, and pronounced heterogeneous in electron density, granules in their cytoplasm than control mast cells (R.A. Rana, unpublished data).

In conclusion, Mpl^{null} mice, in spite of a lower frequency of MPC in their marrow, express higher numbers of mast cells in marrow, dermis, and peritoneum than wild-type littermates.

IMPLICATIONS OF THE BIOLOGICAL EFFECTS OF THROMBOPOIETIN ON MAST CELL DIFFERENTIATION FOR NORMAL AND PATHOLOGICAL HEMATOPOIESIS

Thrombopoietin is one of the most important *in vivo* regulator of megakaryocytic differentiation[53] as Mpl^{null}[68] and TPO^{high}[69] mice have, respectively, decreased and increased levels of megakaryocytes in their marrow and of platelets in the blood. We show here that the two mutations also induce alterations of opposite sign on mast cell numbers in extramedullary sites (FIGS. 5B, 7). In this case, however, the Mpl^{null} mutation increases and the TPO^{high} decreases the number of viable mast cells in extramedullary tissues. As megakaryocytes and platelets represent two cellular elements capable of releasing growth factors in the microenvironment, it could be argued that growth factors released by these cells might indirectly mediate the effects induced by thrombopoietin on mast cell differentiation described here. However, we believe that the fact that mast cells express detectable levels of Mpl argues instead for a direct function of this factor in mast cell differentiation. It remains to be established what could be the physiological role played by this growth

←──────────────────────────────────

FIGURE 5A (*continued*). from *Gata1*^{low} mice. After thrombopoietin treatment, $TUNEL^{pos}$ erythroblasts became detectable in spleen from wild-type mice, whereas the number of $TUNEL^{pos}$ cells in spleen from *Gata1*^{low} mice was greatly reduced by treatment. Because erythroblast apoptosis is linked to the levels of GATA1 expressed by the cells, these results indicate that thrombopoietin affects expression of GATA1 in erythroblasts as well as in megakaryocytes. Original magnification X40 in the left and right panels and X200 in the middle ones. (Data modified by permission from Ref. 62.)

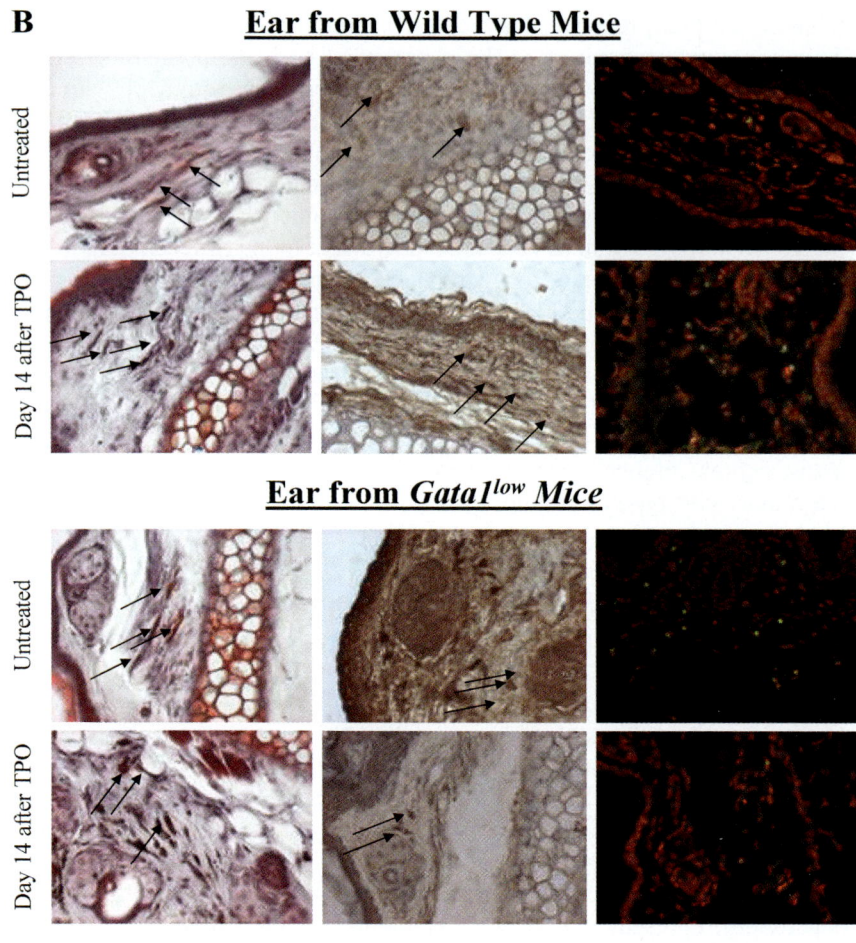

FIGURE 5B. Alcian blue safranin (ABSO), GATA1, and TUNEL staining of ear sections from untreated wild-type and $Gata1^{low}$ littermates and from animals that had been treated 14 days earlier with thrombopoietin (100 µg/kg/day per 5 days), as indicated (the same animals analyzed in FIG. 5A). Alcian blue safranin staining (*left panels*) indicates that the number of mast cells (indicated by *arrows*) in the ear from wild-type mice is increased by thrombopoietin treatment. However, by immunostaining (*middle panels*), GATA1 is detectable in mast cells from wild-type and thrombopoietin-treated $Gata1^{low}$ mice but not in those from untreated $Gata1^{low}$ and thrombopoietin-treated wild-type mice. TUNEL staining parallels the results of GATA1-immunostaining. TUNEL^{pos} cells are detectable in ears from mice with reduced $Gata1^{low}$ expression (both untreated $Gata1^{low}$ and thrombopoietin-treated wild-type mice) but not in those with detectable GATA1 content (untreated wild-type and thrombopoietin-treated $Gata1^{low}$ mice). Therefore, thrombopoietin treatment has the same paradoxical effects on $Gata1$ expression in megakaryocytes and mast cells. Original magnification X40 in all panels.

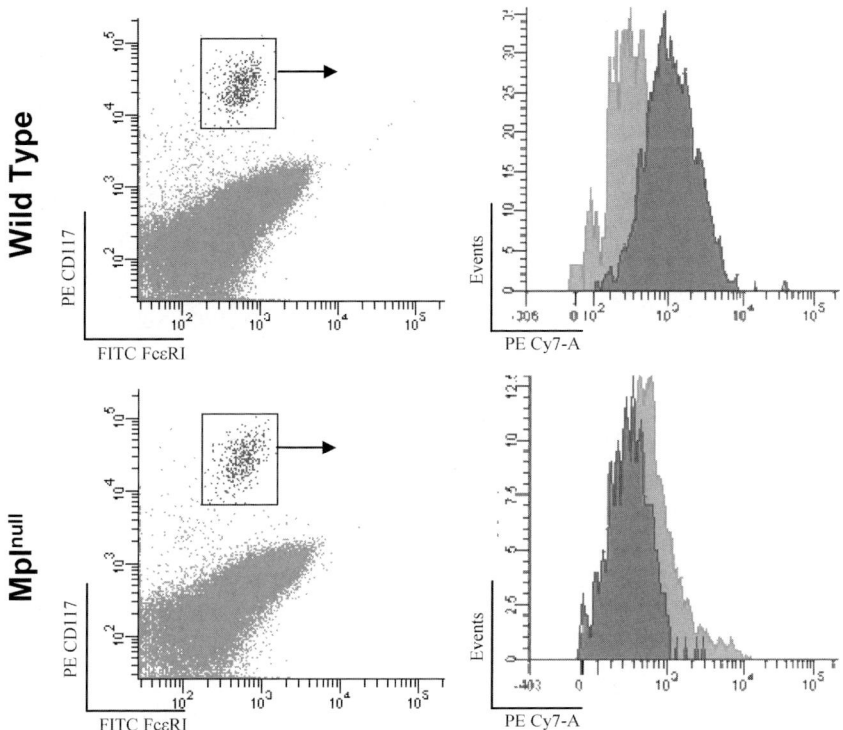

FIGURE 6. Flow cytometry analysis for the expression of MPL on serosal mast cells from wild-type and Mpl^{null} mice, as indicated. The CD117 antibody recognize c-Kit. Cells in the prospective c-Kit^{pos}FcϵRI^{pos} gate indicated by the rectangle were analyzed either with a Cy7A-labeled irrelevant antibody (green area) or with the Cy7A-labeled AMM2 antibody (blue area). Wild-type serosal mast cells express levels of MPL clearly above background. (In color in Annals online.)

factor in normal mastocytopoiesis. We will discuss two not mutually exclusive hypotheses.

The first hypotheses is inspired by the knowledge that thrombopoietin is not only an important regulator of megakaryocytopoiesis, but represents also a regulator of stem/progenitor cell expansion.[67] Originally believed to be a real hormone produced by the liver and delivered to the marrow by the blood stream, thrombopoietin has been recently shown by Suda.[70] to be produced, together with SCF, by osteoblasts on the marrow endosteum. It is currently believed that osteoblasts represent key elements of the hematopoietic stem cell niche.[71] The combination of SCF and thrombopoietin produced by these cells would induce proliferation and detachment of the stem cells from the niche. Once detached, the stem cells would begin their journey into

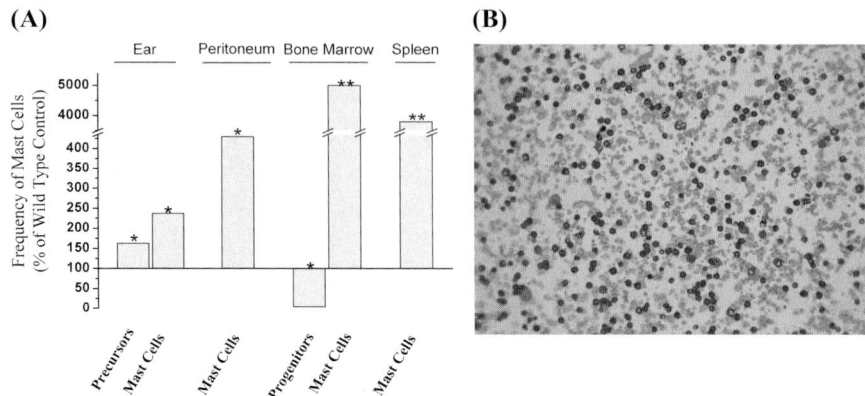

FIGURE 7. (**A**) Impairment of *Mpl* expression (Mpl^{null} mutation) decreases the frequency of mast cell-restricted progenitor (MCP) cells in marrow but increases the frequency of mast cells, and of their precursors, in ear, peritoneum, marrow, and spleen of the animals. Mast cells were defined by histological (alcian blue positivity) criteria on ear sections and by flow cytometry (as c-Kit^{pos}/FcεRI^{pos} cells) in peritoneum, marrow, and spleen cell suspensions. Mast cell precursors were identified by histogical criteria (toluidine blue positive cells) only. MCP were identified by flow cytometry (c-Kit^{pos}/T1/ST2^{pos}, see also FIG. 3). Results are expressed as percent of the values observed in wild-type littermates. * and ** indicate values statistically different ($p < 0.05$ and $p < 0.01$, respectively) from wild-type control. (**B**) A representative toluidine blue staining of peritoneal lavages from Mpl^{null} mice is presented. Compare the frequency of toluidine blue^{pos} cells present in this smear with that present in the smear of peritoneal lavages from wild-type mice presented in FIGURE 4. Original magnification: X40. (In color in Annals online.)

the medulla and, in this process, would differentiate into progenitor cells restricted for the different lineages. Hematopoietic progenitors have the potential to differentiate directly into mast cells without cell division upon interacting with SCF.[8] Some of them must actually do so directly in the marrow. High numbers of mast cells in the marrow, as well as in other organs, lead to the development of fibrosis. A homeostatic mechanism must, then, be in place to limit the number of mast cells produced in the marrow. The fact that the same growth factor, thrombopoietin, would synergize SCF in favoring progenitor cell formation (by inducing their proliferation) and antagonize this growth factor action on mast cell formation (by inducing mast cell apoptosis), represents an extremely efficient mechanism for the organism to achieve this goal.

On the other hand, the observation that thrombopoietin regulates *Gata1*, the transcription factor required for the serosal-specific MC-CPA expression, suggests a more bold hypotheses. One of the consequences of parasite infection is gut bleeding. Bleeding represents an activator of thrombopoietin expression. Indeed, smears of peritoneal lavages from thrombopoietin-treated mice contain high numbers both of megakaryocytes and of mast cells (manuscript in preparation). It would be, then, extremely efficient for an organism to react to parasite-induced bleeding by increasing with the same mechanism, that is, increasing thrombopoietin levels, both megakaryocytes and mast cells present in the gut. In other words, thrombopoietin might represent the physiological stimulus that mediates the selective increase in mucosal mast cell numbers observed after experimentally induced infection with the intestinal worm Nippostrongilus Brasiliensis.[12] Parasite infection, however, by promoting proliferation of mast cell progenitors in the marrow,[11] should increase the number of dermal mast cells as well. The specificity of the physiological response to parasite infection would be ensured by the fact that thrombopoietin, after having favored mast cell differentiation in the marrow, would induce apoptosis of mast cell precursors for the dermal lineage. In other words, thrombopoietin might regulate the switch of lineage between serosal and dermal mast cell differentiation in response to specific external stimuli. Additional support for this hypothesis is provided by the observation that mast cells present in the peritoneal lavages from Mpl^{null} mice express high levels of MMP-7, a MMCP normally expressed by dermal rather than serosal mast cells.

It remains to be discussed whether the mast cell function of thrombopoietin is species-specific and limited to mice or holds true also in men. Two independent studies have recently reported that thrombopoietin induces increased cellular output in BMMC cultures seeded with human $CD34^{pos}$ cells.[72,73] It cannot, therefore, be excluded that thrombopoietin also might affect maturation of human mast cells.

In conclusion, our data suggest a regulatory mechanism for mast cell differentiation according to which thrombopoietin induces apoptosis and alters the differentiation state of mast cell precursors by reducing expression of *Gata1* in these cells. As thrombopoietin is normally present in the bone marrow microenvironment, we propose this growth factor as one of the factors restricting mastocytopoiesis to extramedullary sites.

ACKNOWLEDGMENTS

The authors gratefully acknowledge the assistance of Fabrizio Martelli and Maria Verrucci in performing the experiments and of Carol DiPalermo for editorial assistance. Murine thrombopoietin and the AMM2 anti-Mpl antibody were provided by Kirin (Transfer of material agreement of March 29, 2002). This study was supported by a grant from the National Cancer Institute to the MPD-RC consortium (P01-CA108671-01A2).

REFERENCES

1. GURISH, M.F. & K.F. AUSTEN. 2001. The diverse roles of mast cells. J. Exp. Med. **194:** F1–F5.

2. GALLI, S.J., S. NAKAE & M. TSAI. 2005. Mast cells in the development of adaptive immune responses. Nat. Immunol. **6:** 135–142.
3. KITAMURA, Y., M. YOKOYAMA, H. MATSUDA, et al. 1981. Spleen colony-forming cell as common precursor for tissue mast cells and granulocytes. Nature **291:** 159–160.
4. KIRSHENBAUM, A.S., J.P. GOFF, T. SEMERE, et al. 1999. Demonstration that human mast cells arise from a progenitor cell population that is CD34(+), c-kit(+), and expresses aminopeptidase N (CD13). Blood **94:** 2333–2342.
5. RODEWALD, H.R., M. DESSING, A.M. DVORAK & S.J. GALLI. 1996. Identification of a committed precursor for the mast cell lineage. Science **271:** 818–822.
6. REYNOLDS, D.S., R.L. STEVENS, W.S. LANE, et al. 1990. Different mouse mast cell populations express various combinations of at least six distinct mast cell serine proteases. Proc. Natl. Acad. Sci. USA **87:** 3230–3234.
7. CHEN, C.C., M.A. GRIMBALDESTON, M. TSAI, et al. 2005. Identification of mast cell progenitors in adult mice. Proc. Natl. Acad. Sci. USA **102:** 11408–11413.
8. KITAMURA Y., A. ITO. 2005. Mast cell-committed progenitors. Proc. Natl. Acad. Sci. USA **32:** 11129–11130.
9. JAMUR, M.C., A.C. GRODZKI, E.H. BERENSTEIN, et al. 2005. Identification and characterization of undifferentiated mast cells in mouse bone marrow. Blood **105:** 4282–4289.
10. MARTELLI, F., B. GHINASSI, B. PANETTA, et al. 2005. Variegation of the phenotype induced by the GATA-1low mutation in mice of different genetic background. Blood **106:** 4102–4113.
11. PENNOCK, J.L. & R.K. GRENCIS. 2004. In vivo exit of c-kit+/CD49d(hi)/beta7 +mucosal mast cell precursors from the bone marrow following infection with the intestinal nematode Trichinella spiralis. Blood **103:** 2655–2660.
12. ARIZONO, N., T. KASUGAI, M. YAMADA, et al. 1993. Infection of Nippostrongylus brasiliensis induces development of mucosal-type but not connective tissue-type mast cells in genetically mast cell-deficient Ws/Ws rats. Blood **81:** 2572–2578.
13. KOBAYASHI, T., T. NAKANO, T. NAKAHATA, et al. 1986. Formation of mast cell colonies in methylcellulose by mouse peritoneal cells and differentiation of these cloned cells in both the skin and the gastric mucosa of W/Wv mice: evidence that a common precursor can give rise to both "onnective tissue-type" and "mucosal" mast cells. J. Immunol. **136:** 1378–1384.
14. DAYTON, E.T., P. PHARR, M. OGAWA, et al. 1988. 3T3 fibroblasts induce cloned interleukin 3-dependent mouse mast cells to resemble connective tissue mast cells in granular constituency. Proc. Natl. Acad. Sci. USA **85:** 569–572.
15. NOCKA, K., J. BUCK, E. LEVI & P. BESMER. 1990. Candidate ligand for the c-kit transmembrane kinase receptor: KL, a fibroblast derived growth factor stimulates mast cells and erythroid progenitors. Embo. J. **9:** 3287–3294.
16. DURAND, B., G. MIGLIACCIO, N.S. YEE, et al. 1994. Long-term generation of human mast cells in serum-free cultures of CD34+ cord blood cells stimulated with stem cell factor and interleukin-3. Blood **84:** 3667–3674.
17. ROTTEM, M., S. BARBIERI, J.P. KINET & D.D. METCALFE. 1992. Kinetics of the appearance of Fc epsilon RI-bearing cells in interleukin-3-dependent mouse bone marrow cultures: correlation with histamine content and mast cell maturation. Blood **79:** 972–980.
18. NAKANO, T., T. SONODA, C. HAYASHI, et al. 1985. Fate of bone marrow-derived cultured mast cells after intracutaneous, intraperitoneal, and intravenous transfer

into genetically mast cell-deficient W/W^v mice. Evidence that cultured mast cells can give rise to both connective tissue type and mucosal mast cells. J. Exp. Med. **162:** 1025–1043.
19. MCNEIL, H.P., D.S. REYNOLDS, V. SCHILLER, et al. 1992. Isolation, characterization, and transcription of the gene encoding mouse mast cell protease 7. Proc. Natl. Acad. Sci. USA **89:** 11174–11178.
20. YEE, N.S., I. PAEK & P. BESMER. 1994. Role of kit-ligand in proliferation and suppression of apoptosis in mast cells: basis for radiosensitivity of white spotting and steel mutant mice. J. Exp. Med. **179:** 1777–1787.
21. TSAI, M., T. TAKEISHI, H. THOMPSON, et al. 1991. Induction of mast cell proliferation, maturation, and heparin synthesis by the rat c-kit ligand, stem cell factor. Proc. Natl. Acad. Sci. USA **88:** 6382–6386.
22. MOLLER, C., J. ALFREDSSON, M. ENGSTROM, et al. 2005. Stem cell factor promotes mast cell survival via inactivation of FOXO3a-mediated transcriptional induction and MEK-regulated phosphorylation of the proapoptotic protein Bim. Blood **106:** 1330–1336.
23. SUZUKI, K., H. NAKAJIMA, N. WATANABE, et al. 2000. Role of common cytokine receptor gamma chain (gamma(c))- and Jak3-dependent signaling in the proliferation and survival of murine mast cells. Blood **96:** 2172–2180.
24. BESMER, P. 1991. The kit ligand encoded at the murine Steel locus: a pleiotropic growth and differentiation factor. Curr. Opin. Cell Biol. **3:** 939–946.
25. GHINASSI B., M. SANCHEZ, F. MARTELLI, et al. 2007. The hypomorphic Gata1^{low} mutation alters the proliferation/differentiation potential of the common megakaryocytic-erythroid progenitor. Blood **109:** 1460–1471.
26. NOCKA, K., J.C. TAN, E. CHIU, et al. 1990. Molecular bases of dominant negative and loss of function mutations at the murine c-kit/white spotting locus: W37, Wv, W41 and W. Embo. J. **9:** 1805–1813.
27. KIMURA, Y., N. JONES, M. KLUPPEL, et al. 2004. Targeted mutations of the juxtamembrane tyrosines in the Kit receptor tyrosine kinase selectively affect multiple cell lineages. Proc. Natl. Acad. Sci. USA **101:** 6015–6020.
28. FUKAO, T., T. YAMADA, M. TANABE, et al. 2002. Selective loss of gastrointestinal mast cells and impaired immunity in PI3K-deficient mice. Nat. Immunol. **3:** 295–304.
29. VOSSELLER, K., G. STELLA, N.S. YEE & P. BESMER. 1997. c-kit receptor signaling through its phosphatidylinositide-3′-kinase-binding site and protein kinase C: role in mast cell enhancement of degranulation, adhesion, and membrane ruffling. Mol. Biol. Cell **8:** 909–922.
30. BERROZPE G., I. TIMOKHINA, S. YUKL, et al. 1999. The Wsh, W57, and Ph Kit expression mutations define tissue-specific control elements located between 23 and 154 kb upstream of kit. Blood **94:** 2658–2666.
31. LANTZ, C.S., J. BOESIGER, C.H. SONG, et al. 1998. Role for interleukin-3 in mastcell and basophil development and in immunity to parasites. Nature **392:** 90–93.
32. KOHNO, M., S. YAMASAKI, V.L. TYBULEWICZ & T. SAITO. 2005. Rapid and large amount of autocrine IL-3 production is responsible for mast cell survival by IgE in the absence of antigen. Blood **105:** 2059–2065.
33. KITAURA, J., J. SONG, M. TSAI, et al. 2003. Evidence that IgE molecules mediate a spectrum of effects on mast cell survival and activation via aggregation of the FcepsilonRI. Proc. Natl. Acad. Sci. USA **100:** 12911–12916.

34. LORENTZ, A., D. SCHUPPAN, A. GEBER, et al. 2002. Regulatory effects of stem cell factor and interleukin-4 on adhesion of human mast cells to extracellular matrix proteins. Blood **99:** 966–972.
35. TOWNSEND, J.M., G.P. FALLON, J.D. MATTHEWS, et al. 2000. IL-9-deficient mice establish fundamental roles for IL-9 in pulmonary mastocytosis and goblet cell hyperplasia but not T cell development. Immunity **13:** 573-583.
36. FINOTTO, S., Y.A. MEKORI & D.D. METCALFE. 1997. Glucocorticoids decrease tissue mast cell number by reducing the production of the c-kit ligand, stem cell factor, by resident cells: in vitro and in vivo evidence in murine systems. J. Clin. Invest. **99:** 1721–1728.
37. YAMAGUCHI, M., K. HIRAI, A. KOMIYA, et al. 2001. Regulation of mouse mast cell surface Fc epsilon RI expression by dexamethasone. Int. Immunol. **13:** 843–851.
38. MORII, E., T. JIPPO, T. TSUJIMURA, et al. 1997. Abnormal expression of mouse mast cell protease 5 gene in cultured mast cells derived from mutant mi/mi mice. Blood **90:** 3057–3066.
39. HODGKINSON, C.A., K.J. MOORE, A. NAKAYAMA, et al. 1993. Mutations at the mouse microphthalmia locus are associated with defects in a gene encoding a novel basic-helix-loop-helix-zipper protein. Cell **74:** 395–404.
40. TSAI, F.Y. & S.H. ORKIN. 1997. Transcription factor GATA-2 is required for proliferation/survival of early hematopoietic cells and mast cell formation, but not for erythroid and myeloid terminal differentiation. Blood **89:** 3636–3643.
41. HARIGAE, H., S. TAKAHASHI, N. SUWABE, et al. 1998. Differential roles of GATA-1 and GATA-2 in growth and differentiation of mast cells. Genes Cells **3:** 39–50.
42. MARTIN, D.I., L.I. ZON, G. MUTTER & S.H. ORKIN. 1990. Expression of an erythroid transcription factor in megakaryocytic and mast cell lineages. Nature **344:** 444–447.
43. MIGLIACCIO, A.R., R.A. RANA, M. SANCHEZ, et al. 2003. GATA-1 as a regulator of mast cell differentiation revealed by the phenotype of the GATA-1low mouse mutant. J. Exp. Med. **197:** 281–296.
44. TSUJIMURA, T., E. MORII, M. NOZAKI, et al. 1996. Involvement of transcription factor encoded by the mi locus in the expression of c-kit receptor tyrosine kinase in cultured mast cells of mice. Blood **88:** 1225–1233.
45. MORII, E., T. TSUJIMURA, T. K. JIPPO, et al. 1996. Regulation of mouse mast cell protease 6 gene expression by transcription factor encoded by the mi locus. Blood **88:** 2488–2494.
46. OGIHARA, H., E. MORII, D.K. KIM, et al. 2001. Inhibitory effect of the transcription factor encoded by the mutant mi microphthalmia allele on transactivation of mouse mast cell protease 7 gene. Blood **97:** 645–651.
47. WEISS, M.J. & S.H. ORKIN. 1995. GATA transcription factors: key regulators of hematopoiesis. Exp. Hematol. **23:** 99–107.
48. JIPPO, T., H. MIZUNO, Z. XU, et al. 1996. Abundant expression of transcription factor GATA-2 in proliferating but not in differentiated mast cells in tissues of mice: demonstration by *in situ* hybridization. Blood **87:** 993–998.
49. ZON, L.I., M.F. GURISH, R.L. STEVENS, et al. 1991. GATA-binding transcription factors in mast cells regulate the promoter of the mast cell carboxypeptidase A gene. J. Biol. Chem. **266:** 22948–22953.
50. NISHIYAMA, C., T. YOKOTA, K. OKUMURA & C. RA. 1999. The transcription factors Elf-1 and GATA-1 bind to cell-specific enhancer elements of human high-affinity IgE receptor alpha-chain gene. J. Immunol. **163:** 623–630.

51. MAEDA, K., C. NISHIYAMA, T. TOKURA, et al. 2003. Regulation of cell type-specific mouse Fc epsilon RI beta-chain gene expression by GATA-1 via four GATA motifs in the promoter. J. Immunol. **170:** 334–340.
52. NISHIYAMA, C., T. ITO, M. NISHIYAMA, et al. 2005. GATA-1 is required for expression of Fc{varepsilon}RI on mast cells: analysis of mast cells derived from GATA-1 knockdown mouse bone marrow. Int. Immunol. **17:** 847–856.
53. KAUSHANSKY, K. & J.G. DRACHMAN. 2002. The molecular and cellular biology of thrombopoietin: the primary regulator of platelet production. Oncogene **21:** 3359–3367.
54. SHIVDASANI, R.A., Y. FUJIWARA, M.A. MCDEVITT & S.H. ORKIN. 1997. A lineage-selective knockout establishes the critical role of transcription factor GATA-1 in megakaryocyte growth and platelet development. Embo. J. **16:** 3965–3973.
55. VYAS, P., K. AULT, C.W. JACKSON, et al. 1999. Consequences of GATA-1 deficiency in megakaryocytes and platelets. Blood **93:** 2867–2875.
56. CENTURIONE, L., A. DI BALDASSARRE, M. ZINGARIELLO, et al. 2004. Increased and pathologic emperipolesis of neutrophils within megakaryocytes associated with marrow fibrosis in GATA-1(low) mice. Blood **104:** 3573–3580.
57. VANNUCCHI, A.M., L. BIANCHI, C. CELLAI, et al. 2001. Accentuated response to phenylhydrazine and erythropoietin in mice genetically impaired for their Gata-1 expression (GATA-1low mice). Blood **97:** 3040–3050.
58. VANNUCCHI, A.M., L. BIANCHI, C. CELLAI, et al. 2002. Development of myelofibrosis in mice genetically impaired for GATA-1 expression (GATA-1low mice). Blood **100:** 1123-1132.
59. ZHOU, W, C.F. TOOMBS, T. ZOU, et al. 1997. Transgenic mice overexpressing human cmpl ligand exhibit chronic thrombocytosis and display enhanced recovery from 5fluorouracil or antiplatelet serum treatment. Blood **89:** 1551–1559.
60. YAN, X.Q., D. LACEY, C. FLETCHER, et al. 1995. Chronic exposure to retroviral vector encoded MGDF (mpl-ligand) induces lineage-specific growth and differentiation of megakaryocytes in mice. Blood **86:** 4025–4033.
61. YAN, X.Q., D. LACEY, D. HILL, et al. 1996. A model of myelofibrosis and osteosclerosis in mice induced by overexpressing thrombopoietin (mpl ligand): reversal of disease by bone marrow transplantation. Blood **88:** 402–409.
62. VANNUCCHI, A.M., L. BIANCHI, F. PAOLETTI, et al. 2005. A pathobiologic pathway linking thrombopoietin, GATA-1, and TGF-beta1 in the development of myelofibrosis. Blood **105:** 3493–3501.
63. YANAGIDA, M., Y. IDE, A. IMAI, et al. 1997. The role of transforming growth factor-beta in PEG-rHuMGDF-induced reversible myelofibrosis in rats. Br. J. Haematol. **99:** 739–745.
64. SCHMITT, A., H. JOUAULT, J. GUICHARD, et al. 2000. Pathologic interaction between megakaryocytes and polymorphonuclear leukocytes in myelofibrosis. Blood **96:** 1342–1347.
65. KAUSHANSKY, K. 2003. Etiology of the myeloproliferative disorders: the role of thrombopoietin. Semin. Hematol. **40:** 6–9.
66. MIGLIACCIO, A.R., A.M. VANNUCCHI, G. MIGLIACCIO & R. HOFFMAN. 2006. Molecular advances toward the understanding of the patho-biology of idiopathic myelofibrosis. Curr. Imm. Rev. **2:** 169–186.
67. KAUSHANSKY, K. 2006. Lineage-specific hematopoietic growth factors. N. Engl. J. Med. **354:** 2034–2045.

68. ALEXANDER, W.S., A.W. ROBERTS, N.A. NICOLA, *et al*. 1996. Deficiencies in progenitor cells of multiple hematopoietic lineages and defective megakaryocytopoiesis in mice lacking the thrombopoietic receptor c-Mpl. Blood **87:** 2162–2170.
69. FOX, N., G. PRIESTLEY, T. PAPAYANNOPOULOU & K. KAUSHANSKY. 2002. Thrombopoietin expands hematopoietic stem cells after transplantation. J. Clin. Invest. **110:** 389–394.
70. SUDA, T. 2006. Quiescent stem cells in the niche. 4th ISSCR Annual Meeting. Toronto.
71. SCADDEN, D.T. 2006. The stem-cell niche as an entity of action. Nature **441:** 1075-1079.
72. SAWAI, N., K. KOIKE, H.H. MWAMTEMI, *et al*. 1999. Thrombopoietin augments stem cell factor-dependent growth of human mast cells from bone marrow multipotential hematopoietic progenitors. Blood **93:** 3703–3712.
73. KIRSHENBAUM, A.S., C. AKIN, J.P. GOFF & D.D. METCALFE. 2005. Thrombopoietin alone or in the presence of stem cell factor supports the growth of KIT(CD117)low/ MPL(CD110) +human mast cells from hematopoietic progenitor cells. Exp. Hematol. **33:** 413–421.

Thrombopoietic Cells and the Bone Marrow Vascular Niche

H. G. KOPP[a,b] AND S. RAFII[b]

[a]*Department of Hematology/Oncology, Eberhard-Karls University of Tubingen, 72076 Tubingen, Germany*

[b]*Department of Genetic Medicine, Howard Hughes Medical Institute, Weill Cornell Medical College, New York, New York 10021, USA*

ABSTRACT: Megakaryocytes and platelets have been known to secrete angiogenic growth factors for a long time. However, there is little *in vivo* data on the regulation of angiogenesis by thrombopoietic cells. Both megakaryocytes and platelets are known to carry and release a multitude of both pro- and antiangiogenic mediators. Thus, it remained unknown how the "angiogenic phenotype" of thrombopoietic cells would be determined. Our group established that platelets contribute to angiogenesis as carriers of SDF-1, which is released by platelets in response to stimulation with hematopoietic cytokines. Indeed, even the action of VEGF-A seems to be mediated in part by the release of SDF-1 from stimulated platelets, thereby attracting proangiogenic hematopoietic cells. Moreover, the analysis of murine plasma and serum showed that similar to VEGF-A, SDF-1 is almost exclusively derived from platelets, and only trace amounts are detectable in platelet poor plasma. Because tumor patients' platelets have been shown to contain lower amounts of thrombospondin (Tsp), we generated Tsp-1 and Tsp-2 double knockout mice by crossing the single knockout lines. Interestingly, megakaryocytes and platelets derived from these mice confer a proangiogenic phenotype both in the bone marrow and in reperfusion of ischemic hindlimbs, thereby verifying the hypothesis of pro- and antiangiogenic platelet constituents "in balance."

KEYWORDS: megakaryocytes; platelets; bone marrow vascular niche; SDF-1/CXCL-12

INTRODUCTION

The bone marrow's sinusoidal microvasculature is special in that it represents a discontinuous endothelium, allowing both mobilization and homing of hematopoietic cells including hematopoietic stem/progenitor cells

Address for correspondence: Hans-Georg Kopp, M.D., Otfried-Muller Str. 10, 72076 Tubingen, Germany. Voice: +49-0-7071-2982726; fax: +49-0-7071-293671.
hans-georg.kopp@med.uni-tuebingen.de

(HSCs/HPCs).[1] Bone marrow endothelial cells (BMECs) support *in vitro* propagation of HPCs[2,3] as well as leukemia cells.[4] In fact, bone marrow microvascular density (BMVD) is increased in almost all hematological malignancies.[5] The marrow's vasculature is a dynamic niche, which is almost completely destroyed and regenerates within a few days after myelosuppression with cytostatic agents or radiation.[6] The vascular endothelial receptor Tie2 is a key receptor in this context, linking hematopoietic to angiogenic recovery after myelosuppression.[7] Nevertheless, the functional regulation of BMVD in the steady state remains to be determined. Thrombopoietic cells have been known to modulate angiogenic processes for a long time, but most studies analyzing the proangiogenic effects of platelets have been performed *in vitro* or were merely correlative. Thrombopoietic cells are known to carry and release the most prominent proangiogenic mediator, VEGF-A.[8] However, platelets also contain potentially counteracting antiangiogenic factors, such as PF-4 and thrombospondin (Tsp).[9] Thus, it remained unknown how the "angiogenic phenotype" of platelets would be determined.

MUTUAL DEPENDENCE OF MEGAKARYOCYTES AND BONE MARROW SINUSOIDAL ENDOTHELIUM

Given the typical, close proximity of mature megakaryocytes and sinusoidal endothelial cells in the marrow,[10] the hypothesis that there is a reciprocal influence on survival and/or proliferation is not far fetched. Indeed, inhibition of vascular recovery after administration of myelosuppressive doses of 5-FU by adding the vascular disrupting agent combretastatin-A4-phosphate (CA4P) resulted in a selective inhibition of thrombopoietic recovery, while other lineages displayed timely regeneration.[11] Similar results were obtained when adenoviral vectors encoding the soluble decoy receptor Tie2Fc were injected after 5-FU: by interfering with vascular regeneration and remodeling after myelosuppression, megakaryocyte and sinusoidal endothelial interaction was disturbed, thereby inhibiting platelet release and causing prolonged thrombocytopenia.[7] Vice versa, targeted knockout of the megakaryopoiesis-inhibiting Tsps not only resulted in increased numbers of megakaryocytes and vascular sinusoids in the steady state, but also caused increased survival of both cell types after myelosuppression. Consequently, Tsp-deficient mice displayed enhanced vascular and thrombopoietic recovery after 5-FU.[11]

PLATELETS IN PERIPHERAL REVASCULARIZATION

After induction of hindlimb ischemia by ligation of the superficial femoral artery of mice, vascular disruption and tissue necrosis can be detected in the gastrocnemius muscle followed by reperfusion and regeneration of the ischemic tissue both through arteriogenesis and angiogenesis. We have shown that similar to neoangiogenic vessels in malignant tumors,[12] platelets adhere to neoangiogenic vessels within the ischemic tissue and release Tsp. This was ultimately proven in an experiment, where wild-type platelets were transfused into Tsp-deficient mice after induction of hindlimb ischemia, and provides evidence that Tsp deployed by thrombopoietic cells may play a quantitative role in the process of revascularization after femoral vessel ligation.

SIGNIFICANCE OF SDF-1 RELEASED BY PLATELETS

In ischemic hindlimb studies, we previously established that platelets contribute to angiogenesis by releasing SDF-1 in response to stimulation with hematopoietic cytokines, such as soluble Kit-Ligand (sKitL = stem cell factor), thrombopoietin, granulocyte-monocyte colony stimulating factor, and erythropoietin. Indeed, even the action of VEGF-A was mediated in part by the release of SDF-1 from stimulated platelets, thereby attracting proangiogenic hematopoietic cells.[13] Others have reported similar results in a model of myocardial ischemia, where platelets initiate vascular remodeling by SDF-1-mediated recruitment of bone marrow-derived HPCs.[14] SDF-1 therefore seems to be a key chemokine in angiogenesis and may represent a common element of different proangiogenic signaling pathways.

Importantly, we found platelets to be the main source of serum SDF-1 in mice. Only trace amounts were detected in carefully obtained platelet poor plasma, while serum contained the quantities reported in the literature. Therefore, SDF-1 is similar to VEGF-A in that it is almost exclusively derived from platelets.[15] Moreover, quantitative secretion of SDF-1 upon stimulation with platelet agonists, such as collagen and ADP, is altered in a specific way in Tsp-deficient mice: not only do these platelets secrete larger absolute quantities of SDF-1, but they display a differential increase in SDF-1 secretion as standardized to serotonin in the ^{14}C-serotonin release assay.[11] These results suggest that Tsps are not only some more platelet constituents, but that they play a regulatory role in addition, influencing the platelet release function in a differential manner.

DISCUSSION

Tumor patients' platelets contain higher quantities of VEGF-A, but lower amounts of Tsp-1.[16] However, it remained unclear how exactly this may translate into a proangiogenic tumor or bone marrow environment. Most of the protein content of platelets stems from their mother megakaryocyte, with some protein being produced by mature platelets.[17] In addition, megakaryocytes have also been shown to take up macromolecules, such as Tsp-2, from their extracellular environment.[18] In tumor patients, serum VEGF-A levels do not correlate with tumor mass, but with platelet numbers.[15] Whether increased quantities of both serum and plasma VEGF-A in tumor patients are derived from increased production by megakaryocytes in the marrow or even by uptake of VEGF-A in the tumor neoangiogenic microvasculature by platelets remains unknown.

Our results identify the expression of Tsp-1 and Tsp-2 as a molecular switch in the determination of both BMVD and peripheral revascularization in ischemic limbs. Tsp is secreted by activated platelets in ischemic limb microvasculature and may thereby limit the angiogenic response.[11] Oncogenes like Myc and Ras have been found to mediate the downregulation of Tsp-1 in tumor cells, thereby linking tumorigenicity to angiogenicity.[19]

It is now generally accepted that platelets contribute to metastasis formation, although there are still inconsistencies and missing links between proangiogenic *in vitro* activity of platelets, descriptive analyses of the platelet proteome outweighing potential pro- and antiangiogenic factors, and the 40-year-old knowledge that platelet numbers are correlated with metastasis in malignant tumors.[20] One potential scenario that can be envisioned would be activated platelets secreting SDF-1, thereby recruiting hematopoietic cells, which in turn create the premetastatic niche,[21] enabling tumor cells to settle down, form metastases, and take advantage of an already angiogenic environment.

REFERENCES

1. Kopp, H.G., S.T. Avecilla, A.T. Hooper & S. Rafii. 2005. The bone marrow vascular niche: home of HSC differentiation and mobilization. Physiology (Bethesda) **20:** 349–356.
2. Rafii, S., F. Shapiro, R. Pettengell, *et al.* 1995. Human bone marrow microvascular endothelial cells support long-term proliferation and differentiation of myeloid and megakaryocytic progenitors. Blood **86:** 3353–3363.
3. Rafii, S., R. Mohle, F. Shapiro, *et al.* 1997. Regulation of hematopoiesis by microvascular endothelium. Leuk. Lymphoma **27:** 375–386.
4. Dias, S., K. Hattori, B. Heissig, *et al.* 2001. Inhibition of both paracrine and autocrine VEGF/ VEGFR-2 signaling pathways is essential to induce long-term remission of xenotransplanted human leukemias. Proc. Natl. Acad. Sci. USA **98:** 10857–10862.
5. Moehler, T.M., A.D. Ho, H. Goldschmidt & B. Barlogie. 2003. Angiogenesis in hematologic malignancies. Crit. Rev. Oncol. Hematol. **45:** 227–244.
6. Fliedner, T.M., D. Graessle, C. Paulsen & K. Reimers. 2002. Structure and function of bone marrow hemopoiesis: mechanisms of response to ionizing radiation exposure. Cancer Biother. Radiopharm. **17:** 405–426.
7. Kopp, H.G., S.T. Avecilla, A.T. Hooper, *et al.* 2005. Tie2 activation contributes to hemangiogenic regeneration after myelosuppression. Blood **106:** 505–513.
8. Mohle, R., D. Green, M.A. Moore, *et al.* 1997. Constitutive production and thrombin-induced release of vascular endothelial growth factor by human megakaryocytes and platelets. Proc. Natl. Acad. Sci. USA **94:** 663–668.

9. KISUCKA, J., C.E. BUTTERFIELD, D.G. DUDA, et al. 2006. Platelets and platelet adhesion support angiogenesis while preventing excessive hemorrhage. Proc. Natl. Acad. Sci. USA **103:** 855–860.
10. TAVASSOLI, M. & M. AOKI. 1989. Localization of megakaryocytes in the bone marrow. Blood Cells **15:** 3–14.
11. KOPP, H.G., A.T. HOOPER, M.J. BROEKMAN, et al. 2006. Thrombospondins deployed by thrombopoietic cells determine angiogenic switch and extent of revascularization. J. Clin. Invest. **116:** 3277–3291.
12. VERHEUL, H.M., K. HOEKMAN, F. LUPU, et al. 2000. Platelet and coagulation activation with vascular endothelial growth factor generation in soft tissue sarcomas. Clin. Cancer Res. **6:** 166–171.
13. JIN, D.K., K. SHIDO, H.G. KOPP, et al. 2006. Cytokine-mediated deployment of SDF-1 induces revascularization through recruitment of CXCR4(+) hemangiocytes. Nat. Med. **12:** 557–567.
14. MASSBERG, S., I. KONRAD, K. SCHURZINGER, et al. 2006. Platelets secrete stromal cell-derived factor 1{alpha} and recruit bone marrow-derived progenitor cells to arterial thrombi *in vivo*. J. Exp. Med. **203**(5): 1221–1233. PMID: 16618794 [PubMed-indexed for Medline].
15. VERHEUL, H.M., K. HOEKMAN, S. LUYKX-DE BAKKER, et al. 1997. Platelet: transporter of vascular endothelial growth factor. Clin. Cancer Res. **3:** 2187–2190.
16. GONZALEZ, F.J., A. RUEDA, I. SEVILLA, et al. 2004. Shift in the balance between circulating thrombospondin-1 and vascular endothelial growth factor in cancer patients: relationship to platelet alpha-granule content and primary activation. Int. J. Biol. Markers **19:** 221–228.
17. KIEFFER, N., J. GUICHARD, J.P. FARCET, et al. 1987. Biosynthesis of major platelet proteins in human blood platelets. Eur. J. Biochem. **164:** 189–195.
18. KYRIAKIDES, T.R., P. ROJNUCKARIN, M.A. REIDY, et al. 2003. Megakaryocytes require thrombospondin-2 for normal platelet formation and function. Blood **101:** 3915–3923.
19. WATNICK, R.S., Y.N. CHENG, A. RANGARAJAN, et al. 2003. Ras modulates Myc activity to repress thrombospondin-1 expression and increase tumor angiogenesis. Cancer Cell **3:** 219–231.
20. GASIC, G.J., T.B. GASIC & C.C. STEWART. 1968. Antimetastatic effects associated with platelet reduction. Proc. Natl. Acad. Sci. USA **61:** 46–52.
21. KAPLAN, R.N., R.D. RIBA, S. ZACHAROULIS, et al. 2005. VEGFR1-positive haematopoietic bone marrow progenitors initiate the pre-metastatic niche. Nature **438:** 820–827.

Differential Effects of G Protein–Coupled Receptors on Hematopoietic Progenitor Cell Growth Depend on their Signaling Capacities

XINGKUI XUE,[a,b,*] ZHEN CAI,[b,*] GABRIELE SEITZ,[a] LOTHAR KANZ,[a] KATJA C. WEISEL,[a] AND ROBERT MÖHLE[a]

[a]*Department of Medicine II, University of Tübingen, 72076 Tübingen, Germany*

[b]*Department of Hematology, First Affiliated Hospital, Zhejiang University School of Medicine, Hangzhou 310003, China*

ABSTRACT: We have shown that $CD34^+$ hematopoietic progenitor and stem cells (HPCs) consistently express several G protein–coupled receptors (GPCRs): the chemokine receptor CXCR4, the cysteinyl-leukotriene receptor cysLT1, and receptors for sphingosine 1-phosphate (S1P), particularly S1P1. These GPCRs differentially mediate chemotactic, adhesive, and proliferative responses in HPCs. To elucidate the diversity of the responses observed, we compared their signaling capacities in $CD34^+$ cells. In primary $CD34^+$ progenitors, the strongest effects on calcium signaling (intracellular calcium fluxes) were mediated by cysLT1. Analyses in $CD34^+$ cell lines revealed that calcium signaling induced by cysLT1 was only partially inhibited by pertussis toxin (PTX), while responses induced by CXCR4 and S1P receptors were completely blocked. These findings indicate that cysLT1 signals via Gi and Gq proteins, while CXCR4 and also S1P receptors (e.g., S1P1) only induce Gi protein-mediated effects. By analysis of downstream signaling, we could provide further evidence that combined activation of PTX-insensitive (Gq-mediated) and PTX-sensitive (Gi-mediated) pathways by cysLT1 may explain the strong and broad effects of cysteinyl-leukotrienes in early hematopoietic cells, while signaling of CXCR4 and S1P1 solely depends on Gi proteins, resulting in effects mainly restricted to migration and adhesion.

KEYWORDS: hematopoietic stem cells; G protein–coupled receptors; signal transduction

Address for correspondence: Robert Möhle, M.D., Department of Medicine II, University of Tübingen, Otfried-Müller-Str. 10, 72076 Tübingen, Germany. Voice: 49-7071-2983179; fax: 49-7071-293179.
robert.moehle@med.uni-tuebingen.de
*X.X. and Z.C. contributed equally to this work.

INTRODUCTION

In addition to the chemokine receptor CXCR4,[1,2] hematopoietic progenitor and stem cells (HPCs) express several other G protein–coupled receptors (GPCRs), including the cysteinyl-leukotriene receptor cysLT1[3] and receptors for sphingosine 1-phosphate (S1P), particularly S1P1.[4–6] CysLT1 represents a key regulator in inflammatory processes, such as allergic asthma,[7–9] while receptors for S1P were initially discovered as "endothelial differentiation genes" (EDG receptors),[10] due to their expression and functional properties in endothelial cells.[11] Similar to vascular endothelial growth factor (VEGF), S1P is a major angiogenic factor and essentially required during vasculogenesis.[11] In immature hematopoietic cells, however, these receptors may have additional functions independent of inflammation and endothelial development.[12] One might speculate that receptors for S1P play a particular role in hematopoietic stem cell biology in analogy to the receptors for VEGF, although the function of VEGFR-1[13] and VEGFR-2[14] in early hematopoiesis is still a subject of discussion. While the role of CXCR4 in HPC homing to the bone marrow is well established,[15] the contribution of cysLT1 and S1P1 (and other S1P receptors) to progenitor trafficking still remains elusive. Further studies suggest that CXCR4[16] and particularly cysLT1[17] also contribute to survival and proliferation of HPCs. In contrast, chemotaxis in response to the CXCR4 ligand stromal cell-derived factor-1 (SDF-1) is more pronounced than migration observed in response to leukotriene D4 (LTD4), the most potent ligand of cysLT1.[3] The S1P1 ligand S1P, however, may modulate chemokine (e.g., SDF-1)-dependent migration rather than acting as a chemoattractant itself.[6] Therefore, S1P appears to contribute to HPC homing indirectly. Ubiquitously, high levels of S1P are present in the circulation that may result in sustained activation of S1P1, which supports CXCR4-mediated homing mechanisms.[12] Differential signaling capacities of the three GPCRs expressed in CD34$^+$ hematopoietic progenitor cells could explain the diversity of the effects observed. In order to evaluate signaling of CXCR4, cysLT1, and S1P receptors in immature hematopoietic cells, we used primary CD34$^+$ progenitors and CD34$^+$ hematopoietic cell lines (KG1a, Jurkat) as model systems. The results demonstrate that CXCR4 and S1P receptors only signal via Gi proteins, while cysLT1 activates multiple pathways involving Gi and pertussis-insensitive G proteins (most likely Gq proteins), which may explain the unique proliferative response observed after stimulation of HPCs with cysteinyl-leukotrienes.

MATERIALS AND METHODS

CD34$^+$ Hematopoietic Progenitors and Cell Lines

After informed consent, peripheral blood cells were obtained from healthy donors or patients with nonhematological malignancies during granulocyte colony-stimulating factor (G-CSF)-induced stem cell mobilization according to the guidelines of the ethical committee of the University of Tübingen (project number 268/2003-LP). Mononuclear cells were separated by Ficoll density gradient centrifugation. Enrichment of $CD34^+$ HPCs was performed using immunomagnetic microbeads (MACS system, Miltenyi Biotec, Bergisch Gladbach, Germany) according to the manufacturer's instructions. HPCs were purified to 96–99% as determined by flow cytometry after staining with CD45 and CD34 antibodies. The $CD34^+$ human cell lines KG1a and Jurkat were propagated in RPMI1640 medium with 10% fetal calf serum (FCS), 100 U/mL penicillin and 100 μg/mL streptomycin.

Specific Reagents

The cysLT1 receptor antagonist MK571 (added at a final concentration of 2 μM) was obtained from Biomol (Hamburg, Germany). MK571 specifically binds to cysLT1, but not to other receptors for cysteinyl-leukotrienes (e.g., cysLT2[18]).

Measurement of Intracellular Calcium Mobilization

Measurement of intracellular free Ca^{2+} was performed as described previously with some modifications.[3] Briefly, HPCs were incubated in Hank's balanced salt solution (HBSS, Sigma-Aldrich, München, Germany) containing 10 μM Fluo-3 (Molecular Probes, Leiden, the Netherlands) for 30 min at 37°C. After a 1:5 dilution in HBSS/1% FCS followed by incubation for 40 min, the cells were washed three times, resuspended in HEPES-buffered saline, and incubated for at least 10 min at 37°C. After stimulation with LTD4 (final concentration 1 μM, Paesel & Lorei, Hanau, Germany), SDF-1 (final concentration 100 ng/mL, SDF-1α, R&D Systems, Wiesbaden, Germany), or S1P (final concentration 1 μM, Sigma-Aldrich), the FL-1 fluorescence (FITC channel) was analyzed using a FACSCalibur flow cytometer (Becton-Dickinson, Heidelberg, Germany) in 5-s acquisition intervals.

Cell Lysate Preparation and Western Blotting

Cells were serum-starved for 3 h, resuspended at 2×10^7/mL in serum-free medium, and incubated with 100 nM LTD4, 100 ng/mL SDF-1a, or 100 nM S1P (final concentrations) for the indicated time periods at 37°C. After stimulation, whole cell lysates were prepared by resuspending the pelleted cells in buffer containing 50 mM HEPES (pH 7.5), 10% glycerol, 1% Triton X-100, 1.5 mM $MgCl_2$, 150 mM NaCl, protease inhibitors (Complete Mini,

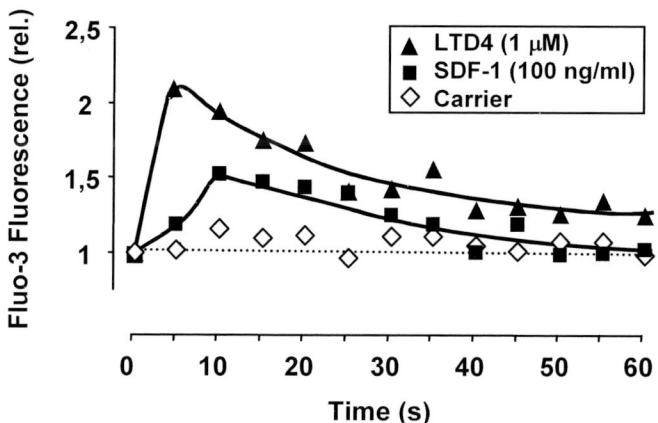

FIGURE 1. LTD4 and SDF-1-induced calcium fluxes in $CD34^+$ hematopoietic progenitor cells. Primary $CD34^+$ hematopoietic progenitor cells were loaded with the calcium-sensitive fluorescent dye Fluo-3 *in vitro* and stimulated with optimal doses of the cysLT1 ligand LTD4 and the CXCR4 ligand SDF-1. Cells were analyzed by flow cytometry (FITC channel) at the indicated time points. Stronger calcium fluxes were observed in response to LTD4 compared to SDF-1.

Roche, Mannheim, Germany), and the phosphatase inhibitors NaF (100 mM), $Na_4P_2O_7$ (10 mM), and activated Na_3VO_4 (1 mM; all from Sigma). Equal amounts of protein (20 μg/lane) were separated on a 10% SDS-polyacrylamide gel, and transferred onto nitrocellulose membrane (Schleicher & Schuell, Dassel, Germany). The blots were probed with phosphospecific polyclonal antibodies against p44/42 MAP kinase (Thr202/Tyr204), Pyk2 (Tyr402), or with control antibodies against the nonphosphorylated forms, respectively (all from Cell Signaling Techn., Beverly, MA). Bands were visualized by ECL staining (Amersham Biosciences, Freiburg, Germany).

RESULTS

Calcium Signaling

In addition to primary $CD34^+$ HPCs, we used $CD34^+$ hematopoietic cell lines that either expressed cysLT1 (KG1a) or CXCR4 and several S1P receptors (Jurkat) as a model system to analyze the effects of GPCR ligands on early hematopoietic cells. In primary $CD34^+$ HPCs, intracellular calcium fluxes induced by the cysLT1 ligand LTD4 were substantially stronger than those observed in response to SDF-1 at an optimal dose of 100 ng/mL (FIG. 1). Also in the $CD34^+$ cell lines, the strongest calcium fluxes were observed in response to LTD4, but direct comparison in the same cells was not possible

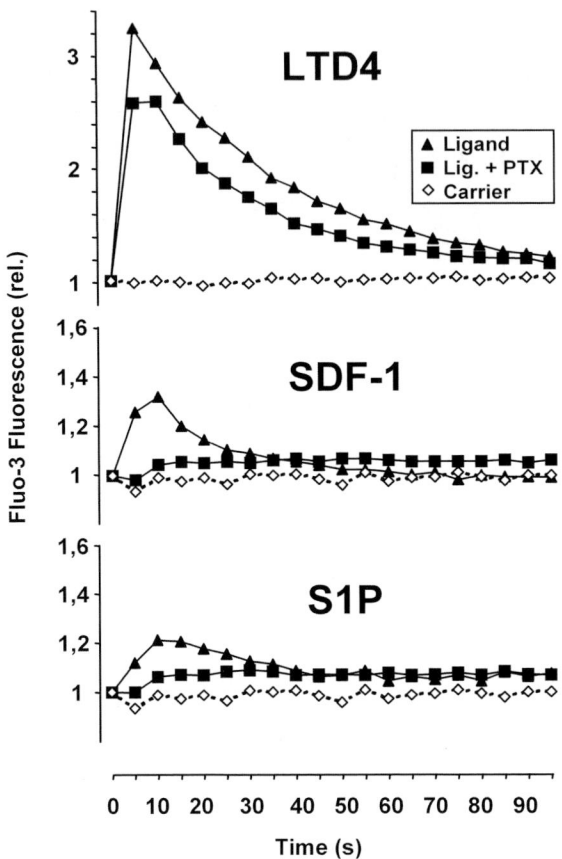

FIGURE 2. LTD4, SDF-1, and S1P-induced calcium fluxes in CD34$^+$ hematopoietic cell lines. CD34$^+$ hematopoietic cell lines (KG1a for analysis of the cysLT1 ligand, Jurkat for analysis of the ligands for CXCR4 and S1P1) were loaded with Fluo-3 and analyzed by flow cytometry at the indicated time points after stimulation. The strong calcium fluxes observed in the cell line KG1a in response to the cysLT1 ligand LTD4 were only partially inhibited by PTX. In contrast, moderate calcium fluxes induced by the CXCR4 ligand SDF-1 and the S1P1 ligand S1P were nearly completely abrogated by pretreatment with PTX.

due to the unavailability of a CD34$^+$ cell line expressing cysLT1, CXCR4, and S1P receptors simultaneously (FIG. 2). Mobilization of intracellular calcium induced by S1P was weak in the cell line (FIG. 2) and virtually undetectable in primary CD34$^+$ cells (data not shown). Pertussis toxin (PTX) slightly reduced LTD4/cysLT1-mediated calcium fluxes, while responses mediated by the other GPCRs were almost completely blocked (FIG. 2). In addition to the data shown in FIGURE 2, we could show that the intracellular calcium release induced by LTD4 was specifically mediated by cysLT1, as it could be blocked with the specific receptor antagonist MK571.

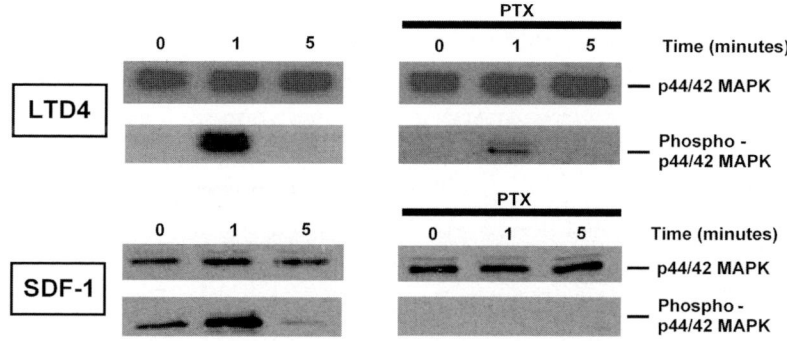

FIGURE 3. Phosphorylation of Erk/MAPK in response to LTD4 and SDF-1. Phosphorylation of p42/44 MAP-kinase (ERK-1 and -2) in response to the cysLT1 ligand LTD4 and the CXCR4 ligand SDF-1 was analyzed by Western blot in CD34$^+$ hematopoietic cell lines (KG1a for analysis of the cysLT1 ligand LTD4, Jurkat for analysis of the CXCR4 ligand SDF-1). Protein was extracted from the cells without stimulation ($t = 0$), 1 and 5 min after addition of the respective GPCR ligand. The strong Erk/MAP-kinase phosphorylation induced by LTD4 was only partially inhibited by PTX, whereas complete inhibition of the SDF-1-induced effect was observed.

GPCR-Induced Phosphorylation of Erk/MAP Kinase in CD34$^+$ Cell Lines

Similar to the effects on calcium signaling, a stronger GPCR-induced MAP kinase phosphorylation was observed in response to the cysLT1 ligand LTD4 compared to the CXCR4 ligand SDF-1 (FIG. 3). PTX completely blocked SDF-1-induced p44/42 Erk/MAPK phosphorylation, while the effect of LTD4 was only partially inhibited. As we could not detect S1P-induced p44/42 MAPK phosphorylation in primary CD34$^+$ cells using varying experimental conditions (data not shown), we did not further analyze the effect of S1P on MAPK in CD34$^+$ cell lines. In addition to the data shown in FIGURE 3, we could also demonstrate that MAPK phosphorylation in response to LTD4 was completely blocked after pretreatment with the specific cysLT1 receptor antagonist MK571.

GPCR-Induced Phosphorylation of Pyk2 in CD34$^+$ Cell Lines

The tyrosine kinase Pyk2 was only weakly activated by SDF-1 and S1P. Preincubation with PTX completely blocked SDF-1- and S1P-induced phosphorylation (FIG. 4). In contrast, a strong signal due to Pyk2 phosphorylation was observed in response to the cysLT1 ligand LTD4. Interestingly, PTX did not inhibit or reduce Pyk2 phosphorylation induced by LTD4, indicating that Gi proteins play a minor role in the signal pathways involved (FIG. 4).

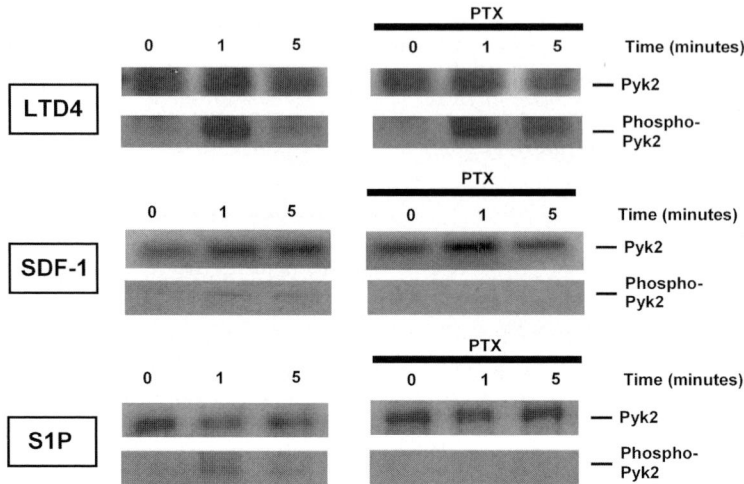

FIGURE 4. Phosphorylation of Pyk2 in response to LTD4, SDF-1, and S1P. Phosphorylation of Pyk2 in response to the cysLT1 ligand LTD4, the CXCR4 ligand SDF-1, and the S1P1 ligand S1P was analyzed by Western blot in CD34$^+$ hematopoietic cell lines (KG1a for analysis of LTD4, Jurkat for analysis of SDF-1 and S1P). Protein was extracted from the cells without stimulation ($t = 0$), 1 and 5 min after addition of the respective GPCR ligand. Only weak phosphorylation was observed after incubation with SDF-1 and S1P, which was completely inhibited by PTX. However, the strong response to the cysLT1 ligand LTD4 in KG1a cells was not influenced by pretreatment of the cells with PTX.

DISCUSSION

In this study, we could demonstrate that GPCRs expressed in hematopoietic progenitor cells may differentially affect cellular functions due to employment of specific signal transduction pathways. Particularly cysLT1, which was found to be highly expressed in CD34$^+$ HPCs[3], mediated strong effects on calcium signaling that were only partially PTX-sensitive and exceeded the responses induced by SDF-1, the ligand of CXCR4. Also in nonhematopoietic cells, such as smooth muscle cells of the airways, which are responsible for bronchoconstriction in allergic asthma, many effects of cysLT1 are mediated by PTX-insensitive, Gq-dependent signaling pathways, including effects that involve the Erk/MAP-kinase pathway.[19] In immature CD34$^+$ hematopoietic cells, cysLT1 is obviously also coupled to both Gi and Gq proteins. This can be concluded from the finding that at optimal ligand concentrations, stronger effects on intracellular calcium fluxes were observed after stimulation of cysLT1 compared to GPCRs that are coupled only to Gi proteins (e.g., CXCR4). We have shown previously that in primary CD34$^+$ HPCs, activation of Erk/MAP-kinase was more pronounced after stimulation of cysLT1 in contrast to CXCR4 and S1P receptors.[20] According to the results shown in this study, incomplete suppression of cysLT1-mediated Erk/MAP-kinase activation by PTX demonstrated involvement of both, Gi and Gq-dependent signaling pathways. As Erk/MAP-kinase is related to cell proliferation,[21] stronger activation could result in increased expansion of HPCs *in vitro*, which was previously observed in liquid cultures supplemented with cysteinyl-leukotrienes.[17,20] In these cultures, repeated addition of the cysLT1 ligand LTD4 was as efficient as classic hematopoietic growth factors (cytokines). Although there are some indications that the SDF-1/CXCR4 axis may contribute to survival of HPCs,[16] at least in our hands, SDF-1 is not an efficient growth factor in expansion cultures *in vitro*.

To further characterize the diversity of effects induced by different GPCRs expressed in CD34$^+$ hematopoietic cells, we explored signaling pathways that might be specifically activated in early hematopoietic cells by Gq proteins coupled to cysLT1. It has been shown that the focal adhesion-related tyrosine kinase Pyk2 links PTX-insensitive, Gq-mediated calcium signaling with key elements of downstream signaling, which results in cell proliferation, that is, the Erk/MAP-kinase pathway.[22] Pyk2 can also link Gi-mediated GPCR signaling with Erk/MAP-kinase, but in this study, we observed only weak activation of Pyk2 by ligands of CXCR4 and S1P1. Furthermore, the presence of PTX did not result in any reduction of Pyk2 phosphorylation in response to the cysLT1 ligand LTD4. These findings strongly suggest that in CD34$^+$ cells, cysLT1-induced Pyk2 activation is predominantly mediated by Gq proteins rather than Gi proteins. Therefore, both Gq and Gi protein–coupled receptor signaling induced by cysLT1 ligands converge on the Erk/MAPK pathway, which could explain the potent proliferative effects of LTD4 observed in primary CD34$^+$ HPCs. In contrast, signaling of CXCR4 and S1P1 is restricted to Gi protein-mediated pathways and may therefore not be sufficient to induce HPC proliferation to an extent that can be compared with classic hematopoietic growth factors.

ACKNOWLEDGMENTS

This work was supported by grants from Deutsche José Carreras Leukämiestiftung (DJCLS R04/13) and Deutsche Krebshilfe (106097). We thank Andreas Lächele for excellent technical assistance.

REFERENCES

1. MÖHLE, R., F. BAUTZ, S. RAFII, *et al.* 1998. The chemokine receptor CXCR-4 is expressed on CD34$^+$ hematopoietic progenitors and leukemic cells and mediates transendothelial migration induced by stromal cell-derived factor-1. Blood **91:** 4523–4530.

2. AIUTI, A., L. TURCHETTO, M. COTA, et al. 1999. Human CD34(+) cells express CXCR4 and its ligand stromal cell-derived factor-1. Implications for infection by T-cell tropic human immunodeficiency virus. Blood **94**: 62–73.
3. BAUTZ, F., C. DENZLINGER, L. KANZ, et al. 2001. Chemotaxis and transendothelial migration of CD34(+) hematopoietic progenitor cells induced by the inflammatory mediator leukotriene D4 are mediated by the 7-transmembrane receptor CysLT1. Blood **97**: 3433–3440.
4. YANAI, N., N. MATSUI, T. FURUSAWA, et al. 2000. Sphingosine-1-phosphate and lysophosphatidic acid trigger invasion of primitive hematopoietic cells into stromal cell layers. Blood **96**: 139–144.
5. WHETTON, A.D., Y. LU, A. PIERCE, et al. 2003. Lysophospholipids synergistically promote primitive hematopoietic cell chemotaxis via a mechanism involving Vav 1. Blood **102**: 2798–2802.
6. KIMURA, T., A.M. BOEHMLER, G. SEITZ, et al. 2004. The sphingosine 1-phosphate receptor agonist FTY720 supports CXCR4-dependent migration and bone marrow homing of human $CD34^+$ progenitor cells. Blood **103**: 4478–4486.
7. LYNCH, K.R., G.P. O'NEILL, Q. LIU, et al. 1999. Characterization of the human cysteinyl leukotriene CysLT1 receptor. Nature **399**: 789–793.
8. SARAU, H.M., R.S. AMES, J. CHAMBERS, et al. 1999. Identification, molecular cloning, expression, and characterization of a cysteinyl leukotriene receptor. Mol. Pharmacol. **56**: 657–663.
9. DENZLINGER, C., C. HABERL & W. WILMANNS. 1995. Cysteinyl leukotriene production in anaphylactic reactions. Int. Arch. Allergy Immunol. **108**: 158–164.
10. HLA, T., M.J. LEE, N. ANCELLIN, et al. 2001. Lysophospholipids–receptor revelations. Science **294**: 1875–1878.
11. LIU, Y., R. WADA, T. YAMASHITA, et al. 2000. Edg-1, the G protein-coupled receptor for sphingosine-1-phosphate, is essential for vascular maturation. J. Clin. Invest **106**: 951–961.
12. SEITZ, G., A.M. BOEHMLER, L. KANZ, et al. 2005. The role of sphingosine 1-phosphate receptors in the trafficking of hematopoietic progenitor cells. Ann. N. Y. Acad. Sci. **1044**: 84–89.
13. RAFII, S., S. AVECILLA, S. SHMELKOV, et al. 2003. Angiogenic factors reconstitute hematopoiesis by recruiting stem cells from bone marrow microenvironment. Ann. N. Y. Acad. Sci. **996**: 49–60.
14. PELOSI, E., M. VALTIERI, S. COPPOLA, et al. 2002. Identification of the hemangioblast in postnatal life. Blood **100**: 3203–3208.
15. LAPIDOT, T. & I. PETIT. 2002. Current understanding of stem cell mobilization: the roles of chemokines, proteolytic enzymes, adhesion molecules, cytokines, and stromal cells. Exp. Hematol. **30**: 973–981.
16. LATAILLADE, J.J., D. CLAY, C. DUPUY, et al. 2000. Chemokine SDF-1 enhances circulating CD34(+) cell proliferation in synergy with cytokines: possible role in progenitor survival. Blood **95**: 756–768.
17. BOEHMLER, A.M., C. DENZLINGER, L. KANZ, et al. 2003. Potential role of cysteinyl leukotrienes in trafficking and survival of hematopoietic progenitor cells. Adv. Exp. Med. Biol. **525**: 25–28.
18. HEISE, C.E., B.F. O'DOWD, D.J. FIGUEROA, et al. 2000. Characterization of the human cysteinyl leukotriene 2 receptor. J. Biol. Chem. **275**: 30531–30536.
19. BILLINGTON, C.K. & R.B. PENN. 2003. Signaling and regulation of G protein-coupled receptors in airway smooth muscle. Respir. Res. **4** (Epub Mar-14-2003): 2.

20. XUE, X., A.M. BOEHMLER, L. KANZ, *et al.* 2006. Differential signaling of G protein-coupled receptors (CXCR4, cysLT1 and S1P1) expressed in hematopoietic progenitor cells [abstract]. Blood **108:** 392a.
21. PEARSON, G., F. ROBINSON, G.T. BEERS, *et al.* 2001. Mitogen-activated protein (MAP) kinase pathways: regulation and physiological functions. Endocr. Rev. **22:** 153–183.
22. DIKIC, I., G. TOKIWA, S. LEV, *et al.* 1996. A role for Pyk2 and Src in linking G-protein-coupled receptors with MAP kinase activation. Nature **383:** 547–550.

Effect of FLT3 Inhibition on Normal Hematopoietic Progenitor Cells

KATJA C. WEISEL, SEDAT YILDIRIM, ERIC SCHWEIKLE, LOTHAR KANZ, AND ROBERT MÖHLE

University of Tuebingen, Medical Center, Department of Hematology, Oncology and Immunology, Tuebingen, Germany

ABSTRACT: Ligand-mediated activation of the FMS-like tyrosine kinase-3 (FLT3) receptor is important for normal proliferation of primitive hematopoietic cells. FLT3 expression in the bone marrow is restricted to $CD34^+$ cells and a subset of dendritic precursors. FLT3, as a member of the type III RTK subfamily, is closely related to c-kit, c-FMS, and $PDGF\alpha/\beta$ and is an unspecific target of tyrosine kinase inhibitors, such as imatinib. Activating mutations of FLT3 play an important role in leukemogenesis and their presence is associated with poor prognosis in acute myeloid leukemia (AML). Targeting the mutation by inhibiting the tyrosine kinase activity of FLT3 is a promising therapeutic option in the treatment of AML patients. CEP-701 (Lestaurtinib), an indocarbazole derivate, is an FLT3 tyrosine kinase inhibitor. In this study, we investigated the effect of FLT3 kinase inhibition on normal hematopoietic stem and progenitor cells *in vitro*. FLT3 inhibition in normal $CD34^+$ cells resulted in a dose-dependent inhibitory effect in cell expansion. In contrast, progenitor cell function remained nearly unaffected. Blocking the FLT3 ligand by a neutralizing antibody partially restored the effects of FLT3 inhibition. These findings might explain hematotoxicity of tyrosine kinase inhibitors such as imatinib.

KEYWORDS: hematopoietic stem cells; FLT3; FL; signal trasduction; CEP-701

INTRODUCTION

Hematopoietic stem cells (HSCs) are characterized by the unique ability of self-renewal and multilineage differentiation. These combined properties are reflected in the ability of an HSC to completely and durably reconstitute hematopoiesis of a myeloablated recipient and maintain it throughout the entire life span.[1,2] The unique expression profiles of hematopoietic cells at various stages of commitment allow cytokines and growth factors to direct the differentiation cascade until development terminates at mature lineage-specific

Address for correspondence: Katja C. Weisel, M.D., University of Tübingen, Department of Hematology, Oncology and Immunology, Otfried-Müller-Strasse 10, 72076 Tübingen, Germany. Voice: +49-7071-2982726; fax: +49-7071-293671.
katja.weisel@med.uni-tuebingen.de

progeny.[3,4] The FMS-like tyrosine kinase-3 (FLT3) receptor is a member of the type III RTK subfamily that also include c-kit, c-FMS, and PDGFα/β[5–8] and plays a key role in normal and malignant hematopoiesis.[9,10] FLT3 expression in the bone marrow is restricted to CD34$^+$ cells and a subset of dendritic precursors.[11] In the mouse system, FLT3 expression is correlated with short-term reconstituting HSC (Lin$^-$Sca-1$^+$c-kit$^+$FLT3$^+$).[12] So far, there is no evidence for its expression in mouse long-term HSC. Furthermore, FLT3 is expressed at high levels in early cell populations with lymphoid and myeloid differentiation potential.[13,14] Ligand-mediated activation of the FLT3 receptor is important for normal proliferation of primitive hematopoietic cells. FLT3 ligand (FL) mRNA is expressed in most hematopoietic and nonhematopoietic tissues, yet detection of the FL protein, as a membrane bound and soluble isoform, is restricted to T lymphocytes and stromal fibroblasts of the bone marrow microenvironment.[15–17] FL by itself does not efficiently induce proliferation of normal myeloid and lymphoid progenitors, but strongly synergizes with other hematopoietic growth factors and interleukins.[18,19] In acute myeloid leukemia (AML) blasts, it was shown that FL is expressed in cells that similarly express the FLT3 receptor, which suggests autocrine and paracrine regulation of the FLT3 response.[20]

Involvement of FLT3 in proliferation of highly undifferentiated hematopoietic cells suggests the oncogenic potential of this signaling pathway. FLT3 is aberrantly expressed on the malignant cells in the majority of patients with AML.[21] So far, two distinct types of FLT3 mutations have been identified: internal tandem duplication (ITD) mutations and point mutations (PM) within the kinase domain.[22]

CEP-701 (Lestaurtinib; Cephalon, West Chester, PA), an indolocarbazole alkaloid is an FLT3 inhibitor that is selective for FLT3 with an IC$_{50}$ of 3 nM for inhibition of phosphorylation of wild-type FLT3, ITD, and PM.[23] The most closely related kinases (PDGFR, FMS, and KIT) are inhibited only at concentrations of 500–1000 nM or greater. Recently, it was shown that CEP-701 is cytotoxic to primary AML and ALL blast samples.[24,25] In addition, CEP-701 showed in a phase 1/2 trial clinical activity in patients with relapsed or refractory AML.[26]

To elucidate the role of FLT3 in normal hematopoiesis, we evaluated the effect of FTL3 inhibition in adult peripheral blood (PB) mobilized HSCs.

METHODS AND RESULTS

FLT3 Inhibition Results in Growth Inhibition of HSCs

To investigate the role of FLT3 inhibition on normal HSCs, PB CD34$^+$ progenitors, highly purified by immunomagnetic absorption, were seeded into cytokine-supplemented (SCF and IL-6) serum-free liquid cultures. In a first step, cell counts after 7 days of liquid culture were evaluated under various concentrations of CEP-701 (DMSO 0, 5, 10, 20, and 50 nM). Compared to the DMSO-control, addition of CEP-701 in a concentration of 5 nM resulted in a more than 50% reduction in cell counts. With a maximum concentration of 50 nM, only a residual cell population was observed after 7 days of culture. Addition of FL in same culture conditions resulted, despite a slight increase in the DMSO control, in an identical effect.

In an MTT assay, the mean IC50 ($\pm$ SD) for CEP-701 on $CD34^+$ cells was 56 ± 19 nM.

Effect of FLT3 Inhibition on Progenitor Cell Function

To investigate the effect of FLT-3 inhibition on hematopoietic progenitor cell function, production of colony forming unit (CFU) and cobblestone area-forming cells (CAFCs) was evaluated. After 7 days of cytokine-supplemented (IL-6 and SCF) liquid culture, defined cell numbers were assayed for colony formation using the standard cytokine-supplemented methylcellulose assay. This assay demonstrated a constant CFU production per input cells independent of concentrations of CEP-701 (DMSO 0, 5, 10, 20, and 50 nM). Absolute numbers of CFC, however, declined with increasing concentrations of CEP-701 in parallel with decreasing absolute cell counts in culture. These results were again independent of additional supplementation with FL during the 7-day liquid culture period. In further detailed analyses, proportions of myeloid (CFU-GM) and erythroid (BFU-E) cultures differed not significantly between the subgroups with various CEP-701 concentrations.

To evaluate the effect of FLT3 inhibition on more primitive progenitor cells, CAFC formation was enumerated. The production of week 5 CAFC in human $CD34^+$ *ex vivo* expansion cultures has been shown to be indicative for the presence and quantity of *in vivo* repopulating HSC as measured using the quantitative NOD/SCID xenotransplant model. For CAFC formation, 1×10^5 cells were seeded after 1 week of liquid culture in presence or absence of CEP-701 for 5 weeks on MS5 stroma. After 5 weeks, CAFC formation was slightly decreased in the cocultures, where input cells were previously exposed to CEP-701.

Immunophenotypic Analysis under FLT3 Inhibition

To furthermore investigate the effect of FLT3 inhibition on the progenitor cell fraction, imunophenotypic analyses were performed. After 1 week of liquid culture in presence of IL-6 and SCF and various concentrations of CEP-701, the relative proportion of $CD34^+$ cells decreased with increased concentrations of CEP-701. In culture condition without CEP-701, $71 \pm 4\%$ cells were still CD34$^+$. In presence of 5 nM CEP-501, this amount decreased to 54 ± 7%. When 50 nM CEP-701 was added to the culture, only 16 ± 4% were still CD34$^+$. In contrast, the relative amount of CD34$^+$/CD38$^-$ progenitor remained relatively stable (13 ± 2%, 13 ± 1%, 11 ± 2%, respectively). However, because of the significant decrease of absolute cell counts in liquid culture with increasing concentration of CEP-701, absolute counts of CD34$^+$/CD38$^-$ cells were also diminished.

Effect of FLT3 Inhibition on Normal Hematopoietic Progenitor Cells Can Be Partially Restored by Addition of FL-Antibody

As described earlier, the effect of FLT3 inhibition on HSCs and progenitor cells was independent of exogenous addition of FL in the liquid culture system. To further specify, whether this effect is due to a continuously FLT3 activation by production of FL of the culture cells and to confirm that the observed effect of FLT3 inhibition is selective, cultures were evaluated with or without addition of an antibody directed against FL. In these experiments, it was demonstrated that the cytotoxic effects of FLT3 inhibition were partially restored by adding a neutralizing antibody against FL to the liquid cultures. This effect was independent of additional supplementation with FL. In the presence of the neutralizing antibody and CEP-701, cell counts after 1 week of coculture did not decrease in the same manner than in cultures with FLT3 inhibition by CEP-701 only.

DISCUSSION

Through a number of recent observations, FLT3 and its ligand have emerged as important regulators of the stem cell pool in mice and humans.[11,27–29] In conjunction with other cytokines, FL stimulates the expansion of CD34$^+$ progenitor cells. Repopulation assays in NOD/SCID mouse model demonstrated that the repopulating efficacy is higher in the CD34$^+$FLT3$^+$ cell fraction, despite CAFC formation representing primitive progenitors are distributed equally between CD34$^+$FLT3$^+$ and CD34$^+$FLT3$^-$ cells.[30] To further investigate the role of FLT3 in normal hematopoiesis, we used a serum-free liquid culture system supplemented with cytokines that had been shown to be effective for *ex vivo* expansion of primitive hematopoietic progenitors.[31] FLT3 inhibition was performed by addition of the small-molecule inhibitor of FLT3, CEP-701, to the cultures. The degree of expansion decreased in a significant and dose-dependent manner in treated compared to untreated cells. To extend these studies on the progenitor cell function, untreated cells and cells after exposure to CEP-701 were assayed for CFC and CAFC production. Absolute

CFC and CAFC counts decreased similar to total cell counts; however, evaluation of relative CFC counts showed that surviving cells after 1 week exposure to CEP-701 showed a constant CFC production. Evaluation of CAFC demonstrated a tendency to a decrease that was in first evaluations not significant. In the liquid culture system, expansion of $CD34^+$ cells was supported by cytokine supplementation with IL-6 and SCF. Effects of FLT3 inhibition were independent of additional supplementation of FL. This observation might indicate that FL is produced by the progenitor cells in culture. To support this observation and to investigate whether the inhibitory effects were selectively induced by FLT3 inhibition and not because of inhibition of further tyrosine kinases, we expanded $CD34^+$ cells under the same conditions outlined but either in the absence or by adding neutralizing monoclonal antibody against FL. Neutralizing FL did partially restore the effects of FLT3 inhibition by CEP-701 even in conditions without supplementation of FL. These observations argue for an autocrine/paracrine activation of FLT3 by its ligand in cultures of $CD34^+$ cells.

The IC50 of CEP-701 on normal hematopoietic cells measured by MTT assay was low (56 nM) compared with the mean through plasma concentration measured in patient's samples (4.4 μM).[26] We hypothesized that the extensive cytokine stimulation by the growth factor cocktail used in this study was one of the reasons for the increased susceptibility of the normal cells to FLT3 inhibition by CEP-701.

Tyrosine kinase inhibition is a successful treatment option in various malignancies. The most widely and successful use of tyrosine kinase inhibition represents the inhibition of the abl-kinase in chronic myeloid leukemia by imatinib. It is known that hematologic side effects of imatinib are dose-dependent, reversible, and include all hematopoietic lineages. Imatinib has known inhibitory effects not only to abl but also to c-kit, PDGF-R, and FLT3. It was already shown that imatinib has inhibitory effects on normal hematopoietic progenitor cells. In the same publication, it was excluded that these effects were mediated by inhibition of c-kit or PDGF-R.[32] Our findings may indicate that the hematologic side effects of imatinib could be mediated by inhibition of FLT3.

ACKNOWLEDGMENT

We thank Steven Trusko, Cephalon Inc. for providing CEP-701.

REFERENCES

1. MOORE, M.A.S. 2003. *In vitro* and *in vivo* hematopoiesis. *In* Atlas of Blood Cells: function and Pathology, 3rd ed. D. Zucker-Franklin, M.F. Greaves, C.E. Grossi & A.M. Marmont, Eds.: 1–38. Arti Grafiche Salea. Milan.

2. KONDO, M., A.J. WAGERS, M.G. MANZ, et al. 2003. Biology of hematopoietic stem cells and progenitors: implications for clinical application. Annu. Rev. Immunol. **21:** 759–806.
3. SHIVDASANI, R.A. & S.H. ORKIN. 1996. The transcriptional control of hematopoiesis. 1996. Blood **87:** 4025–4039.
4. ZHU, J. & S.G. EMMERSON. 2002. Hematopoietic cytokines, transcription factors and lineage commitment. Oncogene **21:** 3295–3313.
5. MATHEWS, L.S. & W.W. VALE. 1991. Expression cloning of an activin receptor, a predicted transmembrane serin kinase. Cell **65:** 973–982.
6. ROSNET, O., M.G. MATTEI, S. MARCHETTO & D. BIRNBAUM. 1991. Isolation and chromosomal localization of a novel FMS-like tyrosine kinase. Genomics **9:** 380–385.
7. ROSNET, O., C. SCHIFF, M.J. PEBUSQUE, et al. 1993. Human FLT3/FLK2 gene: cDNA cloning and expression in hematopoietic cells. Blood **82:** 1110–1119.
8. SMALL, D., M. LEVENSTEIN, E. KIM, et al. 1994. STK-1, the human homolog of Flk-2/Flt-3, is selectively expressed in CD34+ human bone marrow cells and is involved in the proliferation of early progenitor/stem cells. Proc. Natl. Acad. Sci. USA **91:** 459–463.
9. PARCELLS, B.W., A.K. IKEDA, T. SIMMS-WALDRIP, et al. 2006. FLT-like tyrosine kinase 3 in normal hematopoiesis and acute myeloid leukemia. Stem Cells. **24**(5): 1174–1184.
10. GILLILAND, D.G. & J.D. GRIFFIN. 2002. The roles of FLT3 in hematopoiesis and leukemia. Blood **100:** 1532–1542.
11. LYMAN, S.D. & S.E. JACOBSEN. 1998. c-kit ligand and Flt3 ligand: stem/progenitor factors with overlapping yet distinct activities. Blood **91:** 1101–1134.
12. ADOLFSSON, J., R. MANSSON, N. BUZA-VIDAS, et al. 2005. Identification of Flt3+ lympho-myeloid stem cells lacking erythro-megakaryocytic potential. A revised road map for adult blood lineage commitment. Cell **121:** 295–306.
13. ADOLFSSON, J., O.J. BORGE, D. BRYDER, et al. 2001. Upregulation of Flt3 expression within the bone marrow Lin(-)Sca1(+)c-kit(+) stem cell compartment is accompanied by loss of self-renewal capacity. Immunity **15:** 659–669.
14. CHRISTENSEN, J.L. & I.L. WEISSMAN. 2001. Flk-2 is a marker in hematopoietic stem cell differentiation: a simple method to isolate long-term stem cells. Proc. Natl. Acad. Sci. USA **98:** 14541–14546.
15. HANNUM, C., J. CULPEPPER, D. CAMPBELL, et al. 1994. Ligand for FLT3/FLK2 receptor tyrosine kinase regulates growth of haematopoietic stem cells and is encoded by variant RNAs. Nature **368:** 643–648.
16. LYMAN, S.D., L. JAMES, L. JOHNSON, et al. 1994. Cloning of the human homologue of the murine flt3 ligand: a growth factor for early hematopoietic progenitor cells. Blood **83:** 2795–2801.
17. BRASEL, K., S. ESCOBAR, R. ANDERBERG, et al. 1995. Expression of the Flt-3 receptor and its ligand on hematopoietic cells. Leukemia **9:** 1212–1218.
18. LYMAN, S.D., K. BRASEL, A.M. ROUSSEAU & D.E. WILLIAMS. 1994. The Flt3 ligand: a hematopoietic stem cell factor whose activities are distinct from steel factor. Stem Cells **12:** 99–107.
19. RAY, R.J., C.J. PAIGE, C. FURLANGER, et al. 1996. Flt3 ligand supports the differentiation of early B-cell progenitors in the presence of interleukin-11 and interleukin-7. Eur. J. Immunol. **26:** 1504–1510.
20. ZHENG, R., M. LEVIS, O. PILOTO, et al. 2004. FLT3 ligand causes autocrine signaling in acute myeloid leukemia cells. Blood **103:** 267–274.

21. BIRG, F., O. ROSNET, N. CARBUCCIA & D. BIRNBAUM. 1994. The expression of FMS, KIT and FLT3 in hematopoietic malignancies. Leuk. Lymphoma **13:** 223–227.
22. LEVIS, M. & D. SMALL. 2003. ITDoes matter in leukemia. Leukemia **17:** 1738–1752.
23. LEVIS, M., J. ALLENBACH, K.F. TSE, *et al.* 2002. A FLT3-targeted tyrosine kinase inhibitor is cytotoxic to leukemia cells *in vitro* and *in vivo*. Blood **99:** 3885–3891.
24. BROWN, P., S. MESHINCHI, M. LEVIS, *et al.* 2004. Pediatric AML primary samples with FLT3/ITD mutations are preferentially killed by FLT3 inhibition. Blood **104:** 1841–1849.
25. BROWN, P., M. LEVIS, S. SHURTLEFF, *et al.* 2005. FLT3 inhibition selectively kills childhood acute lymphoblastic leukemia cells with high levels of FLT3 expression. Blood **105:** 812–820.
26. SMITH, B.D., M. LEVIS, M. BERAN, *et al.* 2004. Single-agent CEP-701, a novel FLT3 inhibitor, shows biologic and clinical activity in patients with relapsed or refractory acute myeloid leukemia. Blood **103:** 3669–3676.
27. PIACIBELLO, W., F. SANAVIO, L. GARETTO, *et al.* 1997. Extensive amplification and self-renewal of human primitive hematopoietic stem cells from cord blood. Blood **89:** 2644–2653.
28. PETZER, A.L., P.W. ZANDSTRA, J.M. PIRET & C.J. EAVES. 1996. Differential cytokine effects on primitive (CD34+CD38-) human hematopoietic cells: novel responses to FLT3-ligand and thrombopoietin. J. Exp. Med. **18:** 2551–2558.
29. DAO, M.A., C.H. HANNUM, D.B. KOHN & J.A. NOLTA. 1997. FLT3 ligand preserves the ability of human CD34+ progenitors to sustain long-term hematopoiesis in immune-deficient mice after *ex vivo* retroviral transduction. Blood **89:** 446–456.
30. SITNICKA, E., N. BUZA-VIDAZ, S. LARSSON, *et al.* 2003. Human CD34+ hematopoietic stem cells capable of multilineage engrafting NOD/SCID mice express flt3: distinct flt3 and c-kit expression and response patterns on mouse and candidate human hematopoietic stem cells. Blood **102:** 881–886.
31. GAMMAITONI, L., K.C. WEISEL, M. GUNETTI, *et al.* 2004. Elevated telomerase activity, minimal telomere loss in cord blood long-term cultures with extensive stem cell replication. Blood **103:** 4440–4448.
32. BARTOLOVIC, K., S. BALABANOV, U. HARTMANN, *et al.* 2004. Inhibitory effect of imatinib on normal progenitor cells *in vitro*. Blood **103:** 523–529.

The Cdx-Hox Pathway in Hematopoietic Stem Cell Formation from Embryonic Stem Cells

CLAUDIA LENGERKE,[a,b,c] SHANNON McKINNEY-FREEMAN,[a,b,c] OLAIA NAVEIRAS,[a,b,c] FRANK YATES,[a,b,c] YUAN WANG,[a,b,c] DIMPLE BANSAL,[d] AND GEORGE Q. DALEY[a,b,c,d]

[a]*Division of Pediatric Hematology/Oncology, Children's Hospital Boston, Boston, Massachusetts 02115, USA*

[b]*Harvard Stem Cell Institute, Cambridge, Massachusetts 02138, USA*

[c]*Department of Biological Chemistry and Molecular Pharmacology, Harvard Medical School, Boston, Massachusetts 02115, USA*

[d]*Division of Hematology, Brigham and Women's Hospital, Boston, Massachusetts 02115, USA*

ABSTRACT: Embryonic stem cells (ESCs) differentiated *in vitro* will yield a multitude of hematopoietic derivatives, yet progenitors displaying true stem cell activity remain difficult to obtain. Possible causes are a biased differentiation to primitive yolk sac-type hematopoiesis, and a variety of developmental or functional deficiencies. Recent studies in the zebrafish have identified the caudal homeobox transcription factors (cdx1/4) and posterior hox genes (*hoxa9a*, *hoxb7a*) as key regulators for blood formation during embryonic development. Activation of Cdx and Hox genes during the *in vitro* differentiation of mouse ESCs followed by co-culture on supportive stromal cells generates ESC-derived hematopoietic stem cells (HSCs) capable of multilineage repopulation of lethally irradiated adult mice. We show here that brief pulses of ectopic *Cdx4* or *HoxB4* expression are sufficient to enhance hematopoiesis during ESC differentiation, presumably by acting as developmental switches to activate posterior Hox genes. Insights into the role of the Cdx-Hox gene pathway during embryonic hematopoietic development in the zebrafish have allowed us to improve the derivation of repopulating HSCs from murine ESCs.

KEYWORDS: embryonic stem cells; hematopoietic stem and progenitor cells; *HoxB4*; *Cdx4*; engraftment

Address for correspondence: George Q. Daley, M.D., Ph.D., Children's Hospital Boston, 300 Longwood Avenue, Boston, MA 02115. Voice: 617-919-2015; fax: 617-730-0222.
george.daley@childrens.harvard.edu

INTRODUCTION

Embryonic stem cells (ESCs) are derived from the inner cell mass of blastocysts,[1,2] and retain the pluripotent features of early epiblast cells. When reintroduced to the blastocyst, ESCs contribute to all tissues of chimeric mice, including the germ line.[3] *In vitro*, ESCs display self-renewal capacity and differentiate into a multitude of tissues, and thus represent a powerful tool for the study of developmental processes and a promising resource for cell-based therapies.[4] We have focused on the differentiation of ESCs along the hematopoietic lineage, and have explored the role of homeobox genes implicated in blood progenitor development and function on the derivation of ESC-derived hematopoietic precursors.

Under standard conditions, ESCs differentiate into embryoid bodies (EBs) and display robust *in vitro* hematopoietic activity.[5] However, little hematopoietic stem cell (HSC) activity has been reported in transplant assays. We have shown that ectopic expression of *HoxB4* and *Cdx4*, an upstream regulator of the Hox family, promote hematopoietic development from ESCs and enable the generation of engraftable progenitors with adult HSC properties. We show here that a brief window of induction is sufficient to strongly enhance hematopoietic activity, and suggest that the effect is achieved via two mechanisms: (*a*) recapitulation of embryonic patterning, and (*b*) expansion of prepatterned populations by stimulation of stem and progenitor self-renewal and proliferative capacity. We have explored the phenotype of *Cdx4-HoxB4*-modified ESC-derived hematopoietic cells, and compared their *in vitro* colony-forming ability and *in vivo* repopulation capacity. In comparison with adult bone marrow, ESC-derived hematopoietic progenitors show much lower *in vivo* repopulation capacity, despite displaying much stronger *in vitro* clonogenic properties. Homing deficiencies might account for this characteristic, and alternative transplantation protocols, such as intra-bone marrow injection, could benefit ESC-derived HSC transplantation.

MATERIALS AND METHODS

Cell Culture and Transplantation

ESCs were maintained and differentiated according to published protocols.[5] Doxycycline was added to the culture medium from day 3 to day 4 at 0.1 μg/mL and from day 4 to 6 at 0.5 μg/mL to induce *Cdx4* expression, unless otherwise indicated. A total of 10^5 EB cells were plated onto semiconfluent OP9 cells in 6-well dishes and were infected with retroviral supernatants, produced in 293T cells by Fugene (Roche, Basel, Switzerland) cotransfection of viral plasmid MSCV-*HoxB4*-ires-GFP and packaging-defective helper plasmid, pCL-Eco. Infected EB cells were cultured according to published protocols.[5] Blast colony-forming/replating assay and hematopoietic colony formation assay were performed as previously described.[6,7] Six-week- to 3-month-old $Rag2^{-/-}/\gamma c^{-/-}$ female mice were given two doses of 400 cGy-irradiation, separated by 4 h, and were injected via lateral tail vein with 2×10^6 cells in 400 μL of IMDM/2% IFS.[8]

Quantitative Real-Time Polymerase Chain Reaction (PCR)

Cells were harvested in RNA Stat-60 (Tel-Test, Friendswood, TX), and total RNA was isolated. All RNA samples were treated with DNaseI (Ambion Austin, TX). cDNAs were prepared according to the manufacturer's instruction (Superscript II Reagent; Invitrogen, Carlsbad, CA). Real-time PCR was performed in triplicates with SYBR Green reagent kits (Stratagene, La Jolla, CA) on a MX3000P Stratagene PCR machine. Primer sequences and PCR conditions are listed in Reference 8.

Spleen Colony-Forming Assay

ESC-derived hematopoietic progenitor cells (10^5 whole BM or 10^6) were administered retroorbitally, in 200 μL of phosphate-buffered saline (PBS), in irradiated adult $Rag2^{-/-}/\gamma c^{-/-}$ recipients. Spleens were fixed in Bouin's buffer and scored for the colony-forming units of the spleen (CFU-S).

FACS Analysis

Peripheral blood leukocytes, splenocytes, and bone marrow cells were treated with red cell lysis buffer (Sigma-Aldrich, St. Louis, MO). Antibodies were purchased from BD Biosciences Pharmingen (San Diego, CA). Propidium iodide was added to exclude dead cells.

RESULTS

Cdx4 *Enhances Hematopoietic Patterning from Murine ESCs by Activation of Hox Genes*

Recently, the *cdx* (caudal-related homeobox containing transcription factor) gene family has been identified as essential for embryonic blood fate specification in the zebrafish. Mutations in *cdx4* cause a severe, but not complete, deficit in embryonic blood cells,[9] and additional morpholino-mediated knockdown of *cdx1a* causes a complete failure to specify blood.[9,10] On a molecular level, *cdx*-deficient fish display severe perturbations of hox gene expression.[10] Furthermore, ectopic expression of specific posterior hox genes (e.g., *hoxa9a* and

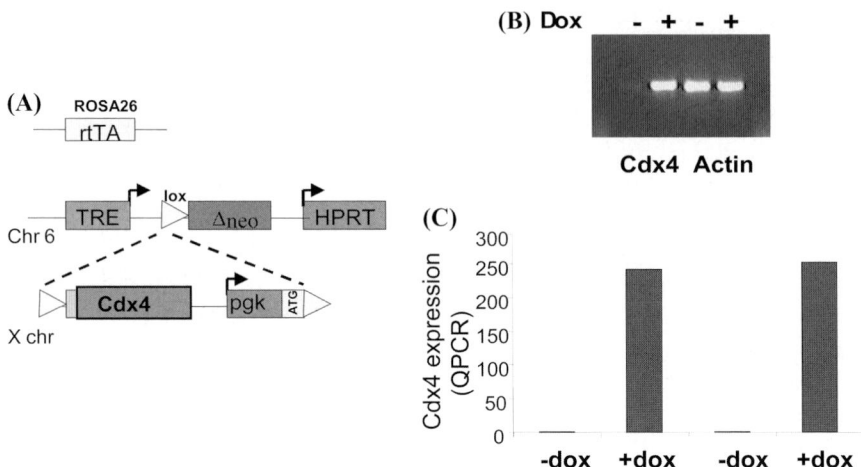

FIGURE 1. Generation of a tetracycline(dox)-*Cdx4* inducible ESC line. (**A**) Schematic representation of integrated expression cassettes. The rtTA is integrated into the constitutive ROSA26 locus on chromosome 6. Cre-mediated recombination of targeting vectors into the homing site on the X chromosome restores resistance to the antibiotic G418 (neo), thereby facilitating efficient isolation of transgenic cells. (**B**) *Cdx4* amplification by PCR. Uninduced ES cells, and induced ES cells (cultured in media supplemented with 1 μg/mL doxycycline). (**C**) Fold increase in *Cdx4* expression after doxycycline addition to differentiating EBs (day 4 of differentiation). QPCR at 4 and 8 h after doxycycline induction (1 μg/mL culture media). Abbreviations: TRE = tetracycline response element; PGK = phosphoglycerokinase promoter; ATG = methionine initiation codon; triangle = lox recognition sequence for Cre recombinase; Δneo = truncated neomycin (G418) resistance gene; dox = doxycycline.

hoxb7a, but not *hoxb8a*) can partially rescue blood formation, supporting the hypothesis that *cdx* genes pattern hematopoiesis by activation of downstream hox targets.[9,10]

We have used the ESC system to interrogate the role of Cdx genes during embryonic hematopoiesis, and engineered a murine ESC line with tetracycline-inducible Cdx4.[5,8,9] *Cdx4* gene expression was strongly induced after only 4 h of incubation with doxycycline containing media. No further upregulation could be observed after another 4 h of doxycycline exposure (FIG. 1).

Ectopic *Cdx4* expression during EB development (day 3–6) promotes the development of hematopoietic colony-forming cells, especially multipotent progenitors, in colony formation assays in semisolid media supplemented with hematopoietic cytokines, and improves the derivation of cells with lymphoid reconstitution potential in transplant assays.[8] Furthermore, ectopic *Cdx4* enhances the numbers of hemangioblastic progenitors in blast colony assays, suggesting an early patterning function during hematopoietic fate specification from the mesodermal germ layer.[8] We show here that a brief exposure of 24 h to

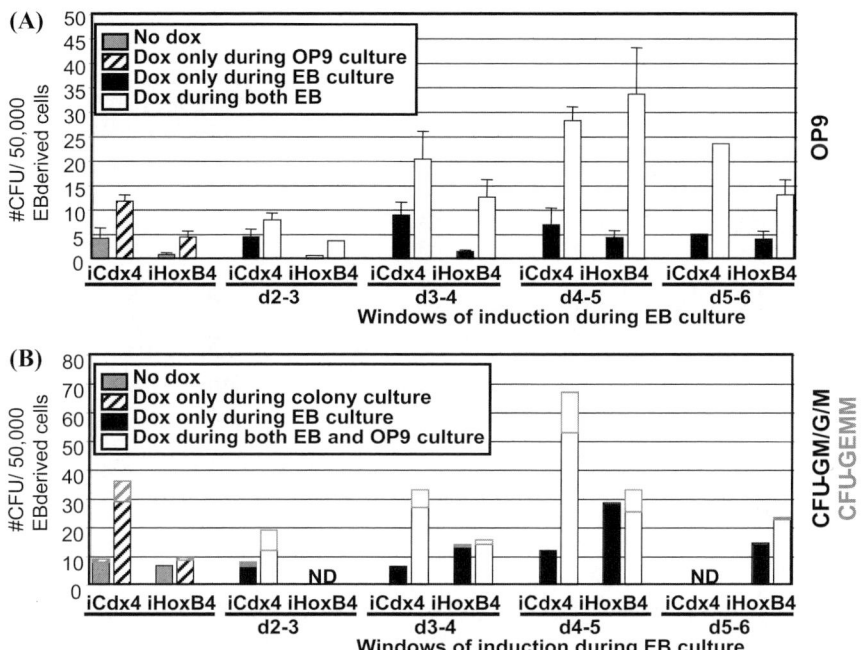

FIGURE 2. Brief pulses of ectopic *Cdx4* or *HoxB4* induction are sufficient to promote hematopoiesis from ESCs. (**A**) Quantification of OP9 colonies. A total of 50,000 EB-derived cells were harvested at day 6 and plated on semiconfluent OP9 cells. Colonies were counted 6 days after plating. Gene induction with doxycycline was performed during specific windows of EB development, or during co-culture on OP9. (**B**) Quantification of hematopoietic colony-forming units (CFU) (GM/G/M, GEMM) in semisolid media supplemented with hematopoietic cytokines. Abbreviations: GM = granulocyte-monocyte/macrophage; G = granulocyte; M = macrophage; GEMM = granulocyte-erythroid-monocyte-macrophage.

ectopic *Cdx4*, as well as to *HoxB4*, is sufficient to strongly augment hematopoietic progenitor formation as assessed by hematopoietic colony-forming assays in semisolid media supplemented with hematopoietic cytokines, and OP9 coculture assays (FIG. 2). The effect was strongest during exposure between day 4 and 5 of EB development. On a molecular level, *Cdx4* expression induces the expression of posterior Hox genes (FIG. 3), suggesting that posterior patterning pathways are conserved between zebrafish embryos and murine ESCs.

CDX4 *and* HoxB4 *Enhance the Clonogenic Potential of CD41+ c-kit + EB-Derived Cells*

CD41 has been identified as the earliest marker of hematopoietic progenitors in the mouse embryo and during *in vitro* EB development,[11] and the

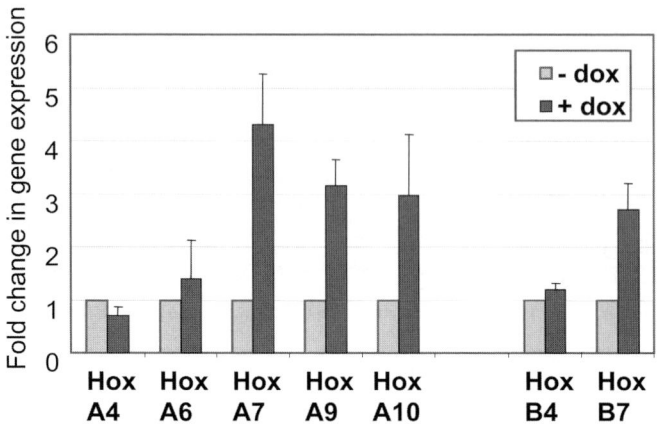

FIGURE 3. Hox gene response after *Cdx4* induction. Quantitative PCR performed on differentiating EBs, 48 h after doxycycline addition to the culture media. Gene expression levels are shown after normalization to actin.

CD41+c-kit+ population is enriched for cells with clonogenic properties in semisolid media assays.[11] We replicated these results under our EB differentiation conditions, and observed that the OP9 colony-initiating cell (OP9-CIC) resides within the CD41+ fraction and is highly enriched in the c-kit+ population (data not shown).

We assessed the impact of ectopic *Cdx4* and *HoxB4* expression during EB differentiation on the frequency and potency of the CD41+c-kit+ population in day 6 EBs. Both genes increased the frequency of these cells and their ability to produce OP9-CIC and hematopoietic progenitors (data not shown).

Hematopoietic Engraftment from Murine ESCs

Ectopic expression of *Cdx4* during EB development and subsequent retroviral infection with *HoxB4* and co-culture on OP9 stroma cells enables the derivation of definitive HSCs capable of repopulating the hematopoietic system of irradiated mice.[8] We compared the *in vitro* and *in vivo* functional characteristics of ESC-derived hematopoietic progenitors with adult bone marrow cells. The ESC-derived *Cdx4-HoxB4* cells expanded on OP9 stroma in the presence of hematopoietic cytokines are predominantly of hematopoietic phenotype (77% CD45+ cells in one representative experiment), and phenotypically resemble cells from the adult bone marrow: 62% display Gr1+, 31% Ter119+, and 16% B220+ (FIG. 4A). Major histocompatibility complex (MHC) gene expression is very low in ESCs and day 6 *Cdx4*-modified EBs, but comparable to adult bone marrow on EB-derived cells expanded on OP9 stroma (FIG. 4A).

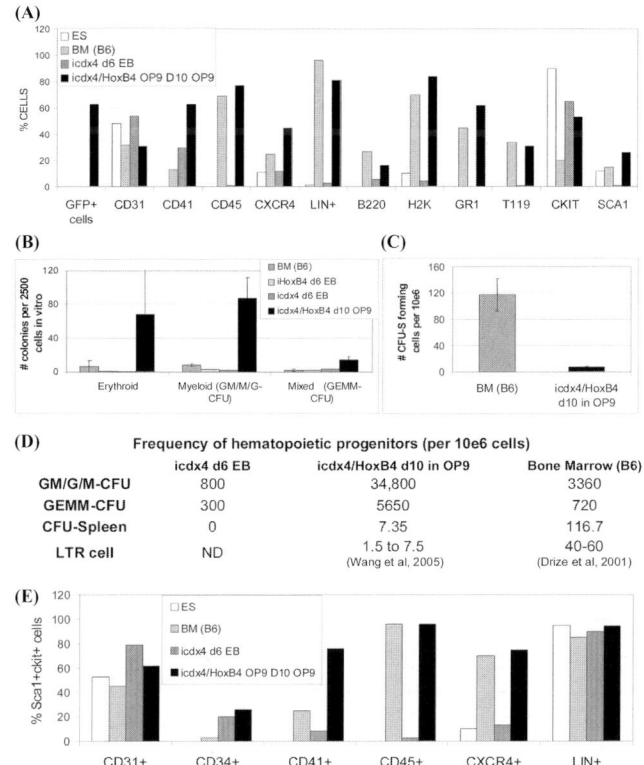

FIGURE 4. ES-derived hematopoietic cells are highly enriched in *in vitro* progenitor activity, but impaired in *in vivo* engraftment potential compared to C57BL/6 bone marrow. (A) Phenotypic analysis of the cells throughout the ES differentiation protocol, compared to adult C57BL/6 bone marrow. (B) Quantification of hematopoietic colony-forming assays in semisolid media supplemented with hematopoietic cytokines for ES-derived hematopoietic cells at the embryoid body stage, or after expansion on OP9 stroma, compared to bone marrow. (C) Quantification of day 12 spleen colony-forming ability from ES-derived hematopoietic cells after expansion on OP9 stroma, compared to bone marrow. (D) Summary of the hematopoietic multipotent progenitor content in bone marrow versus ES-derived hematopoietic cells compared by *in vitro* assay (CFU-GEMM) and *in vivo* assay (CFU-S and long-term repopulation). (Wang, *et al.*[8] and Drize N.J., *et al.* Exp. Hematol. 2001; 29(6): 786–794(E) Phenotypic analysis of the c-kit+Sca-1+ progenitor-containing subset of differentiating ES cells compared to bone marrow. Abbreviations: iHoxB4 d6 EB = embryoid bodies induced with doxycycline for *HoxB4* expression from day 4 to day 6 and tested for CFU on day 6; icdx4 d6 EB = embryoid bodies induced with doxycycline for *cdx4* expression from day 3 to day 6 and tested for CFU on day 6; icdx4/HoxB4 d10 OP9 = embryoid bodies induced with doxycycline for *cdx4* expression from day 3 to day 6, infected with *HoxB4* retrovirus and plated in OP9 stroma in the presence of hematopoietic cytokines for 10 days after which they were tested for CFU or spleen colony activity. BM = adult C57BL/6 bone marrow; LTR = long-term repopulating cell; GM = granulocyte-monocyte/macrophage; G = granulocyte; M = macrophage; GEMM = granulocyte-erythroid-monocyte-macrophage.

ESC-derived hematopoietic progenitors derived by this method display a striking enrichment in early, colony-forming progenitors, compared to adult bone marrow, and form 8- and 10-fold more multipotent (CFU-GEMM) and myeloid colonies, respectively, than adult bone marrow from 2-month-old C57BL/6 mice (FIG. 4 B). The c-kit+Sca-1+ cells constitute 18.5% of ESC-derived, OP9 expanded cells, compared to 1.9% of adult bone marrow cells.

Despite being enriched in hematopoietic progenitors relative to bone marrow, EB-derived cells do not have comparable functional capacity when transplanted *in vivo*. Multilineage short-term hematopoietic repopulation was measured through spleen colony-forming assays (CFU-S). ESC-derived hematopoietic cultures resulted in 16-fold fewer CFU-S on a per cell basis than bone marrow (FIG. 4 C). These results support previous data from our laboratory showing multilineage long-term repopulating HSCs at a frequency of 2–7 clones per million ESC-derived cells.[8] In comparison, classic limiting dilution studies performed on adult bone marrow indicate a frequency of 100–200 clones per million nucleated bone marrow cells.[12]

DISCUSSION

In the vertebrate genome, Hox genes are grouped together in clusters (HoxA, HoxB, HoxC, and HoxD) that are expressed in overlapping domains along the anterior–posterior (AP) body axis. Specific combinations of Hox genes have been hypothesized to specify tissue identities along the AP axis.[13] Hox genes have also been implicated in formation and regulation of blood cells. A, B, and C, but not D cluster Hox genes are transcribed during normal hematopoiesis in primitive subpopulations of the adult bone marrow.[14–16] Consistent with the hypothesis that embryonic pathways are conserved and reactivated under stress in the adult, aberrant Hox expression has been linked to leukemogenesis,[17,18] and Hox genes have been identified in fusion products of chromosomal translocations associated with bone marrow disorders.[19,20]

The profound impact of Hox genes on the development and function of the hematopoietic system is particularly striking in studies involving defects in cofactors, such as *Pbx1*[21] and *Meis1*,[22] or upstream regulators (e.g., mixed-lineage leukemia gene *Mll*),[23] that simultaneously disrupt multiple Hox pathways. Animals bearing deletions in *Pbx1*[21] or *Meis1*[22] die at embryonic stages due to profound hematopoietic defects. *Mll*-deficient ESCs fail to generate adult lymphoid and myeloid cells in chimeric mouse studies, and analysis of fetal liver and aorta-gonado-mesonephros shows a profound impairment in HSC/progenitor development in the absence of *Mll*.[23] In the *in vitro* ESC differentiation system, *Mll* has been shown to be required for hematopoietic development, and its absence to be rescued by ectopic expression of single Hox genes.[24] Presumably due to redundancy among the many Hox gene paralogues, defects in single Hox genes (e.g., *HoxA9*,

HoxC8, HoxB6, HoxA7, HoxB4, HoxB3) display only modest hematopoietic phenotypes.[25–30]

Recently, Cdx genes (*cdx1a, cdx4*) have been identified as upstream regulators of Hox genes and gatekeepers of hematopoietic fate specification in the zebrafish.[9,10] We have exploited the ESC system to study the effect of ectopic *Cdx4* on blood development. *Cdx4* overexpression induces posterior Hox genes of the A and B cluster, and enhances hematopoietic patterning during EB development. Furthermore, in transplant assays, it augments the frequency of cells displaying lymphoid repopulation capacity, suggesting specific enhancement of adult-like definitive hematopoietic progenitors. Similar to other embryonic hematopoietic pathways, we suggest that *Cdx4* plays a role in regulation and homeostasis of committed hematopoietic stem and progenitor cells, possibly by affecting their self-renewal and proliferative capacity. *Cdx4* expression in CD41+c-kit+-sorted ESC-derived cells modifies their hematopoietic profile. A recent study[31] shows that *Cdx4* is expressed in murine adult stem and progenitor hematopoietic cells. Furthermore, enforced overexpression by retroviral transduction of *Cdx4* in adult hematopoietic cells causes acute myeloid leukemia in a bone marrow transplant murine model by dysregulating expression of specific Hox genes. These findings suggest that regulated expression of *Cdx4* is important for the homeostasis of adult hematopoietic cells, and argue for a role for Cdx genes in the adult hematopoietic system.[32,33]

ESC-derived hematopoietic progenitors modified with ectopic *Cdx4* and *HoxB4* are enriched in multilineage progenitor activity *in vitro*, yet their ability to produce multilineage engraftment *in vivo* is strikingly inferior to that of adult bone marrow. This disparity could be explained either by a reduced proliferative capacity of ESC-derived progenitors or by an impaired ability to engraft adult bone marrow. We suggest that the latter might be the case, since improvement of transplant results has been shown by intra-bone marrow transplantation of ESC-derived cells in another system.[34] We are currently investigating the impact of transplant modality on ESC-derived *Cdx4-HoxB4*-modified hematopoietic progenitors.

CONCLUSIONS

Cdx and Hox genes play critical roles in specifying blood development in the zebrafish embryo, and the effect on hematopoietic patterning is conserved and can be exploited to enhance hematopoietic differentiation from murine ESCs *in vitro*. Using ectopic expression of Cdx and Hox gene family members, we have obtained engraftment of ESC-derived hematopoietic progenitors.[5,8] What signals activate Cdx and Hox genes, and how these genes coordinate tissue fates during embryonic development and become pathologically reactivated in leukemia remain key questions for research.

ACKNOWLEDGMENT

This study was supported by grants from the NIH and the NIH Director's Pioneer Award of the NIH Roadmap for Medical Research. G.Q.D. is a recipient of the Burroughs Wellcome Fund Clinical Scientist Award in Translational Research. C.L. was supported by a fellowship from the Mildred Scheel Foundation for Cancer Research. S.M.F. was supported by a postdoctoral fellowship from the American Cancer Society.

REFERENCES

1. MARTIN, G.R. 1981. Isolation of a pluripotent cell line from early mouse embryos cultured in medium conditioned by teratocarcinoma stem cells. Proc. Natl. Acad. Sci. USA **78:** 7634–7638.
2. EVANS, M.J. & M.H. KAUFMAN. 1981. Establishment in culture of pluripotential cells from mouse embryos. Nature **292:** 154–156.
3. BRADLEY, A., M. EVANS, M.H., KAUFMAN et al. 1984. Formation of germ-line chimaeras from embryo-derived teratocarcinoma cell lines. Nature **309:** 255–256.
4. KELLER, G. 2005. Embryonic stem cell differentiation: emergence of a new era in biology and medicine. Genes Dev. **19:** 1129–1155.
5. KYBA, M., R.C. PERLINGEIRO & G.Q. DALEY 2002. HoxB4 confers definitive lymphoid-myeloid engraftment potential on embryonic stem cell and yolk sac hematopoietic progenitors. Cell **109:** 29–37.
6. KENNEDY, M., M. FIRPO, K. CHOI, et al. 1997. A common precursor for primitive erythropoiesis and definitive haematopoiesis. Nature **386:** 488–493.
7. PERLINGEIRO, R.C., M. KYBA, S. BODIE, et al. 2003. A role for thrombopoietin in hemangioblast development. Stem Cells **21:** 272–280.
8. WANG, Y., F. YATES, O. NAVEIRAS, et al. 2005. Embryonic stem cell-derived hematopoietic stem cells. Proc. Natl. Acad. Sci. USA **102:** 19081–19086.
9. DAVIDSON, A.J., P. ERNST, Y. WANG, et al. 2003. Cdx4 mutants fail to specify blood progenitors and can be rescued by multiple hox genes. Nature **425:** 300–306.
10. DAVIDSON, A.J. & L.I. ZON. 2006. The caudal-related homeobox genes cdx1a and cdx4 act redundantly to regulate hox gene expression and the formation of putative hematopoietic stem cells during zebrafish embryogenesis. Dev. Biol. **292:** 506–518.
11. MIKKOLA, H.K., Y. FUJIWARA, T.M. SCHLAEGER, et al. 2003. Expression of CD41 marks the initiation of definitive hematopoiesis in the mouse embryo. Blood **101:** 508–516.
12. SZILVASSY, S.J., R.K. HUMPHRIES, P.M. LANSDORP, et al. 1990. Quantitative assay for totipotent reconstituting hematopoietic stem cells by a competitive repopulation strategy. Proc. Natl. Acad. Sci. USA **87:** 8736–8740.
13. KRUMLAUF, R. 1994. Hox genes in vertebrate development. Cell **78:** 191–201.
14. PINEAULT, N., C.D. HELGASON, H.J. LAWRENCE, et al. 2002. Differential expression of Hox, Meis1, and Pbx1 genes in primitive cells throughout murine hematopoietic ontogeny. Exp. Hematol. **30:** 49–57.

15. Sauvageau, G., P.M. Lansdorp, C.J. Eaves, et al. 1994. Differential expression of homeobox genes in functionally distinct CD34+ subpopulations of human bone marrow cells. Proc. Natl. Acad. Sci. USA **91:** 12223–12227.
16. Giampaolo, A., E. Pelosi, M. Valtieri, et al. 1995. HOXB gene expression and function in differentiating purified hematopoietic progenitors. Stem Cells **13**(Suppl 1): 90–105.
17. Golub, T.R., D.K. Slonim, P. Tamayo, et al. 1999. Molecular classification of cancer: class discovery and class prediction by gene expression monitoring. Science **286:** 531–537.
18. Thorsteinsdottir, U., G. Sauvageau, M.R. Hough, et al. 1997. Overexpression of HOXA10 in murine hematopoietic cells perturbs both myeloid and lymphoid differentiation and leads to acute myeloid leukemia. Mol. Cell. Biol. **17:** 495–505.
19. Slape, C. & P.D. Aplan 2004. The role of NUP98 gene fusions in hematologic malignancy. Leuk. Lymphoma **45:** 1341–1350.
20. Nakamura, T., D.A. Largaespada, M.P. Lee, et al. 1996. Fusion of the nucleoporin gene NUP98 to HOXA9 by the chromosome translocation t(7;11)(p15;p15) in human myeloid leukaemia. Nat. Genet. **12:** 154–158.
21. DiMartino, J.F., L. Selleri, D. Traver, et al. 2001. The Hox cofactor and proto-oncogene Pbx1 is required for maintenance of definitive hematopoiesis in the fetal liver. Blood **98:** 618–626.
22. Hisa, T., S.E. Spence, R.A. Rachel, et al. 2004. Hematopoietic, angiogenic and eye defects in Meis1 mutant animals. EMBO J. **23:** 450–459.
23. Ernst, P., J.K. Fisher, W. Avery, et al. 2004. Definitive hematopoiesis requires the mixed-lineage leukemia gene. Dev. Cell. **6:** 437–443.
24. Ernst, P., M. Mabon, A.J. Davidson, et al. 2004. An Mll-dependent Hox program drives hematopoietic progenitor expansion. Curr. Biol. **14:** 2063–2069.
25. Lawrence, H.J., C.D. Helgason, G. Sauvageau, et al. 1997. Mice bearing a targeted interruption of the homeobox gene HOXA9 have defects in myeloid, erythroid, and lymphoid hematopoiesis. Blood **89:** 1922–1930.
26. Izon, D.J., S. Rozenfeld, S.T. Fong, et al. 1998. Loss of function of the homeobox gene Hoxa-9 perturbs early T-cell development and induces apoptosis in primitive thymocytes. Blood **92:** 383–393.
27. Shimamoto, T., Y. Tang, Y. Naot, et al. 1999. Hematopoietic progenitor cell abnormalities in Hoxc-8 null mutant mice. J. Exp. Zool. **283:** 186–193.
28. Kappen, C. 2000. Disruption of the homeobox gene Hoxb-6 in mice results in increased numbers of early erythrocyte progenitors. Am. J. Hematol. **65:** 111–118.
29. Brun, A.C., J.M. Bjornsson, M. Magnusson, et al. 2004. Hoxb4-deficient mice undergo normal hematopoietic development but exhibit a mild proliferation defect in hematopoietic stem cells. Blood **103:** 4126–4133.
30. Bjornsson, J.M., N. Larsson, A.C. Brun, et al. 2003. Reduced proliferative capacity of hematopoietic stem cells deficient in Hoxb3 and Hoxb4. Mol. Cell. Biol. **23:** 3872–3883.
31. Bansal, D., C. Scholl, S. Frohling, et al. 2006. Cdx4 dysregulates Hox gene expression and generates acute myeloid leukemia alone and in cooperation with Meis1a in a murine modeel. Proc Natl. Acad. Sci. USA **103**(45): 16924–16929.
32. Chase, A., A. Reiter, L. Burci, et al. 1999. Fusion of ETV6 to the caudal-related homeobox gene CDX2 in acute myeloid leukemia with the t(12;13)(p13;q12). Blood **93:** 1025–1031.

33. Rawat, V.P., M. Cusan, A. Deshpande, *et al.* 2004. Ectopic expression of the homeobox gene Cdx2 is the transforming event in a mouse model of t(12;13)(p13;q12) acute myeloid leukemia. Proc. Natl. Acad. Sci. USA **101:** 817–822.
34. Burt, R.K., L. Verda, D.A. Kim, *et al.* 2004. Embryonic stem cells as an alternate marrow donor source: engraftment without graft-versus-host disease. J. Exp. Med. **199:** 895–904.

Differentiation Potential of Histocompatible Parthenogenetic Embryonic Stem Cells

CLAUDIA LENGERKE,[a–c] KITAI KIM,[a–c] PAUL LEROU,[a,c,d] AND GEORGE Q. DALEY[a–c,e]

[a]*Division of Pediatric Hematology/Oncology, Children's Hospital Boston and Dana-Farber Cancer Institute, Boston, Massachusetts 02115, USA*

[b]*Department of Biological Chemistry and Molecular Pharmacology, Harvard Medical School, Boston, Massachusetts 02115, USA*

[c]*Harvard Stem Cell Institute, Cambridge, Massachusetts 02138, USA*

[d]*Division of Newborn Medicine, Brigham and Women's Hospital and Children's Hospital, Harvard Medical School, Boston, Massachusetts 02115, USA*

[e]*Division of Hematology, Brigham and Women's Hospital, Boston, Massachusetts 02115, USA*

ABSTRACT: Embryonic stem cells (ESCs) hold unique promise for the development of cell replacement therapies, but derivation of therapeutic products from ESCs is hampered by immunological barriers. Creation of HLA-typed ESC banks, or derivation of customized ESC lines by somatic cell nuclear transfer, have been envisioned for engineering histocompatible ESC-derived products. Proof of principle experiments in the mouse have demonstrated that autologous ESCs can be obtained via nuclear transfer and differentiated into transplantable tissues, yet nuclear transfer remains a technology with low efficiency. Parthenogenesis provides an additional means for deriving ESC lines. In parthenogenesis, artificial oocyte activation initiates development without sperm contribution and no viable offspring are produced in the absence of paternal gene expression. Development proceeds readily to the blastocyst stage, from which parthenogenetic ESC (pESC) lines can be derived with high efficiency. We have recently shown that when pESC lines are derived from hybrid mice, early recombination events produce heterozygosity at the major histocompatibility complex (MHC) loci in some of these lines, enabling the generation of histocompatible differentiated cells that can engraft immunocompetent MHC-matched mouse recipients. Here, we explore the differentiation potential of murine pESCs derived in our laboratory.

Address for correspondence: George Q. Daley, Children's Hospital Boston, 300 Longwood Avenue, Boston, MA 02115. Voice: 617-919-2015; fax: 617-730-0222.
george.daley@childrens.harvard.edu

Ann. N.Y. Acad. Sci. 1106: 209–218 (2007). © 2007 New York Academy of Sciences.
doi: 10.1196/annals.1392.011

KEYWORDS: parthenogenesis; embryonic stem cells; developmental potential; MHC-matched; Igf2

INTRODUCTION

Mammalian oocytes remain arrested at the metaphase stage of meiosis II until fertilization, when the sperm enters the oocyte and triggers the events leading to the first zygotic cell division. Under certain conditions, spermatozoa will stimulate the oocyte to undergo embryonic development but only the female nucleus contributes genetically to the developing embryo in a process termed *gynogenesis* or *parthenogenesis* (reviewed in Ref. 1). Alternatively, sperm pro-nuclei can be manipulated to produce an embryo with only paternal genomic contribution (termed *androgenesis*). Mammals require genomic contributions from both the oocyte and sperm because expression of both paternally and maternally imprinted genes is required during embryogenesis. Viable offspring have not been reported from uniparental blastocysts without genetical manipulation.[2]

Chimera studies in mice illustrate specific roles for paternal and maternal genes during embryogenesis. When aggregated with normal eight-cell stage embryos, parthenogenetic blastomeres chimerize the embryo proper and extraembryonic mesoderm, but only rarely the extraembryonic endoderm and trophectoderm. In contrast, androgenetic blastomeres contribute strongly to trophectoderm, but almost not at all to tissues of the embryo proper.[3] Consistent with these findings, spontaneous parthenogenesis may result in the formation of ovarian teratomas,[4–6] while androgenesis is responsible for the development of the hyatidiform mole,[7] a condition characterized by uncontrolled growth of trophoblastic tissue in the absence of regular embryonic development.

In the laboratory, species-specific protocols involving exposure to certain chemicals (e.g., alcohol, ionomycine, 6-dimethylaminopurine, cycloheximide), cold shock, and electrical stimulation have been developed for releasing the oocyte from its arrest and stimulating parthenogenetic development in the absence of sperm. Uniparental preimplantation parthenogenetic embryos have been generated in a number of different mammalian species (e.g., mouse, rat, rabbit, pig, goat, cow, monkey, and human) (reviewed in Ref. 8), and, in some cases, embryonic stem cell (ESC) lines have been successfully derived from such blastocysts (termed *parthenogenetic ESCs*, or pESCs).

We have recently generated a large number of pESCs in our laboratory from hybrid mice (C57BL/6 × CBA F1), and subjected these cells to genome-wide polymorphism analysis.[9] We found that early recombination events lead to heterozygosity at the major histocompatibility complex (MHC) loci in a significant number of the derived pESC lines. Therefore parthenogenesis represents an effective method for derivation of ESCs that should be histocompatible with the oocyte donor mice. Though unable to support the development of viable offspring, parthenogenetic blastocysts will yield pESCs that are pluripotent and are able to chimerize all tissues of the embryo proper. We hypothesize similar mechanisms in the human, and envision the possibility of generating histocompatible ESCs and ESC-derived cell therapies for oocyte donors via parthenogenesis.

RESULTS AND DISCUSSION

Generation of MHC-Heterozygous pESC Lines

In a recent study[9] we compared two methods of parthenogenesis for deriving pESCs. The efficiency of ESC line derivation was 65% when meiosis II was interrupted by preventing the extrusion of the second polar body (p(MII)ESCs), and 37% when parthenogenetic development was induced in immature oocytes by interfering with the segregation of paired homologous chromosomes during meiosis I (p(MI)ESCs). To determine if recombination had restored heterozygosity at the MHC loci, we determined the genotypes of the MHC region in the derived pESC lines by PCR amplification followed by allele-specific restriction enzyme digestion or direct sequencing of single nucleotide polymorphisms (SNP). Thirty-three percent of the p(MII) ESC lines ($n = 72$), and 91.3% of the p(MI)ESC ($n = 23$) lines harbored the heterozygous MHC genotype of the hybrid oocyte donor. The frequency of heterozygosity was higher in p(MI)ESC lines than in p(MII)ESC lines. The vast majority of p(MII)ESC lines (84%) harbored a normal karyotype, whereas aneuploidy was common in p(MI)ESC lines, suggesting that activation of immature oocytes caused abnormal chromosomal segregation. The selected ESC lines were named recombinant MHC-matched p(MII/I)ESC lines.

Undifferentiated ESCs do not express HLA antigens, but upon *in vitro* differentiation in embryoid bodies (EBs)[10] MHC antigens are expressed. In our analysis, pESC lines that had not recombined at the MHC locus according to SNP analysis expressed only one of the parental MHC proteins (e.g., H2k^b for C57BL/6), while recombinant MHC-matched pESC lines expressed, as expected, both H2K^b and H2K^k on all cells.[9]

Moreover, in this study,[9] 17 p(MII)ESC lines and 20 p(MI)ESC lines were selected for high-resolution genotyping using a standard panel of 768 mouse markers spaced across the genome. Interestingly, this genotypic analysis showed a high degree of heterozygosity for the p(MII)ESCs. Homozygosity occurred near the centromere, while we observed increasingly heterozygosity in proportion to the genetic distance from the centromere. Overall, p(MII)ESC lines showed a much lower degree of homozygosity than previously assumed (36.9% of loci among the 17 analyzed cell lines), indicating that recombination events prior to meiosis II render the majority of loci in p(MII)ESCs heterozygous. In p(MI)ESCs, two distinct groups were identified: 60% ($n = 12$) showed a predominant pattern of heterozygosity beginning at the centromere followed on some chromosomes by distal regions of homozygosity, while 40%

($n = 8$) showed complete heterozygosity across all loci,[9] suggesting complete genetic identity with the oocyte donor. However, all p(MI)ESC lines showing complete heterozygosity were aneuploid, suggesting a predisposition to abnormal chromosomal segregation by this method.

Differentiation Potential of p(MI) and p(MII) ESC Lines

Parthenogenetic cells have been reported to chimerize tissues of all three germ layers when aggregated with normal fertilized embryos.[3] Recently, primate parthenogenetic ESCs have been isolated and shown to differentiate *in vitro* into multiple tissues, including contractile cardiomyocyte-like cells, smooth muscle, ciliated epithelia, adipocytes, and transplantable dopaminergic neurons.[11,12]

We have compared the *in vitro* differentiation potential of three p(MI)ESC lines and two p(MII)ESC lines derived in our laboratory, using a standard fertilized ESC line (Ainv) as a control. ESCs were cultured on mouse embryonic fibroblasts under standard conditions in serum and LIF containing media and differentiated according to published protocols.[10] Briefly, undifferentiated ESCs were separated from the supportive fibroblasts and aggregated in differentiation media as hanging drops (100 cells/15 μL drop), allowing EB formation. After 2 days, the EBs were collected, transferred to Petri dishes, and incubated for another 4 days.[10] On day 6, EBs were either collected for RNA isolation and RT-PCR analysis, or transferred into semisolid media supplemented with hematopoietic cytokines for hematopoietic colony formation.

All p(MI)ESC lines and p(MII)ESC lines analyzed showed differentiation toward the three germ layers. We determined expression levels for markers of several distinct populations by semiquantitative PCR: hematopoietic (Scl, Runx1, Gata1, beta-major globin), vascular (Flk1), cardiac (NKX2.5), skeletal muscle (MyoD1, Myogenin), endodermal (HNF4, Gata6, AFP), and ectodermal (surface ectodermal: Cytokeratin 17, neuroectodermal: Pax6) (FIG. 1A). Marker expression varied among cell lines and in comparison with

FIGURE 1. Comparison of the *in vitro* developmental potential of EBs derived from p(MI) and p(MII) ESCs. (**A**) Semi-quantitative RT-PCR for markers of the three germ layers. Expression in undifferentiated ESCs and EBs from a fertilized ESC line (Ainv) were used as controls. PCR amplification over 30 cycles (Actin 20 cycles). (**B**) Quantitative PCR analysis. Relative expression levels were calculated after normalization to actin. Data shown represent the averaged values for three p(MI) and two p(MII) ESC lines. Primer sequences are listed in Table 1. (**C**) Quantitative PCR analysis after Igf2 supplementation. The relative change in gene expression after Igf2 supplementation was calculated for each cell line individually, and then averaged for the p(MI) and p(MII) group. All error bars represent the standard deviation.*$P < 0.0005$.

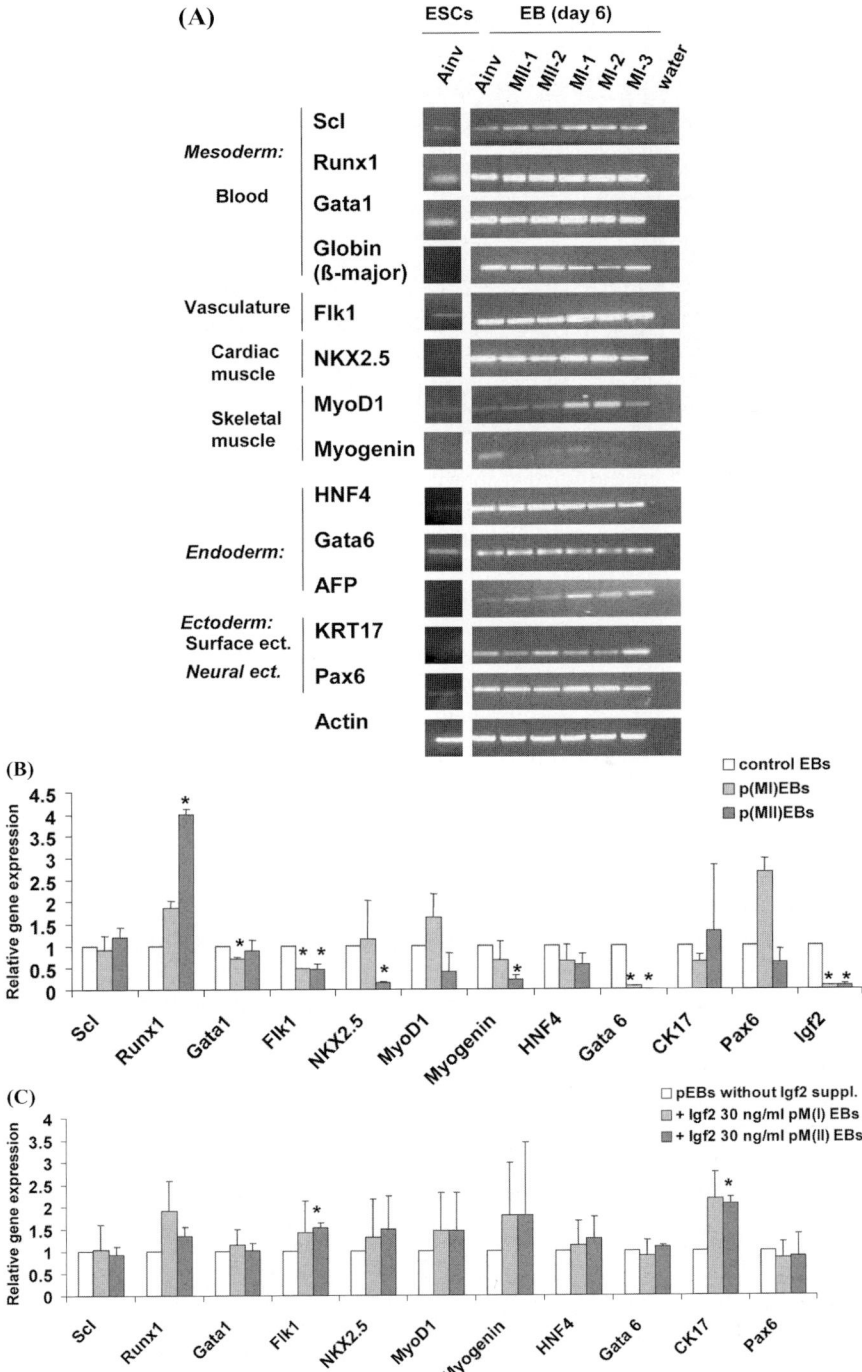

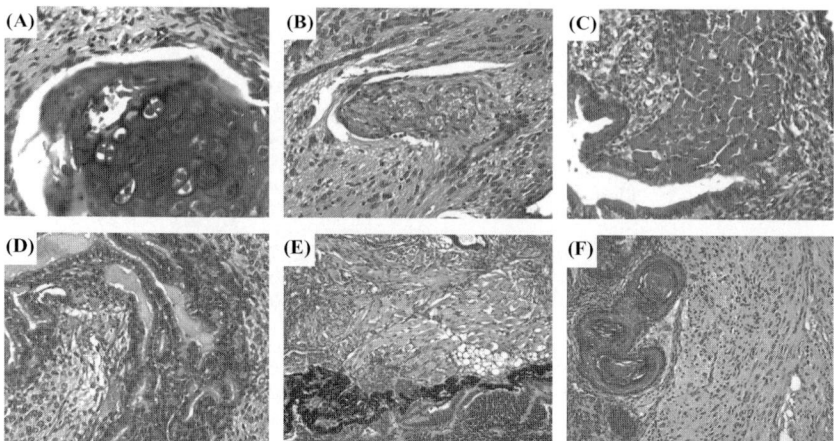

FIGURE 2. Multilineage differentiation in a teratoma from a p(MI) ESC line. (**A**) Bone, cartilage (X160), (**B**) Muscle (X40), (**C**) Pancreas (X40), (**D**) Respiratory epithelium (X40), (**E**) Muscle, retinal pigment epithelium, embryonic brain (X20), (**F**) Skin and brain (X20) (HE-staining).

the fertilized ESC line used as control, but all showed multilineage differentiation potential. We further performed a quantitative PCR analysis, and compared averaged levels of gene expression between p(MI)ESC lines, p(MII)ESC lines, and the fertilized control ESC line. By this analysis, we observed a significant reduction in gene expression levels for Gata1, Flk1, Gata6 in the p(MI) ESC lines, and for Flk1, NKX2.5, Myogenin, Gata6 in the analyzed p(MII) lines (FIG. 1B). The multilineage differentiation potential was also documented in the analysis of teratomas from a p(MI)ESC line (FIG. 2).

As contributions from both maternal and paternal genomes are required for normal embryogenesis, it is probable that the lack of specific paternal genes accounts for the reduced differentiation to particular lineages. Several studies have documented a selection against parthenogenetic cells during embryogenesis. This phenomenon is nonrandom, and particularly affects tissues of mesodermal and endodermal origin (e.g., skeletal muscle, liver, pancreas).[13–16] It remains unclear, however, if specification, terminal differentiation, or survival and proliferation mechanisms are affected in parthenogenetically derived tissues of chimeric mice. *In vitro*, our results show lower levels of Myogenin, but not a significant diminution of MyoD1 in pESCs-derived EBs. While MyoD1 is required for myogenic determination, Myogenin is needed later, during the differentiation of prespecified myoblasts (reviewed in Ref. 17), suggesting that muscle formation from pESCs might be impaired during the terminal differentiation to muscle fibers, rather than during the early specification to muscle fate. Overall, parthenogenetic mouse blastocysts have a significantly smaller cell number than the fertilized control embryos,[18] and general growth retardation

TABLE 1. Primer sequences

Gata6-F	CTTGCGGGCTCTATATGAAACTCCAT
Gata6-R	TAGAAGAAGAGGAAGTAGGAGTCATAGGGACA
KRT17-F	ATCCGAGATTGGTACCAGAAGCAG
KRT17-R	GGTGGCCACAAGGATCTTGTTCTT
NKX2.5-F**	GACAAAGCCGAGACGGATGG
NKX2.5-R**	CTGTCGCTTGCACTTGTAGC
Igf2-F	TCTCATCTCTTTGGCCTTCGCCTT
Igf2-R	GTCCGAACAGACAAACTGAAGCGT
Myogenin-F	ACAATCTGCACTCCCTTACGTCCA
Myogenin-R	TCTCAGTTGGGCATGGTTTCGTCT
MyoD1-F	ATCCCTAAGCGACACAGAACAGGGAA
MyoD1-R	TGCAGTCGATCTCTCAAAGCACCT
SCL-F	TCCCCATATGAGATGGAGATTTC
SCL-R	ATTGATGTACTTCATGGCAAGG
Gata1-F	CATTGGCCCCTTGTGAGGCCAG
Gata1-R	CGCTCCAGCCAGATTCGACCC
Runx1-F	AACCTGAGGTCGTTGAATCTCGCT
Runx1-R	AGCATGGTGGAGGTACTAGCTGA
Globin (beta adult major chain)-F*	CTGACAGATGCTCTCTTGGG
Globin (beta adult major chain)-R*	CACAACCCCAGAAACAGACA
Actin -F	TCTTGGGTATGGAATCCTGTGGCA
Actin -R	ACTCCTGCTTGCTGATCCACATCT

*ref: PNAS, 1999;96:2159; Development, 1998;125:725.**Cell 2006;127(6):1137.

can be contributing to the poor embryo reconstitution with parthenote cells.[13] Specific tissue deficits in parthenote chimeras suggest paternally imprinted genes as key players in formation and/or maintenance of certain tissues, such as striated muscle. Complementary studies using androgenetic ESC lines demonstrate a strong bias toward myogenic differentiation *in vitro* [19] and in teratoma assays.[20]

We hypothesized that introduction of missing paternal genes, or supplementation with their products on the protein level might improve tissue generation from pESCs. Igf2 is a well characterized paternally imprinted gene,[21] that has been implicated in embryonic cell growth,[22] but also in specific differentiation programs, such as myogenesis. Igf2 overexpression has been shown to enhance both differentiation and survival of maturing myoblasts,[23,24] and to promote myogenic differentiation from murine ESCs.[25] Both p(MI)ESC lines and p(MII)ESC lines showed strongly reduced levels of Igf2 (FIG. 1B). We hypothesized that addition of recombinant Igf2 protein might improve the differentiation capacity of p(MI)ESC lines and p(MII)ESC lines, especially along the myogenic lineage. Igf2 (30 ng/mL) (R&D Systems, Minneapolis, MN) was added to the media at the start of ESC differentiation, and samples were analyzed by quantitative PCR at day 6. We observed significant variability among the analyzed lines, but found that the epithelial marker Cytokeratin 17 and the endothelial marker Flk1 showed significant upregulation after Igf2 treatment

(FIG. 1C). The analyzed pESC lines did not show significant enhancement of skeletal muscle markers (MyoD1 and Myogenin) as previously reported in regular (fertilized) ESC lines stably overexpressing Igf2.[25] Further experiments are needed to explore whether the abnormalities of tissue differentiation within pESCs can be abrogated or complemented by culture conditions or genetic modifications. Besides the lack of paternal imprinted genes, excess of maternal imprinted genes, such as H19, may also restrict the development of parthenote cells.[2]

When analyzing the developmental potential of pESCs, it must be taken into account that the process of derivation and of *in vitro* culture of ESCs might affect their imprinting status, and thereby their *in vitro* differentiation and chimerization capacity. Allen *et al.*[26] have shown phenotypic differences in chimera studies between pESCs and parthenote embryo cells. While both pESCs and parthenote cells showed restricted lineage participation in chimeric mice, suggesting similar abnormalities in lineage allocation and differentiation, pESCs did not manifest the same degree of growth disadvantage as was documented for parthenotes. This further suggests that the process of establishing pESC lines in culture may select for epigenetic features that produce more normal growth patterns.

CONCLUSIONS

We have shown in the mouse that pESC lines histocompatible to the oocyte donor can be derived with high efficiency by parthenogenesis.[9] Parthenogenetic ESC lines retain self-renewal capacity, can be extensively cultured *in vitro*, and are pluripotent, giving rise to differentiated cells of all three germ layers. Parthenogenesis has also been documented in the human, where a subset of ovarian teratomas have been demonstrated of parthenote origin, providing *in vivo* evidence for survival and differentiation potential of human parthenote cells.[5] Furthermore, the proof of principle that uniparental cells can be indeed functionally sufficient in the human is provided by a clinical case study of a young boy who was a parthenogenetic chimera. The patient's skin was partially derived from parthenote cells, while the peripheral leukocytes were entirely of parthenote origin, carrying a 46,XX karyotype.[27] As artificial parthenogenetic oocyte activation has been successful in many species including primates, we envision the establishment of protocols for human oocyte activation, and isolation of human pESCs. We thus anticipate efforts to demonstrate the isolation of HLA-matched ESCs from females, and their possible use in modeling disease and developing "autologous" cell therapies. Parthenogenesis provides an alternative and presumably efficient means for derivation of immunologically matched ESC-products, and may enable the establishment of ESC banks composed of cells with broad HLA compatibility.[28]

ACKNOWLEDGMENTS

This study was supported by grants from the NIH and the NIH Director's Pioneer Award of the NIH Roadmap for Medical Research. G.Q.D. is a recipient of the Burroughs Wellcome Fund Clinical Scientist Award in Translational Research. C.L. was supported by a Mildred Scheel Foundation for Cancer Research fellowship. K.K. was supported by the Cooley's Anemia Foundation and is a Special Scholar of the Leukemia and Lymphoma Society. P.L. was partly supported by NIH training grant, "Pathobiology of Newborn and Developmental Diseases" (T32: HD07466).

REFERENCES

1. ROUGIER, N. & Z. WERB. 2001. Minireview: parthenogenesis in mammals. Mol. Reprod. Dev. **59:** 468–474.
2. KONO, T. *et al.* 2004. Birth of parthenogenetic mice that can develop to adulthood. Nature. **428:** 860–864.
3. THOMSON, J.A. & D. SOLTER. 1988. The developmental fate of androgenetic, parthenogenetic, and gynogenetic cells in chimeric gastrulating mouse embryos. Genes Dev. **2:** 1344–1351.
4. LEE, G.H. *et al.* 1997. Genetic dissection of susceptibility to murine ovarian teratomas that originate from parthenogenetic oocytes. Cancer Res. **57:** 590–593.
5. LINDER, D., B.K. MCCAW & F. HECHT. 1975. Parthenogenic origin of benign ovarian teratomas. N. Engl. J. Med. **292:** 63–66.
6. STEVENS, L.C. & D.S. VARNUM. 1974. The development of teratomas from parthenogenetically activated ovarian mouse eggs. Dev. Biol. **37:** 369–380.
7. WAKE, N., N. TAKAGI & M. SASAKI. 1978. Androgenesis as a cause of hydatidiform mole. J. Natl. Cancer Inst. **60:** 51–57.
8. CIBELLI, J.B., K. CUNNIFF & K.E. VRANA. 2006. Embryonic stem cells from parthenotes. Methods Enzymol. **418:** 117–135.
9. KIM, K. *et al.* 2007. Histocompatible embryonic stem cells by parthenogenesis. Science **315**(5811): 482–486. Epub 2006 Dec. 14.
10. KYBA, M., R.C. PERLINGEIRO & G.Q. DALEY. 2002. HoxB4 confers definitive lymphoid-myeloid engraftment potential on embryonic stem cell and yolk sac hematopoietic progenitors. Cell **109:** 29–37.
11. SANCHEZ-PERNAUTE, R. *et al.* 2005. Long-term survival of dopamine neurons derived from parthenogenetic primate embryonic stem cells (cyno-1) after transplantation. Stem Cells **23:** 914–922.
12. VRANA, K.E. *et al.* 2003. Nonhuman primate parthenogenetic stem cells. Proc. Natl. Acad. Sci. USA **100** (Suppl 1): 11911–11916.
13. NAGY, A., M. SASS & M. MARKKULA. 1989. Systematic non-uniform distribution of parthenogenetic cells in adult mouse chimaeras. Development **106:** 321–324.
14. FUNDELE, R. *et al.* 1989. Systematic elimination of parthenogenetic cells in mouse chimeras. Development **106:** 29–35.
15. CLARKE, H.J. *et al.* 1988. The development potential of parthenogenetically derived cells in chimeric mouse embryos: implications for action of imprinted genes. Development **104:** 175–182.

16. NAGY, A. et al. 1987. Prenatal fate of parthenogenetic cells in mouse aggregation chimaeras. Development **101:** 67–71.
17. BUCKINGHAM, M. et al. 2003. The formation of skeletal muscle: from somite to limb. J. Anat. **202:** 59–68.
18. URANGA, J.A. & J. ARECHAGA. 1997. Cell proliferation is reduced in parthenogenetic mouse embryos at the blastocyst stage: a quantitative study. Anat. Rec. **247:** 243–247.
19. MCKARNEY, L.A., M.L. OVERALL & M. DZIADEK. 1997. Myogenesis in cultures of uniparental mouse embryonic stem cells: differing patterns of expression of myogenic regulatory factors. Int. J. Dev. Biol. **41:** 485–490.
20. MANN, J.R. et al. 1990. Androgenetic mouse embryonic stem cells are pluripotent and cause skeletal defects in chimeras: implications for genetic imprinting. Cell **62:** 251–260.
21. DECHIARA, T.M., E.J. ROBERTSON & A. EFSTRATIADIS. 1991. Parental imprinting of the mouse insulin-like growth factor II gene. Cell **64:** 849–859.
22. DECHIARA, T.M., A. EFSTRATIADIS & E.J. ROBERTSON. 1990. A growth-deficiency phenotype in heterozygous mice carrying an insulin-like growth factor II gene disrupted by targeting. Nature **345:** 78–80.
23. STEWART, C.E. et al. 1996. Overexpression of insulin-like growth factor-II induces accelerated myoblast differentiation. J. Cell. Physiol. **169:** 23–32.
24. STEWART, C.E. & P. ROTWEIN. 1996. Insulin-like growth factor-II is an autocrine survival factor for differentiating myoblasts. J. Biol. Chem. **271:** 11330–11338.
25. PRELLE, K. et al. 2000. Overexpression of insulin-like growth factor-II in mouse embryonic stem cells promotes myogenic differentiation. Biochem. Biophys. Res. Commun. **277:** 631–638.
26. ALLEN, N.D. et al. 1994. A functional analysis of imprinting in parthenogenetic embryonic stem cells. Development **120:** 1473–1482.
27. STRAIN, L. et al. 1995. A human parthenogenetic chimaera. Nat. Genet. **11:** 164–169.
28. TAYLOR, C.J. et al. 2005. Banking on human embryonic stem cells: estimating the number of donor cell lines needed for HLA matching. Lancet **366:** 2019–2025.

Hematopoiesis from Human Embryonic Stem Cells

MICKIE BHATIA

Stem Cell and Cancer Research Institute, Michael G. DeGroote School of Medicine, Department of Biochemistry, McMaster University, Hamilton, Ontario L8N 3Z5, Canada

> ABSTRACT: Human embryonic stem cells (hESCs) have the capacity to differentiate into multiple lineages, such as neural and hepatic, and include the ability to form hematopoietic cells. Efforts to develop efficient methods to induce hematopoietic differentiation effectively from hESCs, and generate primitive blood stem cells with multilineage hematopoietic repopulating function, are ongoing in several laboratories. Here, we will discuss the past and current progress of our laboratory in this area, and share our ideas to further this use of hESCs in both experimental and clinical hematology.
>
> KEYWORDS: human embryonic stem cells (hESCs); repopulation; hematopoiesis; development; differentiation; hematopoietic stem cells (HSCs)

Since the initial report by Thomson et al.[1] of the ability to isolate human embryonic stem cells (hESCs) from the intercell mass of human embryos, there has been immense activity to define methods to propagate these cells in culture in order to differentiate hESCs into mature lineages. hESCs will spontaneously differentiate upon removal of conditions that maintain self-renewal, for example, removal of FGF, etc. Under spontaneously differentiating conditions, the propensity of these cells to differentiate into neural lineages has become clear, however, efforts to differentiate to the hematopoietic lineage and generate primitive blood cells that retain repopulating ability have been more arduous. Developing differentiation methods for hESCs of any lineage is central to the importance of these stem cells in cell replacement therapies.

The principle of using hESCs as a cell replacement source is deceptively simple, and fundamentally revolves around the obstacle, or our lack of understanding of human developmental biology, to differentiate these cells in a controlled and precise manner. Previous work using mouse ESCs has demonstrated that bone morphogenic protein-4 (BMP-4) represents a strong

---

Address for correspondence: Dr. M. Bhatia, Stem Cell and Cancer Research Institute (SCC-RI), Michael G. DeGroote School of Medicine, McMaster University, 1200 Main Street West, MDCL 5029, Hamilton, Ontario L8N 3Z5, Canada. Voice: 905-525-9140; ext.: 28687; fax: 905-522-7772.
mbhatia@mcmaster.ca

hematopoietic-inducing factor. Capitalizing on our previous efforts to support adult hematopoietic cells in culture with stem cell factor (SCF) and FLT-3L, we have combined these factors with BMP-4 to treat clusters of hESCs called embryoid bodies (EBs) and demonstrated robust hematopoietic differentiation.[2–4] These conditions work consistently on several hESC lines and represent the optimal conditions our laboratory has identified. It is important to note that the magnitude (exact percentage of hematopoietic cells or progenitors detected by colony-forming unit (CFU) assay) varies among hESC lines. Consistently, all hESC lines that we have tested demonstrate the emergence of hematopoietic cells at day 10 of EB treatment,[5] suggesting that precursor populations within the EB that possess hemogenic properties exist. Detailed investigation and phenotypic analysis of cells comprising EBs prior to day 10 of hematopoietic emergence indicated that cells with endothelial phenotype developed under these conditions and ultimately gave rise to mature hematopoietic cells.[5] Prospective and clonal isolation experiments demonstrated that cells with endothelial phenotype and endothelial potential were exclusively responsible for hematopoietic progenitors derived from hESCs.[5] We had defined these cells as $CD45^-$ to denote the absence of hematopoietic marker expression and expression of cell-surface PECAM-1, VE-cadherin, and Flk-1 as $CD45^{neg}PFV$ cells. $CD45^{neg}PFV$ cells could be differentiated into hematopoietic lineages whereas no other cell in the EB without this phenotype possessed hematopoietic potential. Since endothelial lineages could also be derived from clonal starts of $CD45^{neg}PFV$ cells, this unique population seems to possess hemangioblastic properties.[5] This series of studies provided us with biological insights of hESC differentiation toward hematopoiesis and, more importantly, provided a phenotypic and timed road map of hematopoietic development from which we have expanded our current work. (See proposed working model FIG. 1.)

The ability to isolate $CD45^{neg}PFV$ cells and differentiate them to hematopoietic progenitors provided an approach to engineer and transplant a focused population of hematopoietic cells from hESC cultures.[3,6] This allowed additional studies where we examined the *in vivo* repopulating potential of resulting hematopoietic progenitors derived from $CD45^{neg}PFV$ cells. Initially, primitive $CD34^+CD45^+$ cells derived from hESCs were transplanted into NOD-SCID mice by both intravenous and interfemoral injection.[6] Although intravenous injection did not lead to any detectable reconstitution after greater than 6 weeks in NOD-SCID mice, interfemoral injection allowed multilineage human engraftment to be detected up to 10 weeks in several NOD-SCID recipients. This provided initial observations to suggest that human SCID repopulating cells (SRCs) can be derived from hESCs similar to adult sources of umbilical cord blood (CB) or bone marrow (BM).[6] However, upon further characterization of hESC-derived SRCs, we observed a limited proliferative and migratory capacity compared to adult sources of SRCs. Molecular profiling of ES-derived SRCs to adult sources (both purified for the $CD34^+$ $CD38^-$ $CD45^+$ phenotype) provided surprising differences in both homing molecules

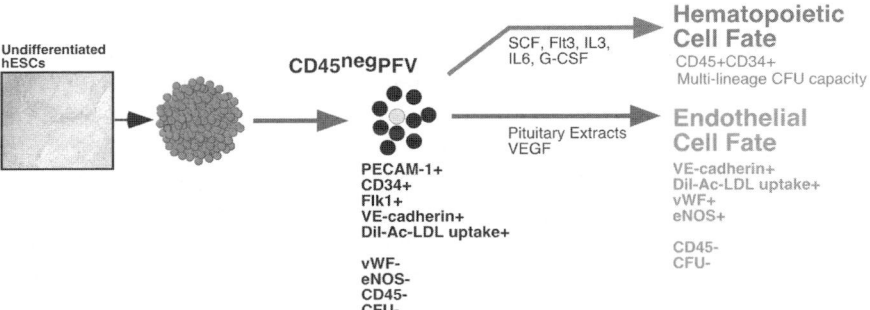

FIGURE 1. Working model outlining our current understanding of hematopoietic cell fate specification from human embryonic stem cells (hESCs). Functional hemogenic capacity emerges between 7 and 10 days of human embryoid bodies (hEB) development. The presence of PECAM-1, CD34, Flk-1, VE-cadherin, uptake of LDL, and absence of CD45 or hematopoietic progenitor function are together used to define endothelial-like cells. A unique subset of CD45^{neg} hEB cells lacking hematopoietic progenitor capacity, but possessing endothelial-like properties, can be isolated from hEBs prior to the emergence of hematopoietic cell fate. Despite functional Dil-Ac-LDL uptake, expression of Flk-1, VE-cadherin, and CD34, the endothelial-like cells lack mature endothelial properties, such as eNOS or vWF factor expression. Hemogenic capacity of hEBs is consistent among hESC lines tested to date, and is exclusive to endothelial-like CD45^{neg}PFV cells derived from these lines. Upon VEGF and pituitary hormone stimulation, CD45^{neg}PFV cells are able to produce mature endothelial cells that continued to express endothelial markers, and upregulate eNOS and vWF expression. This system provides the foundation and cellular road map to further define hematopoietic and endothelial development to evaluate the role of signaling pathways known to affect adult HSCs.

associated with hematopoietic stem cells (HSCs) and expression of HOX gene clusters known to be important in the self renewal of adult HSCs.[6] Taken together, these initial studies were interpreted by our lab as an inability to activate essential genetic programs in hESCs to derive bonafide somatic-like HSCs. We believe this inability leads to the atypical *in vivo* behavior of hESC-derived SRCs. More globally, these studies in the hematopoietic system that use rigorous *in vivo* assays underscore the importance of both molecular and functional comparisons between cell lineages (of any kind) derived from hESCs to their adult/somatic counterparts.

Our continued efforts to derive HSCs from hESCs will be informed by our studies in umbilical CB- and BM-derived HSCs capable of repopulating NOD-SCID mice. Over the past several years we have demonstrated that the Wnt, Hh, and Notch pathways are capable of governing the proliferative capacity of HSCs, and instructing self-renewal and differentiation ability more potently than cytokines traditionally identified. Our findings in the past year indicate that the Wnt signaling pathway is active in hESCs and can be dissected for functional effects of cononical versus noncononical Wnt that have unique effects on hematopoietic differentiation from hESCs. Furthermore, unlike Wnt signals, the Notch signaling pathway is dormant in untreated hESCs, but can be activated upon Notch ligand presentation and seems to depend on the downstream target of Notch signaling HES-1 to induce hemogenic precursor development ($CD45^{neg}PFVs$). Interestingly, when the Hh pathway is downregulated, hematopoietic differentiation from hESCs is augmented. Although the role of Hh in this regard may seem contradictory to that described in adult HSCs, recent evidence from our laboratory suggests that continued Hh signals in HSCs lead to stem cell exhaustion, suggesting that downregulation may be critical when initially specifying hematopoiesis from undifferentiated hESCs. Since use of these pathways further augments hematopoiesis from hESCs, we will examine the repopulating capacity of these resulting hematopoietic progenitors to ascertain whether the increased "quantity" of hematopoietic cells derived from ESCs amounts to increased quality of these hematopoietic cells to possess SRC activity similar to that derived from umbilical CB or BM.

REFERENCES

1. THOMSON, J.A. et al. 1998. Embryonic stem cell lines derived from human blastocysts. Science **282**: 1145–1147.
2. CHADWICK, K. et al. 2003. Cytokines and BMP-4 promote hematopoietic differentiation of human embryonic stem cells. Blood **102**: 906–915.
3. MENENDEZ, P. et al. 2004. Retroviral transduction of hematopoietic cells differentiated from human embryonic stem cell-derived CD45(neg)PFV hemogenic precursors. Mol. Ther. **10**: 1109–1120.
4. WANG, L. et al. 2005. Human embryonic stem cells maintained in the absence of mouse embryonic fibroblasts or conditioned media are capable of hematopoietic development. Blood **105**(12): 4598–4603.
5. WANG, L. et al. 2004. Endothelial and hematopoietic cell fate of human embryonic stem cells originates from primitive endothelium with hemangioblastic properties. Immunity **21**: 31–41.
6. WANG, L. et al. 2005. Generation of hematopoietic repopulating cells from human embryonic stem cells independent of ectopic HOXB4 expression. J. Exp. Med. **201**: 1603–1614.

Emergence of Human Angiohematopoietic Cells in Normal Development and from Cultured Embryonic Stem Cells

ELIAS T. ZAMBIDIS,[a] LIDIA SINKA,[b] MANUELA TAVIAN,[b] VENTA JOKUBAITIS,[c] TEA SOON PARK,[d] PAUL SIMMONS,[e] AND BRUNO PÉAULT[d]

[a]*Kimmel Comprehensive Cancer Center at Johns Hopkins, Baltimore, Maryland 21231, USA*

[b]*INSERM U602, Hôpital Paul Brousse, Villejuif 94807, France*

[c]*Australian Stem Cell Centre, Clayton, Victoria 3800, Australia*

[d]*Stem Cell Research Center at Children's Hospital of Pittsburgh of UPMC, Pittsburgh, Pennsylvania 15213, USA*

[e]*The University of Texas Health Science Center, Houston, Texas 77030, USA*

ABSTRACT: Human hematopoiesis proceeds transiently in the extraembryonic yolk sac and embryonic, then fetal liver before being stabilized in the bone marrow during the third month of gestation. In addition to this classic developmental sequence, we have previously shown that the aorta-gonad-mesonephros (AGM) embryonic territory produces stem cells for definitive hematopoiesis from 27 to 40 days of human development, through an intermediate *blood-forming endothelium* stage. These studies have relied on the use of traditional markers of human hematopoietic and endothelial cells. In addition, we have recently identified and characterized a novel surface molecule, BB9, which typifies the earliest founders of the human angiohematopoietic system. BB9, which was initially identified with a monoclonal antibody raised to Stro-1$^+$ bone marrow stromal cells, recognizes in the adult the most primitive Thy-1$^+$ CD133$^+$ Lin$^-$, non-obese diabetic—severe combined immunodeficiency disease (NOD–SCID) mouse engrating hematopoietic stem cells (HSCs). In the 3- to 4-week embryo, BB9 expression typifies a subset of splanchnopleural mesodermal cells that migrate dorsally and colonize the ventral aspect of the aorta where they establish a population of hemogenic endothelial cells. We have indeed confirmed that hematopoietic potential in the human embryo, as assessed by long-term culture-initiating cell

Address for correspondence: Bruno Péault, Children's Hospital of Pittsburgh of UPMC, 3460 Fifth Avenue, 3302 Rangos Research Center, Pittsburgh, PA 15213. Voice: 412-692-6526; fax: 412-692-5837.

bruno.peault@chp.edu

(LTC-IC) and SCID mouse reconstituting cell (SRC) activities, is confined to BB9-expressing cells. We have further validated these results in the model of human embryonic stem cells (hESCs) in which we have modeled, through the development of hematopoietic embryoid bodies (EBs), primitive and definitive hematopoieses. In this setting, we have documented the emergence of BB9$^+$ hemangioblast-like clonogenic angiohematopoietic progenitors that currently represent the earliest known founders of the human vascular and blood systems.

KEYWORDS: hematopoietic stem cell; embryonic stem cell; hemangioblast; embryo; blood vessel

INTRODUCTION

Emerging hematopoiesis adapts to the fast-changing anatomy of the mammalian embryo, proceeding transiently in the extraembryonic yolk sac, then in the liver before stabilizing in the fetal thymus and bone marrow. In the yolk sac, hematopoietic stem cells (HSCs) and endothelial cells emerge simultaneously from extraembryonic mesoderm, leading to joint formation of blood and blood vessels from a putative common stem cell, the *hemangioblast*.[1,2] It was much later demonstrated that a second phase of *de novo* HSC production takes place within the aorta-gonad-mesonephros (AGM) region of the embryo proper (reviewed in Refs. 3,4). HSCs arising in the AGM sprout from the ventral aspect of the dorsal aorta and vitelline artery, where discrete subsets of vascular endothelial cells exhibit transient blood-forming activity.[5–7] Besides animal models, this developmental sequence has been also closely scrutinized in the human embryo and fetus, in which the abundance of cell lineage markers and robust functional assays have allowed to precisely document incipient angiohematopoiesis from the third week of gestation.[8–16]

Pluripotent mouse embryonic stem cells (mESCs) form cellular clusters termed *embryoid bodies* (EBs) that can also differentiate into hematopoietic and endothelial progenitors and recapitulate embryonic hematopoiesis through a hemangioblast intermediate, even producing yolk sac-like blood islands.[17–19] A clonogenic *hemangioblast* was first characterized as a bipotential yolk sac-type progenitor termed the *blast colony-forming cell* (BL-CFC) derived from mESC[20] and mouse embryos.[21] These VEGF-responsive progenitors arise in yolk sac blood islands and mouse EBs as flk-1/KDR$^+$ (VEGF-R2$^+$) mesoderm cells.[22] The derivation of multipotent human hemangioblasts from human embryonic stem cells (hESCs) has not yet been reported, and remains a sought-after goal, which should allow the unlimited supply of human hematopoietic and vascular cells for transplantation. Since human embryonic and fetal tissues are usually difficult to procure, the basic cellular and genetic mechanisms of human angiohematopoietic cell lineage incipience and development could be investigated using hESC models. We herein report our most recent attempts to understand human developmental hematopoiesis through the parallel exploration of normal embryos and cultured embryonic stem cells. We have notably taken advantage of the identification by some of us of BB9, a novel cell-surface antigen that typifies the most primitive HSCs at adult stages, and a population of candidate hemangioblasts in the emergent human angiohematopoietic system.

BB9, a Novel Marker of Human Early HSCs

In human adult bone marrow, the BB9 antibody exhibits reactivity with a subpopulation of $CD34^+$ cells with the phenotypic characteristics of HSCs, namely low to undetectable levels of CD38, and coexpression of CD90 and CD133.[23] Similarly, BB9 expression is highest on $CD34^+CD133^+CD90^+CD38^-$ cells in fetal liver and umbilical cord blood (UCB). The most direct evidence for BB9 expression by HSCs comes from non-obese diabetic—severe combined immunodeficiency disease (NOD–SCID) mouse transplant studies where $CD34^+BB9^+$ but not $CD34^+BB9^-$ UCB cells successfully engrafted and sustained long-term multilineage haemopoiesis in this most rigorous surrogate assay for human HSC activity (Jokubaitis *et al.*; Pers. Oberv.)

Development of Hematopoiesis in the Human Embryo, from the Splanchnopleura to the Liver Rudiment

As in all other mammals, human hematopoiesis begins in the yolk sac before proceeding to intraembryonic blood-forming organs, upon colonization of the rudiments of the latter by blood-borne HSCs. This pathway set the basis for the prevailing idea that the YS was also the unique provider of HSC, and accordingly no intrinsic hematopoietic potential was observed in any other tissue of the embryo. Evidence for an intraembryonic emergence of hematopoietic progenitors was obtained in birds, mice, and from our own work conducted in the human embryo. In the mouse, the AGM region, and its anlage, the paraaortic splanchnopleura (P-Sp), are endowed with hematopoietic potential.[4] Indeed, cells expressing markers of hematopoietic progenitors are present in the dorsal aorta.[3] In the human embryo, some of us have identified a dense population of $CD34^+$ blood cells adhering to the ventral side of the aortic endothelium,[15,24,25] which display a cell-surface and molecular phenotype typifying primitive hematopoietic progenitors ($CD45^+$, $CD34^+$, $CD31^+$, $CD38^-$, negative for lineage markers, $GATA-2^+$, $GATA-3^+$, $c-myb^+$, $SCL/TAL1^+$, $c-kit^+$, $flk-1/KDR^+$).[8,10,11] Hematopoietic cell clusters appear precisely in the embryo between the 27th and 40th days of development, only 2–3 days before hepatic colonization by $CD34^+$ progenitors.[11] In order to define the origin of these aortic progenitors we have set up a miniaturized *in vitro* system for culturing tissue rudiments dissected from human embryos. In that setting, we detected the existence of a blood-forming potential intrinsic to the human intraaortic splanchnopleura as early as day 19 of development, that is, 2 days before the onset of blood circulation. We further showed that the yolk sac only generates progenitors with limited, myeloid developmental ability, whereas precursors emerging autonomously in the presumptive aortic territory are endowed with multilineage lymphomyeloid potential.[13] These results established that the development of the human blood system is characterized by two independent waves of HSC emergence. The first one occurs in the yolk sac and generates ephemeral progenitors with restricted developmental potential. The second one takes place in the intraembryonic splanchnopleura and gives rise to multipotent HSCs; these intraembryonic progenitors are responsible for the definitive colonization of the liver and therefore are at the origin of all blood cells in the fetus and adult (reviewed in Refs. 15,24,25).

In order to track the earliest forerunners of this intraembryonic blood-forming activity, we have used a novel monoclonal antibody, BB9, which typifies the most primitive human HSCs at adult stages[23] (see also paragraph above). We show that during human development, BB9 is expressed in all blood-forming tissues: P-Sp, yolk sac, fetal liver, and bone marrow. Starting from day 19 of gestation, BB9 is present on cells in yolk sac blood islands. From this stage until 26 days of gestation, BB9 expression identifies rare $CD34^-CD45^-$ cells, in the embryo proper, concentrated in the hemogenic portion of the P-Sp. From the 27th day of gestation, when hematopoiesis occurs inside the embryo, BB9 stains precisely the HSCs that emerge aggregated on the ventral side of the dorsal aorta and surrounding endothelial cells. This pattern is consistent with the hypothesis of a $BB9^+CD34^-CD45^-$ hemangioblastic precursor migrating from the P-Sp toward the ventral aorta, to give rise to $BB9^+CD34^+CD45^+$ hematopoietic progenitors and surrounding $BB9^+CD34^+CD45^-$ endothelial cells. Along this line, preliminary results obtained by culturing cells sorted from 24- to 26-day human embryos indicate that $BB9^+$, but not $BB9^-$ cells, are endowed with hematopoietic ability. Furthermore, functional hematopoiesis assays performed on fetal liver and bone marrow cells have highlighted a direct relationship between expression of BB9 and hematopoietic incipience in these organs, too. Recent functional analyses carried out both *in vitro* and *in vitro* on purified $BB9^+$ cells are in agreement with these observations. Indeed, in fetal liver and bone marrow, long-term culture-initiating cell (LTC-IC) and SCID mouse reconstituting cell (SRC) abilities are restricted to BB9-expressing cells. Altogether, these data define BB9 as a novel marker of very primitive human mesodermal precursors endowed with hemogenic and, possibly, angiogenic potentialities and persists at the surface of human HSCs at embryonic, fetal, and adult stages (Sinka *et al.*, manuscript in preparation).

Hematopoiesis Ontogeny in the Model of hESCs

As mentioned in the paragraphs above, HSCs first emerge during the early weeks of human gestation within the extraembryonic yolk sac, and secondarily within the truncal arteries of the embryo in the AGM region. This second hematopoietic cell wave gives rise to definitive, adult-type HSCs. In both YS blood islands and embryonic aorta, HSCs develop within immediate vicinity of vascular endothelial cells, which secondarily arise from rare mesodermal hemangioblasts. These developmental processes have been elusive, in part because reliable markers of early human hematoendotheliogenesis were lacking. As described in the first paragraph above, we have recently identified a novel marker of undifferentiated HSCs using a new monoclonal antibody, BB9.[23] The pattern of human embryonic BB9 expression is consistent with the hypothesis that CD34⁻ hemangioblasts emigrate dorsally from the p-Sp, and subsequently colonize the ventral aorta to give rise to CD34⁺ hemogenic endothelial cells. Thus, mesodermal BB9⁺CD34⁻ cells may mark the elusive human hemangioblast in the developing human YS and AGM. The rarity of human embryonic tissue makes the detailed analysis of these BB9⁺ AGM/YS progenitors challenging, although such studies are currently in progress in our laboratories. Differentiated hESCs may provide an alternative, more abundant supply of these progenitors for detailed biologic study.

Differentiation of mESCs[17,18] and hESCs[26–29] in suspension culture results in the spontaneous formation of cellular clusters termed *EBs*, containing ectoderm, endoderm, and mesoderm (including hematopoietic progenitors).[30–33] We and others[33–36] have demonstrated hematopoietic differentiation from hESCs. Based on the sequential expression of hematoendothelial genes, surface markers, and emergence of primitive and definitive CFC, our group first demonstrated that the human embryoid body (hEB) differentiation system mimics the early phases of human yolk sac blood development with mesodermal-hemangioblast differentiation, followed by primitive/definitive erythromyeloid hematopoiesis.[16,36] Despite the potential for generating adult-type, transplantable HSCs from hESCs, this goal has thus far been elusive. Our group has suggested that this failure is due to the fact that hESCs produce blood similar to that found in the human yolk sac, and is therefore too immature to provide adult-type hematopoietic reconstitution. A detailed developmental biologic approach will be necessary for understanding how to further mature these yolk sac-like progenitors into adult-type transplantable HSCs.

Our main hypothesis is that mesoderm commitment to hemangioblasts occurs during this hEB differentiation. These hemangioblasts represent common progenitors for human primitive and definitive hematopoiesis, and also vascular endothelium. We have recently improved our original methods[36] for differentiating hEBs into hematopoietic progenitors using culture conditions supplemented with the mesodermal morphogens BMP4 and VEGF (Zambidis *et al.*, manuscript in preparation). Kinetic analysis of differentiating hEB cells generates hematopoietic progenitors representing primitive and definitive hematopoiesis over a 4-week time course. CFC assays with hematopoietic growth factors (SCF, IL-3, IL-6, GM-CSF, G-CSF, erythropoietin) of day 7–12 hEB cells generate organized, mesodermal-hematoendothelial (MHE) colonies, and a robust wave of primitive and definitive erythropoiesis correlated with increased expression of *SCL/TAL1, GATA1, GATA2, EKLF, PU.1,* and *CDX4*. Increases in mRNA levels of these transcription factors coincided with a similarly increasing expression of the hematoendothelial cell markers *CD31, CD34,* and *KDR/flk-1*.

Interestingly, MHE clusters bear striking resemblance to classic histologic descriptions of normal intravascular human yolk sac.[37] Erythroid progenitors arose in two waves: a primitive wave characterized by CFU with scant adult hemoglobin (HbA) but abundant embryonic and fetal hemoglobins (HbE, HbF), and a definitive wave characterized by CFU expressing increased amounts of HbA, and decreased amounts of embryonic and fetal globins.

Expression levels for hematopoietic genes and markers peaked at days 6–10 of hEB differentiation, suggesting that a putative coordinated developmental watershed event in hEB hematoendothelial commitment occurs during this time, and predicts the genesis of a hemangioblast at this stage. This hypothesis is corroborated by the burst of *CDX4* expression accompanying SCL/TAL1 peak expression at day 9 of hEB development, which in turn coincided with the peak of primitive erythroid CFC and subsequent emergence of definitive CFC. Using novel serum-free methodologies, we revealed this progenitor after developing an *in vitro* hemangioblastic blast-colony-forming cell (BL-CFC) assay that detects single clonogenic hEB cells, which could secondarily give rise to adherent endothelial, or primitive and definitive hematopoietic progenitors. Under either serum-free or OP9 bone marrow stromal co-cultures, we demonstrated that BL-CFC colonies contain a common progenitor for both primitive and definitive erythromyelopopiesis and also adherent endothelial cells. Hemangioblastic BL-CFC progenitor formation is correlated directly with the onset of BB9 hEB expression, but prior to the onset of CD34 expression. FACS analysis of isolated, pooled BL-CFC colonies revealed that they are constituted of $BB9^+CD34^{-/lo}CD45^-$ hEB cells.

To further test the definitive hematopoietic potential of FACS-purified $BB9^+CD34^+CD45^-$ or $BB9^+CD43^+CD45^-$ hEB cells, we co-cultured sorted cells on a fetal liver stromal line in the presence of early-acting hematopoietic growth factors (Flt3L, TPO, SCF, IL-6). Sorted hEB cells quickly expanded into mononuclear, cobblestone-forming blasts under these conditions and produced a $CD34^+CD45^+CD38^-$ phenotype. CFC analysis of these cells revealed robust GEMM-CFC, GM-CFC, G-CFC, and M-CFC potentials that were indistinguishable from those of CD34-positive-enriched cord blood CFC controls. $BB9^+CD34^+$ hEB cells also gave rise to definitive-type $CD56^+$ putative NK cells and $CD19^+$ B lymphocytes on OP9 stromal layers. These data collectively suggest that $BB9^+$ hEB cells contain the elusive common hemangioblastic progenitor for endothelium, primitive yolk sac hematopoiesis, and most importantly, can further mature into definitive

AGM-type $CD34^+CD45^+CD38^-$ progenitors after exposure to a stromal microenvironmental niche.

CONCLUSION

Few concepts in developmental biology are at the same time as ancient and as minimally documented, experimentally, than that of an angiohematopoietic stem cell, or hemangioblast. Whereas hemangioblast-like progenitors have been typified with markers among mouse-dissociated embryonic stem cells or extraembryonic tissues, the clonal development of blood islands and surrounding endothelial cells in the yolk sac has been recently disputed, using multiple color tags to trace the progeny of individual cells.[38] While it may appear presumptuous to explore on human cells and tissues the existence of a developmental intermediate that has not yet been fully characterized in the convenient mouse model, we have pursued the study of the incipient human angiohematopoietic system on the following grounds: (a) we have access to human embryos from the third week of development; (b) a large panoply of reliable markers of human vascular cells and blood cells is available; (c) human angiohematopoiesis can be tested functionally in diverse culture and *in vivo* systems; and (d) hESCs can provide robust surrogate models of human histogenesis. We have, indeed, faithfully reproduced primitive and definitive human hematopoiesis in embryonic stem cell-derived EBs, and confirmed in this setting that hematopoietic cells emerge from blood-forming endothelium, a conclusion reached previously using progenitor cells sorted from normal human embryos. The relevance of the embryonic stem cell model to normal development is further illustrated by our most recent studies of the BB9 cell-surface antigen, which at this point appears as a promising candidate marker of human hemangioblasts. This assumption was first based on the anatomic distribution, from the yolk sac and splanchnopleura to the aortic hemogenic endothelium and derived HSCs, and functional properties of BB9-expressing cells. However, we are currently uncovering that BB9 also marks emerging clonogenic angiohematopoietic progenitor cells in hEBs. If confirmed by clonal analyses currently in progress, these experiments could reveal the existence of the first univocal hemangioblast marker.

ACKNOWLEDGMENTS

This work was supported in part by grants from the University of Pittsburgh and McGowan Institute for Regenerative Medicine, by grant Avenir from Inserm, and by grant 4814 from the Association pour la Recherche sur le Cancer to M.T. E.Z. is supported by NIH grant K08 HL077595. We thank Rose Perry for her assistance with the preparation of this report.

REFERENCES

1. SABIN, F.R. 1920. Studies on the origin of blood vessels and of red blood corpuscles as seen in the living blastoderm of chicks during the second day of incubation. Carnegie Inst. Wash. Pub. n°272, Contrib. Embryol. **9:** 214.
2. MURRAY, P.D.F. 1932. The development '*in vitro*' of blood of the early chick embryo. Proc. Roy. Soc. Lond. **11:** 497–521.
3. GODIN, I. & A. CUMANO. 2002. The hare and the tortoise: an embryonic haematopoietic race. Nat. Rev. Immunol. **2:** 593–604.
4. DZIERZAK, E. 2003. Ontogenic emergence of definitive hematopoietic stem cells. Curr. Opin. Hematol. **10:** 229.
5. JAFFREDO, T., R. GAUTIER, A. EICHMANN & F. DIETERLEN-LIEVRE. 1998. Intraaortic hemopoietic cells are derived from endothelial cells during ontogeny. Development **125:** 4575–4583.
6. NISHIKAWA, S.I., S. NISHIKAWA, H. KAWAMOTO, *et al.* 1998. *In vitro* generation of lymphohematopoietic cells from endothelial cells purified from murine embryos. Immunity **8:** 761–769.
7. NORTH, T.E., M.F. DE BRUIJN, T. STACY, *et al.* 2002. Runx1 expression marks long-term repopulating hematopoietic stem cells in the midgestation mouse embryo. Immunity **16:** 661–672.
8. TAVIAN, M., L. COULOMBEL, D. LUTON, *et al.* 1996. Aorta-associated CD34 +hematopoietic cells in the early human embryo. Blood **87:** 67–72.
9. CHARBORD, P., M. TAVIAN, L. HUMEAU & B. PÉAULT. 1996. Early ontogeny of the human marrow from long bones: an immunohistochemical study of hematopoiesis and its microenvironment. Blood **87:** 4109–4119.
10. LABASTIE, M.C., F. CORTÉS, P.H. ROMÉO, *et al.* 1998. Molecular identity of hematopoietic precursor cells emerging in the human embryo. Blood **92:** 3624–3635.
11. TAVIAN, M., M.F. HALLAIS & B. PÉAULT. 1999. Emergence of intraembryonic hematopoietic precursors in the pre-liver human embryo. Development **126:** 793–803.
12. CORTÉS, F., C. DEBACKER, B. PÉAULT & MC. LABASTIE. 1999. Differential expression of KDR/VEGFR-2 and CD34 during mesoderm development of the early human embryo. Mech. Dev. **83:** 161–164.
13. TAVIAN, M., C. ROBIN, L. COULOMBEL, B. PÉAULT. 2001. The human embryo, but not its yolk sac, generates lympho-myeloid stem cells. mapping multipotent hematopoietic cell fate in intraembryonic mesoderm. Immunity **15:** 487–495.
14. OBERLIN, E., M. TAVIAN, I. BLAZSEK & B. PÉAULT. 2002. Blood-forming potential of vascular endothelium in the human embryo. Development **129:** 4147–4157.
15. TAVIAN, M. & B. PÉAULT. 2005. Embryonic development of the human hematopoietic system. Int. J. Dev. Biol. **49:** 243–250.
16. ZAMBIDIS, E.T., E. OBERLIN, M. TAVIAN & B. PÉAULT. 2006. Blood-forming endothelium in human ontogeny: lessons from *in utero* developmental and embryonic stem cell culture. Trends Cardiovasc Med. **16**(3): 95–101.
17. WILES, M.V., & G. KELLER. 1991. Multiple hematopoietic lineages develop from embryonic stem (ES) cells in culture. Development **111:** 259–267.

18. KELLER, G., M. KENNEDY, T. PAPAYANNOPOULOU & M.V. WILES. 1993. Hematopoietic commitment during embryonic stem cell differentiation in culture. Mol. Cell. Biol. **13:** 473–486.
19. VITTET, D., M.H. PRANDINI, R. BERTHIER, et al. 1996. Embryonic stem cells differentiate in vitro to endothelial cells through successive maturation steps. Blood **88:** 3424–3431.
20. CHOI, K., M. KENNEDY, A. KAZAROV, et al. 1998. A common precursor for hematopoietic and endothelial cells. Development **125:** 725–732.
21. HUBER, T.L., V. KOUSKOFF, FEHLING et al. 2004. Haemangioblast commitment is initiated in the primitive streak of the mouse embryo. Nature **432:** 625–630.
22. KABRUN, N., H.J. BUHRING, K. CHOI, et al. 1997. Flk-1 expression defines a population of early embryonic hematopoietic precursors. Development **124:** 2039–2048.
23. RAMSHAW, H.S., D. HAYLOCK, B. SWART, et al. 2001. Monoclonal antibody BB9 raised against bone marrow stromal cells identifies a cell-surface glycoprotein expressed by primitive human hemopoietic progenitors. Exp. Hematol. **29:** 981–992.
24. PÉAULT, B. & M. TAVIAN. 2003. Hematopoietic stem cell emergence in the human embryo and fetus. Ann. N. Y. Acad. Sci. **996:** 132–140.
25. TAVIAN, M. & B. PÉAULT. 2005. The changing cellular environments of hematopoiesis in human development in utero. Exp. Hematol. **33:** 1062–1069.
26. THOMSON, J.A., J. ITSKOVITZ-ELDOR, S.S. SHAPIRO, et al. 1998. Embryonic stem cell lines derived from human blastocysts. Science **282:** 1145–1147.
27. REUBINOFF, B.E., M.F. PERA, C.Y. FONG, et al. 2000. Embryonic stem cell lines from human blastocysts: somatic differentiation in vitro. Nat. Biotechnol. **18:** 399–404.
28. AMIT, M., M.K. CARPENTER, M.S. INOKUMA, et al. 2000. Clonally derived human embryonic stem cell lines maintain pluripotency and proliferative potential for prolonged periods of culture. Dev. Biol. **227:** 271–278.
29. PERA, M.F., B. REUBINOFF & A. TROUNSON. 2000. Human embryonic stem cells. J. Cell. Sci. **113:** 5–10.
30. ITSKOVITZ-ELDOR, J., M. SCHULDINER, D. KARSENTI, et al. 2000. Differentiation of human embryonic stem cells into embryoid bodies comprising the three embryonic germ layers. Mol. Med. **6:** 88–95.
31. ZHANG, S.C., M. WERNIG, I.D. DUNCAN, et al. In vitro differentiation of transplantable neural precursors from human embryonic stem cells. Nat. Biotechnol. **19:** 1129–1133.
32. KEHAT, I., D. KENYAGIN-KARSENTI, M. SNIR, et al. 2001. Human embryonic stem cells can differentiate into myocytes with structural and functional properties of cardiomyocytes. J. Clin. Invest. **108:** 407–414.
33. KAUFMAN, D.S. & JA. THOMSON. 2002. Human ES cells-haematopoiesis and transplantation strategies. J. Anat. **200:** 243–248.
34. KAUFMAN, D.S., E.T. HANSON, R.L. LEWIS, et al. 2001. Hematopoietic colony-forming cells derived from human embryonic stem cells. PNAS **98:** 10716–10721.
35. CHADWICK, K., L. WANG, L. LI, et al. 2003. Cytokines and BMP-4 promote hematopoietic differentiation of human embryonic stem cells. Blood **102:** 906–915.

36. ZAMBIDIS, E.T., B. PÉAULT, T.S. PARK, *et al.* 2005. Hematopoietic differentiation of human embryonic stem cells progresses through sequential hemato-endothelial, primitive, and definitive stages resembling human yolk sac development. Blood **106:** 860–870.
37. BLOOM, W. & G.W. BARTELMEZ. 1940. Hematopoiesis in young human embryos. Am. J. Anat. **67:** 21–53.
38. UENO, H. & I.L. WEISSMAN. 2006. Clonal analysis of mouse development reveals a polyclonal origin for yolk sac blood islands. Dev. Cell **11:** 519–533.

Epigenetic Control of Hematopoietic Stem Cell Aging

The Case of Ezh2

GERALD DE HAAN AND ALICE GERRITS

Department of Cell Biology, Section Stem Cell Biology, University Medical Center Groningen, University of Groningen, 9713 AV Groningen, the Netherlands

ABSTRACT: **Hematopoietic stem cells have potent, but not unlimited, selfrenewal potential. The mechanisms that restrict selfrenewal are likely to play a role during aging. Recent data suggest that the regulation of histone modifications by Polycomb group genes may be of crucial relevance to balance selfrenewal and aging. We provide evidence for the involvement of one of these Polycomb group genes, Ezh2, in aging of the hematopoietic stem cell system.**

KEYWORDS: **stem cells; aging; polycomb; Ezh2; senescence**

The molecular mechanisms underlying cellular aging in mammals have remained largely unknown. Ever since Hayflick described his seminal experiments in which it was shown that the proliferative potential of normal mammalian cells is limited,[1] the process of cellular senescence and its potential relevance for *in vivo* aging has been extensively debated.[2,3] Over the last two decades ample speculations have been put forward as to how relevant these *in vitro* observations are for *in vivo* cell turnover. It is evident that during aging the integrity and functioning of a wide variety of tissues are gradually deteriorating, ultimately resulting in a spectrum of disease-initiating events. A special case of such tissue dysfunctioning is the development of cancer, including leukemia, which is a most prominent age-related disorder. Most notably, the issue has been raised whether cellular senescence is an obligatory step in tumor formation. If this were the case, one would expect to observe an increased frequency of senescent cells in precancerous/premalignant lesions. Some of these cells would somehow escape the stringent antiproliferative pressure and these clones initiate the full-blown tumor. A second prediction would be that delaying senescence leads to proliferative disorders, increasing the pool of

Address for correspondence: Gerald de Haan, Department of Cell Biology, Section Stem Cell Biology, University Medical Center Groningen, University of Groningen, Antonius Deusinglaan 1, 9713 AV Groningen, the Netherlands. Voice: +31503632722; fax: +31503637445.
g.de.haan@med.umcg.nl

actively cycling cells. This large population of proliferating cells is then likely to be vulnerable to tumorigenesis.

Recently, a series of papers using a variety of *in vivo* tumor models have clearly indicated that tumor formation is indeed preceded by a phase of cellular senescence. For example, it was shown that oncogenic Ras initially induces senescent features in relatively benign adenomas. These senescent markers disappeared when some of these premalignant lesions progressed into full-blown adenocarcinomas.[4] Similar findings on senescent markers were reported in nonlethal cancers that developed upon Pten inactivation in the prostate.[5] Only in the absence of p53, the senescent phenotype was overcome and tumors progressed rapidly. Of particular relevance is the finding that a deficiency in the histone methyltransferase Suv39h1 accelerates the development of lymphomas after transduction of hematopoietic cells with senescence-inducing oncogenic Ras.[6]

We propose that somatic stem cells, which can be found in most adult tissues, play a crucial role in the aging process. Somatic stem cells are characteristically defined by their ability to self-renew at the single cell level.[7–9] If self-renewal of adult stem cells is impaired during aging, tissue homeostasis cannot be maintained. In contrast, if self-renewal of adult stem cells is increased beyond normal levels, excessive cell proliferation, potentially resulting in tumor formation will follow (FIG. 1).

It is not easy to define the elusive concept of senescence/aging as it relates to hematopoietic stem cells. Senescence is generally described as a process of irreversible proliferative arrest.[10] There is a strong genetic component to this phenomenon, and cells from mice senesce much faster, but also escape from senescence much easier than human cells. A description of the molecular events leading to senescence is far from complete, and is undoubtedly complex. Essentially, cellular stress induced by a variety of events results in the activation of the p53 and p16/Rb pathways, proteins that possess potent tumor-suppressive activity as they suppress proliferation.[11] As hematopoietic stem cells are typically quantified using functional *in vitro* and *in vivo* assays, senescent stem cells can never be detected in any routine assay as they would not produce offspring. Therefore, in the strictest sense of the word, a senescent stem cell ceases to be a stem cell, and by definition cannot exist. For the purpose of this article, we define stem cell aging/senescence operationally as a loss of stem cell functioning and/or quality with time.

Hematopoietic stem cells at the single level can reconstitute the entire blood-forming tissue after transplantation into an appropriately conditioned recipient.[7,8,12–14] Self-renewal of hematopoietic stem cells is most convincingly demonstrated by their ability to regenerate hematopoiesis after repeated serial transplantations. Clonal analyses have shown that unique stem cell clones can persist to function after serial transplantation.[15] Also, using such a serial transplantation approach, it has been demonstrated that hematopoietic stem cells can outlive their original donor.[16] This would suggest that hematopoietic

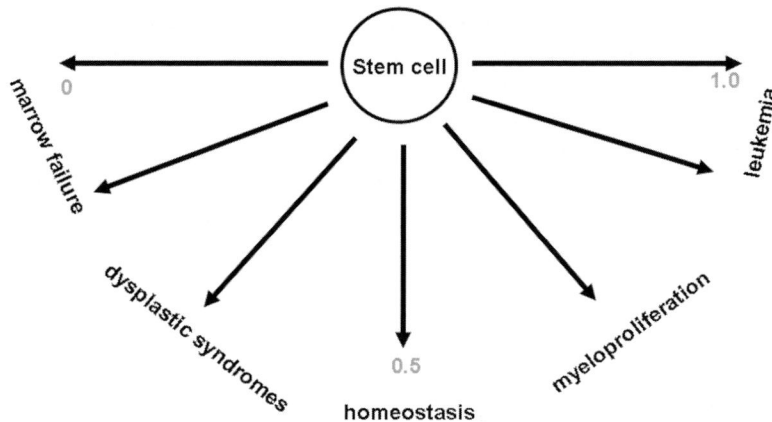

FIGURE 1. Selfrenewal probabilities of hematopoietic stem cells are carefully regulated. During normal homeostasis this probability is 0.5, indicating that on average half of the progeny of the stem cell compartment continues to possess stem cell characteristics, whereas the other half becomes differentiated. If the selfrenewal probability decreases to values lower than 0.5, the inevitable outcome is marrow failure and exhaustion of the system. In contrast, if the selfrenewal probability is higher than 0.5, hyperproliferation is the result.

stem cells do not age. However, a large body of experimental evidence clearly indicates that upon serial transplant, the quality of the transplanted stem cells to repopulate the hematopoietic system decreases substantially.[17–20] Similarly, stem cells isolated from fetal tissue are much more potent than those from adults, and stem cells from young adults are more potent than those from old donors.[21,22]

Why old (in terms of proliferative history) stem cells cease to function as they did when they first developed in the embryo remains completely unknown. It has been argued that accumulating DNA mutations, including loss of telomeric repeats, affect stem cell functioning.[23] In agreement, stem cells that lack an appropriate DNA repair machinery typically perform quite poorly.[24] Genes associated with Fanconi anemia, a disease characterized by life-threatening bone marrow failure, all play an important role in the DNA machinery.[25] In addition, telomerase deficient animals show pronounced phenotypes in rapidly proliferating tissues, including bone marrow.[26] Patients suffering from dyskeratosis congenita, who are essentially telomerase deficient, display major bone marrow insufficiencies.[27] In contrast, overexpression of telomerase does not protect from hematopoietic stem cell exhaustion, although it does maintain telomere length.[28] To what extent accumulating DNA mutations occur in stem cells and how this could potentially affect normal aging of the hematopoietic system is unknown.

In an extensive series of experiments we and others have shown that in mice hematopoietic stem cells display a marked reduction in quantity and/or quality during aging.[17,29–33] However, hematopoietic stem cells from some strains of inbred mice showed much more obvious aging phenotypes than those from other strains. We reported that the rate of hematopoietic cell cycling among eight different inbred strains of mice correlated negatively with their mean life span.[30] Mouse strains with high hematopoietic cell turnover (including AKR/J and DBA/2) had a shorter mean life span than strains with slowly cycling cells (such as C57BL/6 and 129/SvJ). A potential explanation of this phenomenon could be that increased cell turnover (which we measured in the hematopoietic system, but which may be prevalent in all mitotic tissues) is associated with more rapid decline of stem cell functioning, which would negatively impact on organismal life span. Initially, our studies were focused on a description of hematopoietic cell turnover during normal aging, primarily to assess whether stem cell aging is controlled by cell-intrinsic or cell-extrinsic factors.[29,30,34,35] All experiments indicated that the decrease of stem cell quantity or quality was due to cell-intrinsic genetic or epigenetic factors. As marked differences were observed in hematopoietic cell cycling kinetics among different inbred strains of mice, we embarked on a screen to detect the genomic loci that were associated with variation in stem cell turnover and stem cell numbers. We phenotyped a large panel of recombinant inbred BXD (offspring of C57BL/6 and DBA/2 strains) and found strong evidence of a locus on chromosome 11, which regulated cell turnover,[36] and a locus on chromosome 18 involved in stem cell frequency.[34]

To identify the causative genes, we have recently performed an extensive "genetical genomics" profiling study (stem cell samples from 30 different BXD strains × ~12,000 probesets).[37] Genetical genomics combines the power of detailed gene expression patterns of thousands of genes with sophisticated quantitative trait genetic analyses using fully genotyped samples.[38–41] Putative transcriptional networks become immediately visible in an unbiased way that does not require any prior biological information. We found a large cluster of polymorphic and differentially expressed genes on chromosome 11 that is associated with variation in cell cycling.[42] Similarly, we searched for genes that were transcriptionally regulated by the chromosome 18 locus that we had shown to increase stem cell frequency. Interestingly, we demonstrated that *Enhancer of Zeste-2* (*Ezh2*), a Polycomb group protein (PcG) involved in histone methylation and deacetylation,[43] was highly differentially expressed among the 30 strains and its transcript abundance was controlled by the chromosome 18 locus. *Ezh2* expression levels were high in primitive hematopoietic stem cells, and rapidly decreased upon differentiation.[44] We constructed retroviral vectors to assess the consequences of enforced overexpression of *Ezh2* in stem cells. Strikingly, whereas normal hematopoietic stem cells rapidly cease to function after repeated serial transplantation, *Ezh2* overexpression completely prevented exhaustion of long-term repopulating potential.[44] Even after three serial transplantations, a regime that normally eliminates essentially all stem cell activity, *Ezh2* overexpressing stem cells reconstituted lethally irradiated recipients just fine. This qualifies Ezh2 as a stem cell senescence-preventing gene, and indeed, it is the first gene that has been described to have such potential.

Our observation that enforced overexpression of the chromatin modifying gene Ezh2 results in prevention of stem cell exhaustion, provides a novel insight into the molecular pathways regulating this process. To date, two distinct Polycomb repressive complexes, (PRC) 1 and 2, have been identified that play a major role in chromatin remodeling. Mammalian PRC1 contains Cbx, Mph, Ring, Bmi1, and Mel18 and is thought to be important in the maintenance of gene repression. The second complex, PRC2, contains Ezh2, Eed, and Su(z)12 and is thought to be involved in the initiation of gene repression by methylating lysine 27 on Histone H3, thus leading to compaction of chromatin. PcG complexes are targeted to cis-regulatory Polycomb response elements (PREs) by DNA-binding transcription factors.[43] We hypothesize that chromatin modifications play a key role in maintaining a stem cell-specific transcriptome. During self-renewing cell replication complex chromatin modifications must be faithfully copied to both daughter cells. It appears very plausible that minor changes in the copied chromatin structure are associated with stem cell turnover during aging, resulting in a stem cell-incompatible transcriptome, which is reflected as stem cell senescence. This senescence program safeguards stem cells from malignant transformation induced by oncogenic events. It remains possible that potentially oncogenic events (i.e., mutations) result in malignancies only in "permissive" cells in which the senescent program is less stringently maintained.

Our data document that potent stem cell senescence-preventing genes do exist. It is likely that the identification and functional characterization of such genes will provide important information on the molecular mechanisms that affect cellular aging, an area of research of which our knowledge is rudimentary at best. It is evident that modulation of senescence-preventing genes will be required to fully exploit the use of stem cells in future regenerative therapies.

REFERENCES

1. HAYFLICK, L. & P. MOORHEAD. 1961. The serial cultivation of human diploid cell strains. Exp. Cell Res. **25**: 585–621.
2. WRIGHT, W.E. & J.W. SHAY. 2001. Cellular senescence as a tumor-protection mechanism: the essential role of counting. Curr. Opin. Genet. Dev. **11**: 98–103.
3. CAMPISI, J. 2005. Senescent cells, tumor suppression, and organismal aging: good citizens, bad neighbors. Cell **120**: 513–522.
4. COLLADO, M., J. GIL, A. EFEYAN, *et al.* 2005. Tumour biology: senescence in premalignant tumours. Nature **436**: 642.

5. CHEN, Z., L.C. TROTMAN, D. SHAFFER, et al. 2005. Crucial role of p53-dependent cellular senescence in suppression of Pten-deficient tumorigenesis. Nature **436**: 725–730.
6. BRAIG, M., S. LEE, C. LODDENKEMPER, et al. 2005. Oncogene-induced senescence as an initial barrier in lymphoma development. Nature **436**: 660–665.
7. EMA, H., K. SUDO, J. SEITA, et al. 2005. Quantification of self-renewal capacity in single hematopoietic stem cells from normal and Lnk-deficient mice. Dev. Cell **8**: 907–914.
8. OSAWA, M., K. HANADA, H. HAMADA & H. NAKAUCHI. 1996. Long-term lympho-hematopoietic reconstitution by a single CD34-low/negative hematopoietic stem cell. Science **273**: 242–245.
9. POTTEN, C.S. & M. LOEFFLER. 1990. Stem cells: attributes, cycles, spirals, pitfalls and uncertainties. Lessons for and from the crypt. Development **110**: 1001–1020.
10. CAMPISI, J. 1997. The biology of replicative senescence. Eur. J. Cancer **33**: 703–709.
11. BEAUSEJOUR, C.M., A. KRTOLICA, F. GALIMI, et al. 2003. Reversal of human cellular senescence: roles of the p53 and p16 pathways. EMBO J. **22**: 4212–4222.
12. DYKSTRA, B., J. RAMUNAS, D. KENT, et al. 2006. High-resolution video monitoring of hematopoietic stem cells cultured in single-cell arrays identifies new features of self-renewal. Proc. Natl. Acad. Sci. USA **103**: 8185–8190.
13. SIEBURG, H.B., R.H. CHO, B. DYKSTRA, et al. 2006. The hematopoietic stem compartment consists of a limited number of discrete stem cell subsets. Blood **107**: 2311–2316.
14. CAMARGO, F.D., S.M. CHAMBERS, E. DREW, et al. 2006. Hematopoietic stem cells do not engraft with absolute efficiencies. Blood **107**: 501–507.
15. JORDAN, C.T. & I.R. LEMISCHKA. 1990. Clonal and systemic analysis of long-term hematopoiesis in the mouse. Genes Dev. **4**: 220–232.
16. HARRISON, D.E. 1972. Normal function of transplanted mouse erythrocyte precursors for 21 months beyond donor life spans. Nat. New Biol. **237**: 220–222.
17. CHEN, J., C.M. ASTLE & D.E. HARRISON. 1999. Development and aging of primitive hematopoietic stem cells in BALB/cBy mice. Exp. Hematol. **27**: 928–935.
18. VAN ZANT, G., B.P. HOLLAND, P.W. ELDRIDGE & J.J. CHEN. 1990. Genotype-restricted growth and aging patterns in hematopoietic stem cell populations of allophenic mice. J. Exp. Med. **171**: 1547–1565.
19. MAUCH, P., L.E. BOTNICK, E.C. HANNON, et al. 1982. Decline in bone marrow proliferative capacity as a function of age. Blood **60**: 245–252.
20. KAMMINGA, L.M., R. VAN OS, A. AUSEMA, et al. 2005. Impaired hematopoietic stem cell functioning after serial transplantation and during normal aging. Stem Cells **23**: 82–92.
21. REBEL, V.I., C.L. MILLER, C.J. EAVES & P.M. LANSDORP. 1996. The repopulation potential of fetal liver hematopoietic stem cells in mice exceeds that of their liver adult bone marrow counterparts. Blood **87**: 3500–3507.
22. HARRISON, D.E., R.K. ZHONG, C.T. JORDAN, et al. 1997. Relative to adult marrow, fetal liver repopulates nearly five times more effectively long-term than short-term. Exp. Hematol. **25**: 293–297.
23. GEIGER, H. & G. VAN ZANT. 2002. The aging of lympho-hematopoietic stem cells. Nat. Immunol **3**: 329–333.
24. PRASHER, J.M., A.S. LALAI, C. HEIJMANS-ANTONISSEN, et al. 2005. Reduced hematopoietic reserves in DNA interstrand crosslink repair-deficient Ercc1-/- mice. EMBO J. **24**: 861–871.

25. BLOM, E., H.J. VAN DE VRUGT, J.P. DE WINTER, *et al.* 2002. Evolutionary clues to the molecular function of fanconi anemia genes. Acta Haematol. **108:** 231–236.
26. LEE, H.W., M.A. BLASCO, G.J. GOTTLIEB, *et al.* 1998. Essential role of mouse telomerase in highly proliferative organs. Nature **392:** 569–574.
27. MITCHELL, J.R., E. WOOD & K. COLLINS. 1999. A telomerase component is defective in the human disease dyskeratosis congenita. Nature **402:** 551–555.
28. ALLSOPP, R.C., G.B. MORIN, J.W. HORNER, *et al.* 2003. Effect of TERT overexpression on the long-term transplantation capacity of hematopoietic stem cells. Nat. Med. **9:** 369–371.
29. DE HAAN, G. & G. VAN ZANT. 1999. Dynamic changes in mouse hematopoietic stem cell numbers during aging. Blood **93:** 3294–3301.
30. DE HAAN, G., W. NIJHOF & G. VAN ZANT. 1997. Mouse strain-dependent changes in frequency and proliferation of hematopoietic stem cells during aging: correlation between lifespan and cycling activity. Blood **89:** 1543–1550.
31. HARRISON, D.E. & C.M. ASTLE. 1982. Loss of stem cell repopulating ability upon transplantation. Effects of donor age, cell number, and transplantation procedure. J. Exp. Med. **156:** 1767–1779.
32. SUDO, K., H. EMA, Y. MORITA & H. NAKAUCHI. 2000. Age-associated characteristics of murine hematopoietic stem cells. J. Exp. Med. **192:** 1273–1280.
33. MORRISON, S.J., A.M. WANDYCZ, K. AKASHI, *et al.* 1996. The aging of hematopoietic stem cells. Nat. Med. **2:** 1011–1016.
34. DE HAAN, G. & G. VAN ZANT 1997. Intrinsic and extrinsic control of hemopoietic stem cell numbers: mapping of a stem cell gene. J. Exp. Med. **186:** 529–536.
35. KAMMINGA, L.M., I. AKKERMAN, E. WEERSING, *et al.* 2000. Autonomous behavior of hematopoietic stem cells. Exp. Hematol. **28:** 1451–1459.
36. DE HAAN, G. & G. VAN ZANT. 1999. Genetic analysis of hemopoietic cell cycling in mice suggests its involvement in organismal life span. FASEB J. **13:** 707–713.
37. BYSTRYKH, L., E. WEERSING, B. DONTJE, *et al.* 2005. Uncovering regulatory pathways that affect hematopoietic stem cell function using 'genetical genomics'. Nat. Genet. **37:** 225–232.
38. JANSEN, R.C. & J. NAP. 2001. Genetical genomics: the added value from segregation. Trends Genet. **17:** 388–391.
39. JANSEN, R.C. 2003. Studying complex biological systems using multifactorial perturbation. Nat. Rev. Genet. **4:** 145–151.
40. BREM, R.B., G. YVERT, R. CLINTON & L. KRUGLYAK. 2002. Genetic dissection of transcriptional regulation in budding yeast. Science **296:** 752–755.
41. SCHADT, E.E., S.A. MONKS, T.A. DRAKE, *et al.* 2003. Genetics of gene expression surveyed in maize, mouse and man. Nature **422:** 297–302.
42. DE HAAN, G., L.V. BYSTRYKH, E. WEERSING, *et al.* 2002. A genetic and genomic analysis identifies a cluster of genes associated with hematopoietic cell turnover. Blood **100:** 2056–2062.
43. LUND, A.H. & M. VAN LOHUIZEN. 2004. Polycomb complexes and silencing mechanisms. Curr. Opin. Cell. Biol. **16:** 239–246.
44. KAMMINGA, L.M., L.V. BYSTRYKH, A. DE BOER, *et al.* 2006. The Polycomb group gene Ezh2 prevents hematopoietic stem cell exhaustion. Blood **107:** 2170–2179.

Telomere Length in Human Natural Killer Cell Subsets

QIN OUYANG,[a,b] GABRIELA BAERLOCHER,[a,c] IRMA VULTO,[a] AND PETER M. LANSDORP[a,d]

[a]*Terry Fox Laboratory, BC Cancer Research Centre, Vancouver, British Columbia, Canada V5Z 1L3*

[b]*Child and Family Research Institute, University of British Columbia, Vancouver, British Columbia, Canada*

[c]*Department of Hematology, University of Bern, 3010 Bern, Switzerland*

[d]*Department of Medicine, University of British Columbia, Vancouver, British Columbia, Canada V5Z 4E3*

> ABSTRACT: Natural killer (NK) cells are cytotoxic cells that play a critical role in the innate immune response against infections and tumors. In the elderly, the cytotoxic function of NK cells is often compromised. Telomeres progressively shorten with each cell division and with age in most somatic cells eventually leading to chromosomal instability and cellular senescence. We studied the telomere length in NK cell subsets isolated from peripheral blood using "flow FISH," a method in which the hybridization of telomere probe in cells of interest is measured relative to internal controls in the same tube. We found that the average telomere length in human NK cells decreased with age as was previously found for human T lymphocytes. Separation of adult NK cells based on CD56 and CD16 expression revealed that the telomere length was significantly shorter in $CD56^{dim}CD16^+$ (mature) NK cells compared to $CD56^{bright}CD16^-$ (immature) NK cells from the same donor. Furthermore, sorting of NK cells based on expression of activation markers, such as NKG2D and LFA-1, revealed that NK cells expressing these markers have significantly shorter telomeres. Telomere fluorescence was very heterogeneous in NK cells expressing CD94, killer inhibitory receptor (KIR), NKG2A, or CD161. Our observations indicate that telomeric DNA in NK cells is lost with cell division and with age similar to what has been observed for most other hematopoietic cells. Telomere attrition in NK cells is a plausible cause for diminished NK cell function in the elderly.
>
> KEYWORDS: telomere length; NK cells; flow cytometry; flow FISH

Address for correspondence: Dr. Peter M. Lansdorp, Terry Fox Laboratory, BC Cancer Agency, 675 West 10th Avenue, Vancouver, BC, Canada, V5Z 1L3. Voice: 604-675-8135; fax: 604-877-0712.
plansdor@bccrc.ca

INTRODUCTION

Telomeres are specialized structures at the end of chromosomes, which consist of repetitive TTAGGG sequences and a number of associated proteins.[1,2] The length of telomere repeats varies between chromosomes and between species. In most human cells, the length of telomere repeats is remarkably heterogeneous[3] ranging from a few to 20 kb depending on the tissue type, the age of the donor, and the replicative history of the cells. The length of telomere repeats was found to gradually decrease with cell divisions (50–200 bp per cell division in normal somatic cells) leading to chromosomal instability and cellular senescence upon replication *in vitro* and with aging *in vivo*.[4,5] Human telomerase is a reverse transcriptase enzyme that has the ability to extend telomeres by adding single-stranded telomeric DNA to the 3' ends of chromosomes.[6,7] Transfection of the telomerase reverse transcriptase gene into various cells can result in the elongation of telomere length and extension of the *in vitro* replicative life span.[8–13]

A correlation between telomere shortening and life span has been found in cells of the immune system. Effective immune responses in humans rely on a variety of circulating cell types and require rapid expansion of specific T cells or B cells. Previous studies have shown that the average telomere length of memory CD4 and CD8 T lymphocytes from the same donor is significantly shorter than that of naïve T cells.[14,15] In contrast, memory B cells ($CD19^+CD27^+$) have almost 2 kb longer telomeres compared to naïve B cells ($CD19^+CD27^-$)[16] and activated B cells appear to prevent telomere shortening by the upregulation of telomerase activity.[17,18]

Over an individual's lifetime, successful host defense relies on the concerted actions of innate and adaptive immunity. Natural killer (NK) cells are an important component of the innate immune system, which can recognize and induce lysis of a variety of target cells including virally infected cells, tumor cells, and allogeneic cells. NK cells can also provide an early source of cytokines and chemokines without prior sensitization.[19,20] NK cells comprise approximately 10–15% of all circulating lymphocytes and are defined phenotypically by their expression of CD56, an isoform of neural adhesion molecule with unknown function, as well as CD16, and lack of expression of CD3.[21] Density of surface expression of CD56 and CD16 can be used to classify functionally and developmentally distinct NK cell subsets.[22,23] $CD56^{bright}CD16^-$ NK cells presenting immature NK cells mediate low or no cytotoxicity, proliferate in response to IL-2, and produce high levels of inflammatory cytokines. In contrast, $CD56^{dim}CD16^+$ NK cells presenting mature NK cells are potent mediators of cytotoxicity, which have a distinct cytokine and chemokine profile.[22] Furthermore, NK cells express different cell-surface receptors creating a diverse NK cell repertoire, which exhibits specificity in the immune response.[24] A balance between the NK activating and inhibitory receptors on the cell surface controls the cytolytic activity of NK cells, such as human killer cell Ig-like receptors (KIR), member of the C-type-lectin like receptors, CD94-NKG2A heterodimers, and NKG2D.[24] Peripheral blood NK cells are also equipped with a panel of cell-surface molecules that have been documented to participate in the binding of NK cells to endothelial cells, such as lymphocyte function-associated Ag 1 (LFA-1), CD44, CD2, and CD31.[25,26]

It has been reported that the number of NK cells increases and that modifications of NK cell cytolytic activity occur with advancing age.[27–30] A recent study observed an age-associated loss of telomere length and reduction of telomerase activity in $CD16^+$ NK cells of octogenarians.[31] However, little is known about the telomere biology in the NK cell subsets in terms of the expression of different surface molecules and activating or inhibitory receptors. To further characterize the telomere length heterogeneity in peripheral blood cells of normal individuals, we analyzed the length of telomere repeat sequences in NK cell subpopulations that were purified on the basis of phenotypic properties using the partially automated flow FISH technique.[32]

MATERIALS AND METHODS

Subjects

Peripheral blood mononuclear cells (PBMCs) were obtained with informed consent from eight healthy male and female individuals by density centrifugation using Ficoll-Hypaque (Pharmacia, Uppsala, Sweden). The mononuclear cells were frozen in tissue culture medium containing 20% fetal bovine serum (FBS) and 10% dimethylsulfoxide (DMSO). Following thawing, cells were washed and resuspended in phosphate-buffered saline (PBS) containing 2% FBS (staining buffer) for staining or sorting.

Flow Cytometry and Cell Sorting

Leukocytes were stained with fluorescent-labeled monoclonal antibodies against human CD3, CD4, CD8, CD19, CD56, CD16 (all purchased from Becton Dickinson, San Jose, CA), Vα24, and CD161 (Immunotech, Marsielle, France). After labeling, $CD3^+$, $CD4^+CD3^+$, $CD8^+CD3^+$, $CD19^+CD3^-$, $CD16^+CD3^-$, $CD56^+CD3^-$, $Vα24^+$, and $Vα24^+CD161^+$ cells were sorted on a FACSVantage (Becton Dickinson) cell sorter and immediately cryopreserved in 10% DMSO, 20% FBS, and 70% RPMI (freezing medium) for analysis by flow FISH at a later time point.

Human NK Cell Enrichment and NK Cell Subsets Sorting

Human NK cells were isolated from normal donor leukocytes using NK cell enrichment cocktail (StemCell Technologies, Vancouver, BC) according to the manufacturer's directions. Enriched NK cells were stained with monoclonal antibodies against human CD56 (Becton Dickinson), CD16 (Becton Dickinson), CD94 (Immunotech), KIR (NKB1, Becton Dickinson), NKG2A (Immunotech), NKG2D (R&D Systems, Inc., Minneapolis, MN), LFA-1 (Becton Dickinson), and CD161 (Becton Dickinson). After labeling, $CD56^{dim}CD16^+$, $CD56^{bright}CD16^-$, positive and negative cells for CD94, KIR, NKG2A, NKG2D, LFA-1, and CD161 antibodies were sorted and immediately cryopreserved in freezing medium for further flow FISH analysis.

Telomere Length Analysis by Fluorescence In Situ Hybridization and Flow Cytometry (Flow FISH)

The average length of telomere repeats at chromosome ends in individual peripheral blood leukocytes was measured by automated flow FISH as previously described.[14,32,33] Briefly, sorted T cells, B cells, NK cells, and natural killer T (NKT) cells were hybridized with or without 0.3 μg/mL telomere-specific fluorescein isothiocyanate (FITC)-conjugated $(CCCTAA)_3$ PNA probe, washed, and counterstained with 0.01 μg/mL LDS 751 (Exciton Chemical Co. Inc., Dayton, OH). To convert the fluorescence measured in sample cells hybridized with the FITC-labeled telomere PNA probe into kilobases of telomere repeats, fixed bovine thymocytes with known telomere length (internal control) were processed simultaneously with each sample.[32] FITC-labeled fluorescent beads were used to correct for daily shifts in the linearity of the flow cytometer and due to possible fluctuations in the laser intensity and alignment. Flow cytometric data collection was performed on a Becton Dickinson FACSCalibur and the data analysis was performed using CellQuestPro (Becton Dickinson).

Statistical Analysis

All statistical procedures were performed with SPSS™ for Windows™ v.6.1 (SPSS inc., Chicago, IL). The Mann–Whitney U test was used to determine statistically significant differences of the telomere length comparisons between B cells and T cells or NK cells, $CD56^{dim}CD16^+$ and $CD56^{bright}CD16^-$, $NKG2D^+$ and $NKG2D^-$, and $LFA-1^+$ and $LFA-1^-$ NK cells in this study. P values less than 0.05 were considered significant.

RESULTS

Telomere Length in Sorted T, B, and NK Cells

The median telomere length in sorted T cells ($CD3^+$), B cells ($CD19^+CD3^-$), and NK cells ($CD16^+CD3^-$ and $CD56^+CD3^-$) of seven

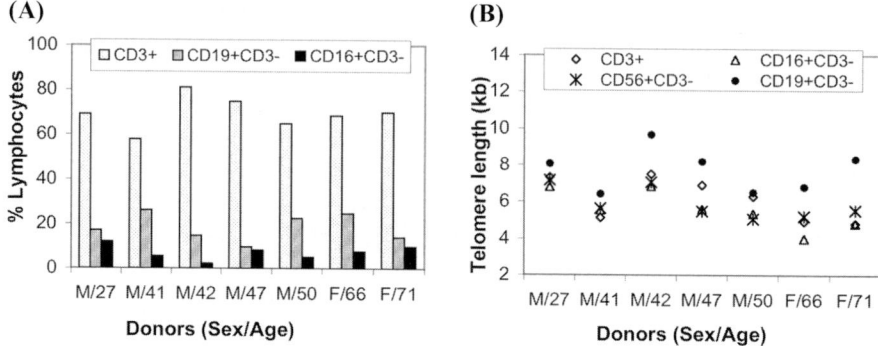

FIGURE 1. Telomere length in sorted T (CD3$^+$), B (CD19$^+$CD3$^-$), and NK (CD16$^+$CD3$^-$) cells from healthy individuals measured by flow FISH. (**A**) Proportion of T cells, B cells, and NK cells among lymphocytes. (**B**) Telomere length (kb) in sorted T, B, and NK (CD16$^+$CD3$^-$ and CD56$^+$CD3$^-$) cells from seven individual donors. The difference of telomere length between B cells and T cells ($P < 0.05$) and NK cells ($P < 0.05$) is statistically significant.

individuals was measured by automated flow FISH. The proportion of T cells, B cells, and NK cells in density separated mononuclear blood cells from those individuals is shown in FIGURE 1 A. The median telomere length in T cells, B cells, and the NK cell populations is shown in FIGURE 1 B. In all individuals B cells had longer telomeres compared to T cells and NK cells. The mean telomere length in B cells was 7.7 ± 0.6 kb, in T cells 6.1 ± 0.6 kb, and in NK cells 5.9 ± 0.4 kb. The most pronounced difference in telomere length between B cells and T cells or NK cells was observed in donor F/71 (3.5 kb and 2.8 kb) and resulted from the relatively longer telomeres in B cells in this donor. The difference in median telomere length between B cells and T cells (or NK cells) was statistically significant ($P < 0.05$). B cells showed the least decline in telomere length with age and B cells from old donors F/66 and F/71 had longer telomeres (resp. 6.8 and 8.3 kb) than B cells from some of the younger donors as M/41 and M/50 (resp. 6.4 and 6.5 kb, FIG. 1B).

Telomere Length in Sorted NK Cell Subpopulations

The proportion of CD56^{dim}CD16$^+$ mature NK cells and CD56^{bright}CD16$^-$ immature NK cells in an enriched NK cell population was analyzed by flow cytometry (FIG. 2 A). The majority of the enriched NK cells from eight individuals were CD56^{dim}CD16$^+$ mature NK cells (mean: 69.6%, range: 62.1–80.9%), and only a small proportion (mean: 3.7%, range: 1.3–7.1%) were CD56^{bright}CD16$^-$ immature NK cells (FIG. 2 C, upper panel). An example of the fluorescence histograms obtained by flow FISH analysis of CD56$^+$CD16$^+$ and CD56$^+$CD16$^-$

populations is shown in FIGURE 2B. The telomere fluorescence of mature CD56⁺CD16⁺ NK cells was more heterogeneous compared to the more immature CD56^{bright}CD16⁻ immature NK cells. This difference in fluorescence histograms most likely represents a more diverse replication history of the mature NK cells, similar to the more heterogeneous telomere fluorescence in memory compared to naïve T cells observed previously.[14] All the eight donors had longer telomeres in CD56⁺CD16⁻ immature (6.3 ± 0.6 kb) than in CD56⁺CD16⁺ mature NK cells (5.5 ± 0.6 kb) and this difference was statistically significant ($P < 0.05$). The maximum telomere length difference for mature and immature NK cells was 1.5 kb (donor M/47, FIG. 2 C, lower panel). To further analyze the telomere length in more specialized NK cell subpopulations, the enriched NK cells were sorted on the basis of CD94, KIR, NKG2A, NKG2D, CD161, and LFA-1 expression. CD94 was expressed on most of the cells (mean: 71.3%, range: 65.6–76.4%) as was NKG2D (mean: 76.6%, rang: 60–94.1%) and LFA-1 (mean: 86.4%, range: 78.1–95%). NKG2A was expressed on a more variable proportion of the cells (mean: 43.2%, range: 27.4–53.1%) and KIR and CD161 were expressed on only a minority (resp. mean: 24.4%, range: 0–47.1% and mean: 17.2%, range: 8.4–23%) of the enriched NK cells (FIG. 2 D).

The telomere length was significantly longer in NKG2D⁺ (5.5 ± 0.6 kb) compared to NKG2D⁻ (5.9 ± 0.7 kb) NK cells from the same donor ($n = 8, P < 0.05$). This was also true for LFA-1⁺ (5.6 ± 0.5 kb) compared to LFA-1⁻ (6.5 ± 0.6 kb) NK cells ($n = 8, P < 0.05$). However, the telomere length was more heterogeneous in CD94⁺ (mean: 5.4 kb, range: 3.0–7.9 kb), CD94⁻ (mean: 5.9 kb, range: 4.4–7.5 kb), NKG2A⁺ (mean: 5.9 kb, range: 4.7–7.2 kb), NKG2A⁻ (mean: 5.3 kb, range: 3.4–7.2 kb), KIR⁺ (mean: 5.1 kb, range: 2.6–7.2 kb), KIR⁻ (mean: 5.9 kb, range: 5.0–7.3kb), CD161⁺ (mean: 6.0 kb, range: 4.7–7.7 kb), and CD161⁻ (mean: 6.2 kb, range: 4.0–8.3 kb) NK cell populations (FIG. 2D) and the expression of these markers was not significantly different. The difference between cells that do or do not express these markers was not significant.

Telomere Length in Sorted Human NKT Cells

We also compared the telomere length in sorted NKT (Vα24⁺CD161⁺) cells and Vα24⁺ T cells. The proportion of Vα24⁺ T cells (mean: 0.62%, range: 0.14–0.89%) within lymphocytes was higher than NKT cells (mean: 0.29%, range: 0.11–0.71%) for all seven individuals (FIG. 3, *upper panel*). The Vα24⁺ T cells (6.2 ± 0.6 kb) had significantly longer telomeres than NKT cells (4.9 ± 0.8 kb) ($n = 7, P < 0.05$) (FIG. 3, *lower panel*).

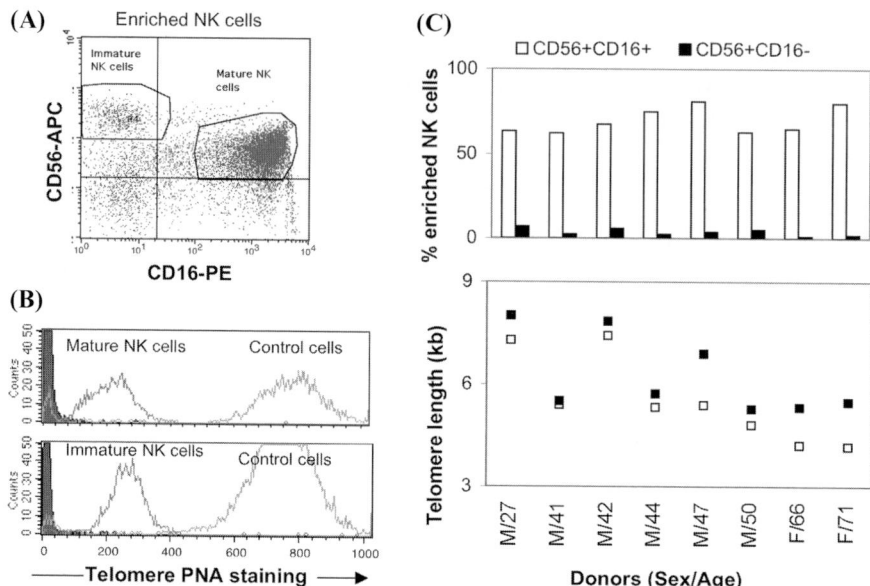

FIGURE 2. Percentages and telomere length of NK cell subsets within enriched NK cell populations. (**A**) The enriched NK cells were stained with anti-CD56-APC and CD16-PE and the $CD56^+CD16^+$ (mature NK) and $CD56^+CD16^-$ (immature NK) cells in the indicated windows were sorted. (**B**) Representative staining obtained by flow FISH analysis of $CD56^+CD16^+$ and $CD56^+CD16^-$ populations relative to the bovine thymocytes with known telomere length used as control cells. (**C**) Proportion of $CD56^+CD16^+$ (mature NK, *open boxes*) and $CD56^+CD16^-$ (immature NK, *black boxes*) NK cells and corresponding telomere length. Cells were sorted from enriched NK cell populations from eight healthy donors ($P < 0.05$). (**D**) Telomere length in sorted NK cell subpopulations measured by flow FISH. Proportions and telomere length of NK subpopulations ($CD94^+$, $CD94^-$, KIR^+, KIR^-, $NKG2A^+$, $NKG2A^-$, $CD161^+$, $CD161^-$, $NKG2D^+$, $NKG2D^-$, $LFA-1^+$, and $LFA-1^-$) within enriched NK cells are shown in upper and middle lower panels as well as upper middle and lowest panels, respectively. Telomere length from the same donor was significantly longer in $NKG2D^+$ than $NKG2D^-$ NK population ($P < 0.05$), so was in $LFA-1^+$ than $LFA-1^-$ NK cells ($P < 0.05$).

DISCUSSION

In this study, we measured the telomere length in NK cell subpopulations using the automated flow FISH method, in which hybridized fluorescently labeled telomere probe in cells of interest is measured relative to internal control cells (bovine thymocytes) in the same tube.[32,33] We observed a marked heterogeneity in the telomere length in adult peripheral NK cell subpopulations. The pattern of telomere length in NK cells is similar to what has been described previously in T cells.[34,35] In each of eight donor samples studied, the $CD56^{dim}CD16^+$ mature NK cells were found to have shorter telomeres than the $CD56^{bright}CD16^-$ immature NK cells. Furthermore, we showed that NK cells carrying NKG2D,

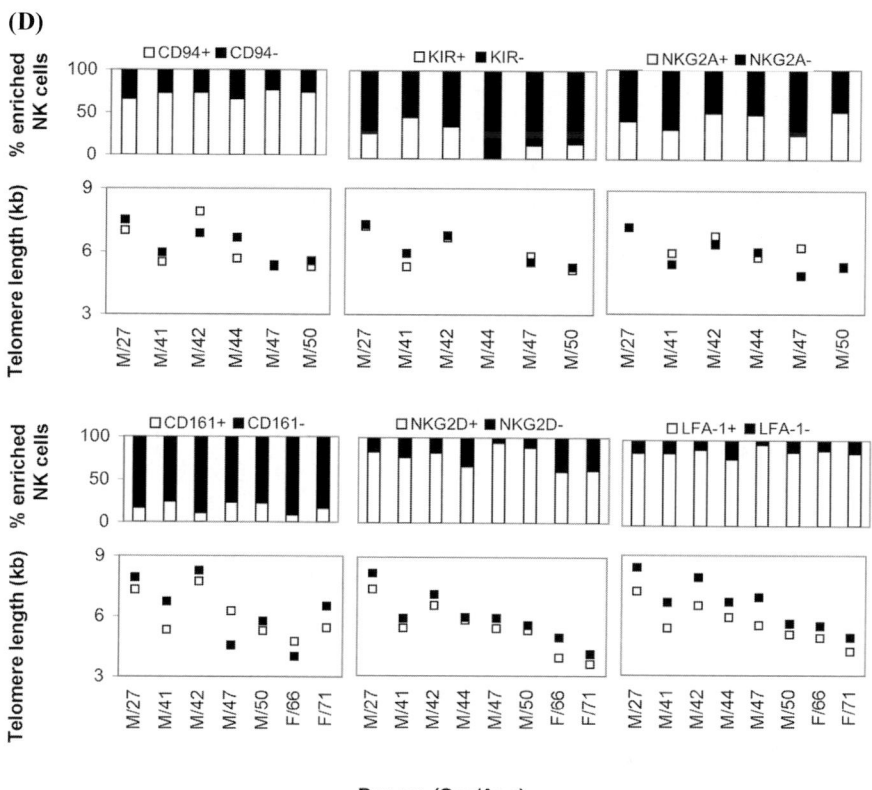

FIGURE 2. Continued.

a NK cell-activating receptor, have significantly shorter telomeres compared to NK cells that do not express this receptor. Moreover, NK cells expressing the adhesion molecule LFA-1 (which may potentiate the cytolytic NK-effector function) also had shorter telomeres than the LFA-1-negative NK cells.

Our data on telomere length analysis of subpopulations of NK cells supports the hypothesis that cells with the greatest proliferative potential have the longest telomeres. $CD56^{dim}CD16^+$ mature NK cells appear to be more terminally differentiated than the $CD56^{bright}CD16^-$ immature NK cells in terms of their higher cytotoxicity against NK-sensitive targets and distinct cytokine and chemokine profile.[22] During normal aging NK cells show a phenotypic and functional shift toward cells with a more mature NK cell phenotype.[36] $CD56^{bright}$ NK cells are considered to be immature NK precursor cells with a potential to differentiate into $CD56^{dim}$ mature NK cells.[37] The observed differences in telomere length between $CD56^{dim}CD16^+$ mature and $CD56^{bright}CD16^-$ immature NK cells are compatible with this notion.

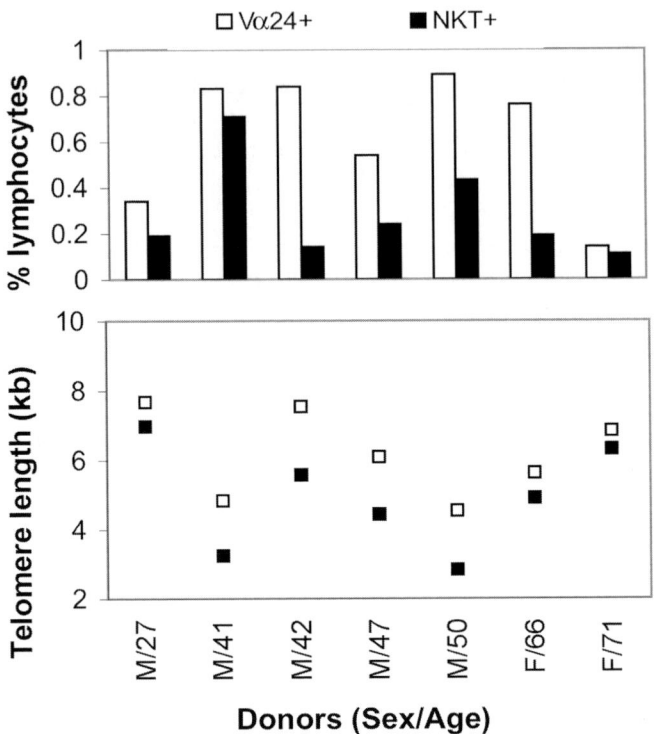

FIGURE 3. Telomere length in sorted NKT cells measured by flow FISH. (*upper panel*) Proportion of NKT cells and Vα24+ T cells within lymphocytes. (*lower panel*) Telomere length in kilobases of sorted NKT and Vα24+ T cells. The Vα24+ T cells have a significantly longer telomere length than NKT cells ($P < 0.05$).

It has been reported that the high density of adhesion molecule LFA-1 may increase the cytolytic-effector function of NK cells,[21] which can be regulated by L-selectin.[38] We observed that LFA-1+ NK cells have shorter telomeres. This observation could reflect differences in replicative history or function of the NK cells expressing high levels of LFA-1 during the early immune response to foreign pathogens.[38]

NK cells are known to be activated by many tumor cells and virus-infected cells.[39,40] NK cells express a large number of different cell-surface receptors that deliver either activating or inhibitory signals, which are involved in the proper regulation of NK cell functions.[41–43] However, the interactions between these two receptor types, which dictate the balance between rest and activation in NK cells, have not yet been clearly defined. Engagement of NKG2D on NK cells can trigger cytolytic activity and can also elicit cytokine and chemokine production.[39,44,45] NK cells expressing NKG2D have shorter telomeres than those that do not, suggesting that activated NK cells typically have replicated more frequently *in vivo* than unactivated NK cells. Unactivated NK cells from mice have been shown to often respond poorly to target cells.[46] However, the telomere length profile is strikingly heterogeneous in NK cells carrying inhibitory and activating NK receptors, such as CD94, NKG2A, KIR, or expressing the surface molecule CD161. Most likely, these receptors are expressed variably among different NK cells (e.g., KIR is not expressed on NK cells of individual M/44 at all).

In seven individuals studied, the sorted B cells have longer telomeres than T cells and NK cells from the same donor. These results are consistent with previous findings from baboons[47] and humans showing that germinal center B cells and certain memory B cells may have longer telomeres, which is most likely due to the high level of telomerase in the germinal center.[16,18]

Vα24 NKT cells are innate immune cells constituting a lymphocyte lineage sharing characteristics of both T cells including a semi-invariant TCR with uniform usage of α-chain variable gene segment 24 (Vα24), preferentially paired with β-chain variable gene segment 11(Vβ11), and NK cells including the expression of the NK cell marker CD16,[48,49] and play immunoregulatory roles in autoimmunity and tumor immunity.[50] By using flow FISH, we found that NKT cells have shorter telomeres than Vα24$^+$ T cells in all seven individuals and shorter than CD3$^+$ T cells and CD161$^+$ NK cells in six of seven individuals (except the oldest individual F/71). The number of Vα24 NKT cells has been reported to be reduced in elderly humans[51] and in patients with systemic lupus erythematosus,[52] and rheumatoid arthritis.[53] The murine Vα14 NKT cells are lost during acute infection with lymphocytic choriomeningitis virus.[54] Compared to T and NK cells, NKT cells are unique in acquiring a memory-activated phenotype before birth,[55] which may account in part for the shorter telomeres in NKT cells.

Taken together, we conclude that NK cell subpopulations with a mature phenotype have shorter telomeres compared to those with an immature phenotype. Furthermore, NK cells with an effector phenotype or expressing activating receptors also have relatively short telomeres. Our observations support the concept that telomere length dynamics in NK cells are similar to that observed in T cells and granulocytes. At advanced age such telomere attrition could result in the disappearance of long-lived, mature cytotoxic NK cells and compromise immunity against foreign cells, virally infected cells, and tumor cells.

ACKNOWLEDGMENTS

This work was supported by NIH Grant AI29524, the Swiss National Science Foundation (Grant number: 31-53774-98); the Bernese Cancer League and the National Cancer Institute of Canada with funds from the Terry Fox Run.

REFERENCES

1. MOYZIS, R.K., J.M. BUCKINGHAM, L.S. CRAM, et al. 1988. A highly conserved repetitive DNA sequence, (TTAGGG)n, present at the telomeres of human chromosomes. Proc. Natl. Acad. Sci. USA **85:** 6622–6626.
2. SMOGORZEWSKA, A. & T. DE LANGE. 2004. Regulation of telomerase by telomeric proteins. Annu. Rev. Biochem. **73:** 177–208.
3. LANSDORP, P.M., N.P. VERWOERD, F.M. VAN DE RIJKE, et al. 1996. Heterogeneity in telomere length of human chromosomes. Hum. Mol. Genet. **5:** 685–691.
4. HARLEY, C.B., A.B. FUTCHER & C.W. GREIDER. 1990. Telomeres shorten during ageing of human fibroblasts. Nature **345:** 458–460.
5. HASTIE, N.D., M. DEMPSTER, M.G. DUNLOP, et al. 1990. Telomere reduction in human colorectal carcinoma and with ageing. Nature **346:** 866–868.
6. GREIDER, C.W. & E.H. BLACKBURN. 1985. Identification of a specific telomere terminal transferase activity in Tetrahymena extracts. Cell **43:** 405–413.
7. LINGNER, J., T.R. HUGHES, A. SHEVCHENKO, et al. 1997. Reverse transcriptase motifs in the catalytic subunit of telomerase. Science **276:** 561–567.
8. BODNAR, A.G., M. OUELLETTE, M. FROLKIS, et al. 1998. Extension of life-span by introduction of telomerase into normal human cells. Science **279:** 349–352.
9. HOOIJBERG, E., J.J. RUIZENDAAL, P.J. SNIJDERS, et al. 2000. Immortalization of human $CD8^+$ T cell clones by ectopic expression of telomerase reverse transcriptase. J. Immunol. **165:** 4239–4245.
10. LUITEN, R.M., J. PENE, H. YSSEL & H. SPITS. 2003. Ectopic hTERT expression extends the life span of human $CD4^+$ helper and regulatory T-cell clones and confers resistance to oxidative stress-induced apoptosis. Blood **101:** 4512–4519.
11. ROTH, A., H. YSSEL, J. PENE, et al. 2003. Telomerase levels control the lifespan of human T lymphocytes. Blood **102:** 849–857.
12. RUFER, N., M. MIGLIACCIO, J. ANTONCHUK, et al. 2001. Transfer of the human telomerase reverse transcriptase (TERT) gene into T lymphocytes results in extension of replicative potential. Blood **98:** 597–603.
13. VAZIRI, H. & S. BENCHIMOL. 1998. Reconstitution of telomerase activity in normal human cells leads to elongation of telomeres and extended replicative life span. Curr. Biol. **8:** 279–282.
14. RUFER, N., W. DRAGOWSKA, G. THORNBURY, et al. 1998. Telomere length dynamics in human lymphocyte subpopulations measured by flow cytometry. Nat. Biotechnol. **16:** 743–747.
15. WENG, N.P., L.D. PALMER, B.L. LEVINE, et al. 1997. Tales of tails: regulation of telomere length and telomerase activity during lymphocyte development, differentiation, activation, and aging. Immunol. Rev. **160:** 43–54.
16. MARTENS, U.M., V. BRASS, L. SEDLACEK, et al. 2002. Telomere maintenance in human B lymphocytes. Br. J. Haematol. **119:** 810–818.
17. BATLIWALLA, F.M., N. RUFER, P.M. LANSDORP & P.K. GREGERSEN. 2000. Oligoclonal expansions in the CD8(+)CD28(−) T cells largely explain the shorter telomeres detected in this subset: analysis by flow FISH. Hum. Immunol. **61:** 951–958.
18. WENG, N.P., L. GRANGER & R.J. HODES. 1997. Telomere lengthening and telomerase activation during human B cell differentiation. Proc. Natl. Acad. Sci. USA **94:** 10827–10832.
19. LANIER, L.L. 1998. NK cell receptors. Annu. Rev. Immunol. **16:** 359–393.

20. MORETTA, L., M.C. MINGARI, D. PENDE, et al. 1996. The molecular basis of natural killer (NK) cell recognition and function. J. Clin. Immunol. **16:** 243–253.
21. ROBERTSON, M.J., M.A. CALIGIURI, T.J. MANLEY, et al. 1990. Human natural killer cell adhesion molecules. Differential expression after activation and participation in cytolysis. J. Immunol. **145:** 3194–3201.
22. COOPER, M.A., T.A. FEHNIGER & M.A. CALIGIURI. 2001. The biology of human natural killer-cell subsets. Trends Immunol. **22:** 633–640.
23. TAKEDA, K., M.J. SMYTH, E. CRETNEY, et al. 2001. Involvement of tumor necrosis factor-related apoptosis-inducing ligand in NK cell-mediated and IFN-gamma-dependent suppression of subcutaneous tumor growth. Cell Immunol. **214:** 194–200.
24. MCQUEEN, K.L. & P. PARHAM. 2002. Variable receptors controlling activation and inhibition of NK cells. Curr. Opin. Immunol. **14:** 615–621.
25. BERMAN, M.E., Y. XIE & W.A. MULLER. 1996. Roles of platelet/endothelial cell adhesion molecule-1 (PECAM-1, CD31) in natural killer cell transendothelial migration and beta 2 integrin activation. J. Immunol. **156:** 1515–1524.
26. BIANCHI, G., M. SIRONI, E. GHIBAUDI, et al. 1993. Migration of natural killer cells across endothelial cell monolayers. J. Immunol. **151:** 5135–5144.
27. FACCHINI, A., E. MARIANI, A.R. MARIANI, et al. 1987. Increased number of circulating Leu 11+ (CD 16) large granular lymphocytes and decreased NK activity during human ageing. Clin. Exp. Immunol. **68:** 340–347.
28. MARIANI, E., G. RAVAGLIA, P. FORTI, et al. 1999. Vitamin D, thyroid hormones and muscle mass influence natural killer (NK) innate immunity in healthy nonagenarians and centenarians. Clin. Exp. Immunol. **116:** 19–27.
29. PAWELEC, G., Y. BARNETT, R. FORSEY, et al. 2002. T cells and aging, January 2002 update. Front Biosci. **7:** d1056–d1183.
30. SOLANA, R. & E. MARIANI. 2000. NK and NK/T cells in human senescence. Vaccine **18:** 1613–1620.
31. MARIANI, E., A. MENEGHETTI, I. FORMENTINI, et al. 2003. Different rates of telomere shortening and telomerase activity reduction in CD8 T and CD16 NK lymphocytes with ageing. Exp. Gerontol. **38:** 653–659.
32. BAERLOCHER, G.M. & P.M. LANSDORP. 2003. Telomere length measurements in leukocyte subsets by automated multicolor flow-FISH. Cytometry A. **55:** 1–6.
33. BAERLOCHER, G.M., J. MAK, T. TIEN & P.M. LANSDORP. 2002. Telomere length measurement by fluorescence *in situ* hybridization and flow cytometry: tips and pitfalls. Cytometry **47:** 89–99.
34. RUFER, N., T.H. BRUMMENDORF, S. KOLVRAA, et al. 1999. Telomere fluorescence measurements in granulocytes and T lymphocyte subsets point to a high turnover of hematopoietic stem cells and memory T cells in early childhood. J. Exp. Med. **190:** 157–167.
35. WENG, N.P., B.L. LEVINE, C.H. JUNE & R.J. HODES. 1995. Human naive and memory T lymphocytes differ in telomeric length and replicative potential. Proc. Natl. Acad. Sci. USA **92:** 11091–11094.
36. KRISHNARAJ, R. & A. SVANBORG. 1992. Preferential accumulation of mature NK cells during human immunosenescence. J. Cell Biochem. **50:** 386–391.
37. NAGLER, A., L.L. LANIER, S. CWIRLA & J.H. PHILLIPS. 1989. Comparative studies of human FcRIII-positive and negative natural killer cells. J. Immunol. **143:** 3183–3191.

38. FREY, M., N.B. PACKIANATHAN, T.A. FEHNIGER, et al. 1998. Differential expression and function of L-selectin on CD56bright and CD56dim natural killer cell subsets. J. Immunol. **161:** 400–408.
39. DIEFENBACH, A., E.R. JENSEN, A.M. JAMIESON & D.H. RAULET. 2001. Rae1 and H60 ligands of the NKG2D receptor stimulate tumour immunity. Nature **413:** 165–171.
40. MOSER, J.M., A.M. BYERS & A.E. LUKACHER. 2002. NK cell receptors in antiviral immunity. Curr. Opin. Immunol. **14:** 509–516.
41. LANIER, L.L. 2005. NK cell recognition. Annu. Rev. Immunol. **23:** 225–274.
42. LOPEZ-BOTET, M., M. LLANO, F. NAVARRO & T. BELLON. 2000. NK cell recognition of non-classical HLA class I molecules. Semin. Immunol. **12:** 109–119.
43. RAVETCH, J.V. & L.L. LANIER. 2000. Immune inhibitory receptors. Science **290:** 84–89.
44. COSMAN, D., J. MULLBERG, C.L. SUTHERLAND, et al. 2001. ULBPs, novel MHC class I-related molecules, bind to CMV glycoprotein UL16 and stimulate NK cytotoxicity through the NKG2D receptor. Immunity **14:** 123–133.
45. JAMIESON, A.M., A. DIEFENBACH, C.W. MCMAHON, et al. 2002. The role of the NKG2D immunoreceptor in immune cell activation and natural killing. Immunity **17:** 19–29.
46. GLAS, R., L. FRANKSSON, C. UNE, et al. 2000. Recruitment and activation of natural killer (NK) cells *in vivo* determined by the target cell phenotype. An adaptive component of NK cell-mediated responses. J. Exp. Med. **191:** 129–138.
47. BAERLOCHER, G.M., J. MAK, A. ROTH, et al. 2003. Telomere shortening in leukocyte subpopulations from baboons. J. Leukoc. Biol. **73:** 289–296.
48. DELLABONA, P., E. PADOVAN, G. CASORATI, et al. 1994. An invariant V alpha 24-J alpha Q/V beta 11 T cell receptor is expressed in all individuals by clonally expanded CD4-8- T cells. J. Exp. Med. **180:** 1171–1176.
49. PORCELLI, S., C.E. YOCKEY, M.B. BRENNER & S.P. BALK. 1993. Analysis of T cell antigen receptor (TCR) expression by human peripheral blood CD4-8- alpha/beta T cells demonstrates preferential use of several V beta genes and an invariant TCR alpha chain. J. Exp. Med. **178:** 1–16.
50. GODFREY, D.I., K.J. HAMMOND, L.D. POULTON, et al. 2000. NKT cells: facts, functions and fallacies. Immunol. Today **21:** 573–583.
51. DELAROSA, O., R. TARAZONA, J.G. CASADO, et al. 2002. Valpha24+ NKT cells are decreased in elderly humans. Exp. Gerontol. **37:** 213–217.
52. OISHI, Y., T. SUMIDA, A. SAKAMOTO, et al. 2001. Selective reduction and recovery of invariant Valpha24JalphaQ T cell receptor T cells in correlation with disease activity in patients with systemic lupus erythematosus. J. Rheumatol. **28:** 275–283.
53. YANAGIHARA, Y., K. SHIOZAWA, M. TAKAI, et al. 1999. Natural killer (NK) T cells are significantly decreased in the peripheral blood of patients with rheumatoid arthritis (RA). Clin. Exp. Immunol. **118:** 131–136.
54. HOBBS, J.A., S. CHO, T.J. ROBERTS, et al. 2001. Selective loss of natural killer T cells by apoptosis following infection with lymphocytic choriomeningitis virus. J. Virol. **75:** 10746–10754.
55. VAN DER VLIET, H.J., N. NISHI, T.D. DE GRUIJL, et al. 2000. Human natural killer T cells acquire a memory-activated phenotype before birth. Blood **95:** 2440–2442.

Flt3 in Regulation of Type I Interferon-Producing Cell and Dendritic Cell Development

NOBUYUKI ONAI,[a,b] AYA OBATA-ONAI,[a,b] MICHAEL A. SCHMID,[a] AND MARKUS G. MANZ[a,c]

[a]*Institute for Research in Biomedicine, CH-6500 Bellinzona, Switzerland*

[b]*Department of Immunology, University of Akita, School of Medicine, Akita 010-8543, Japan*

[c]*Department of Hematology, Oncology, and Immunology, Eberhard-Karls-University Medical School, 42076 Tübingen, Germany*

ABSTRACT: Flt3-ligand is a nonredundant cytokine in type I interferon-producing cell (IPC) and dendritic cell (DC) development. We demonstrated that IPC and DC differentiation potential is confined to Flt3$^+$-hematopoietic progenitor cells, that Flt3-ligand drives development along both lymphoid and myeloid developmental pathways from Flt3$^+$-progenitors to Flt3$^+$-IPCs and -DCs, and that *in vivo* pharmacologic inhibition of Flt3-signaling leads to disruption of IPC and DC development in spite of consecutive Flt3-ligand upregulation in treated animals. We here summarize our recent findings that overexpression of human *Flt3* in Flt3$^-$ and Flt3$^+$ hematopoietic progenitors rescues and enhances their IPC and DC differentiation potential, respectively. Based on these data, we propose an instructive, demand-regulated, cytokine-driven IPC and DC regeneration model, where high Flt3-ligand levels initiate a self-sustaining, Flt3-STAT3 and -PU.1-mediated IPC and DC differentiation program in Flt3$^+$-hematopoietic progenitor cells.

KEYWORDS: Flt3; hematopoiesis; dendritic cells

INTRODUCTION

Hematopoiesis is regarded as a unidirectional, multilinear process where hematopoietic stem cells (HSCs) differentiate into mature hematopoietic cells by progressive loss of developmental options and restriction to one lineage, and where regeneration and expansion of specific lineages are largely regulated extrinsically by different hematopoietic cytokines.[1] However, it is unclear

Address for correspondence: Markus G. Manz, M.D., Institute for Research in Biomedicine (IRB), Via Vincenzo Vela 6, CH-6500 Bellinzona, Switzerland. Voice: +41-91-820-0361; fax: +41-91-820-0312.

manz@irb.unisi.ch

whether under physiologic conditions cytokines are capable to instruct HSCs and multipotent precursors to differentiate with lineage-restricted progenitors (extrinsic determination) or, if alternatively, HSCs and subsequent progenitors commit to lineage-restricted progenitors by intrinsic differentiation programs (intrinsic determination), and restricted progenitors are consecutively stimulated by hematopoietic cytokines, produced upon demand.[2,3] HSCs as well as multiple developmental intermediates with limited cellular expansion potential and restriction to specific mature cell types have been identified in both mice and men. These include myeloid progenitors, as clonal common myeloid progenitors (CMPs) that give rise to either granulocyte/macrophage progenitors (GMPs) or megakaryocyte/erythrocyte progenitors (MEPs),[4,5] and clonal common lymphoid progenitors (CLPs),[6,7] which produce the respective mature cell types.

Access to lineage developmental options and readiness to receive lineage-permissive and -instructive signals might be determined by relative expression levels of diverse transcription factors and cytokine receptors.[8,9] Indeed, experimental deletion or overexpression of single transcription factors is sufficient to reprogram committed progenitors or mature cells to alternative hematopoietic lineages[10] *Pax5*-deficient pre-B cells lose B cell differentiation potential and mature into T and myelomonocytic cells, however, reexpression of *Pax5* restores B cell commitment[10,11]; ectopic expression of *GATA-1* instructs HSCs and CMPs, and converts CLPs and GMPs to the megakaryocyte/erythrocyte lineage, respectively[12]; and enforced expression of *C/EBP*α and *C/EBP*β in B cells leads to macrophage differentiation.[13] Furthermore, it has been shown that *GM-CSF receptor* expression and stimulation with the cognate ligand redirect CLPs and early T cell progenitors to myeloid lineage outcomes.[14–16] The latter proves that, at least in these experimental settings, hematopoietic lineage instruction can be mediated extrinsically by cytokines.

Dendritic cells (DCs) are regulators of innate and adaptive immune responses, involved in initiation of immunity as well as in maintenance of self-tolerance.[17–19] In addition, they are cells of the hematopoietic system and are replenished from hematopoietic stem and progenitor cells.[1] In mice, multiple DC subsets that differ in maturation state, phenotype, location, and in some functions were identified.[20] For simplicity, here we will grossly divide these into $CD11c^+B220^+$ natural type I interferon-producing cells (IPCs, also called plasmacytoid cells or plasmacytoid pre-DCs) and $CD11c^+B220^-$ "conventional" DCs, consisting of $CD11c^+CD8\alpha^-CD4^-CD11b^+$, $CD11c^+CD8\alpha^-CD4^+CD11b^+$, and $CD11c^+CD8\alpha^+CD4^-CD11b^-$ subpopulations.[20] While initially it was suggested that IPCs as well as conventional $CD11c^+CD8\alpha^+$ DCs are derived from lymphoid committed progenitors,[20] it was demonstrated later that any of the IPCs and conventional DCs can be generated via lymphoid and myeloid progenitors.[21–26] Specifically, all IPCs and conventional DCs are generated by mouse CMPs, GMPs, CLPs, and pro-T1 cells, while IPC and DC differentiation potential is lost once definitive MEPs, or B cell commitment occurs.[21–26] Thus, in contrast to other hematopoietic lineages, IPC and DC potentials are conserved along lymphoid and myeloid developmental pathways.

Flt3, a receptor tyrosine kinase with homology to c-Kit (the receptor for stem cell factor) and c-fms (the receptor for M-CSF),[27] has a nonredundant role in steady-state differentiation of IPCs and DCs *in vivo*: *Flt3-ligand* (*Flt3L*)-deficient mice and mice with hematopoietic system restricted deletion of *STAT3*, a transcription factor activated in the Flt3-signaling cascade, as well as mice that are treated with Flt3 tyrosine kinase inhibitors, show massively reduced IPCs and DCs.[28–30] On the other hand, injection or conditional expression of Flt3L in mice increases IPCs and DCs.[31–33] Furthermore, Flt3L as a single cytokine is capable to induce differentiation of IPCs and DCs in mouse bone marrow cell cultures.[34]

Flt3 is expressed in mouse short-term HSCs and multipotent progenitors,[35,36] in most CLPs and CMPs, and at lower levels on fractions of GMPs and pro-T1 cells, as well as on mature steady-state IPCs and DCs, while it is downregulated on pro-B cells, further downstream T cell progenitors, and absent on MEPs.[32,37] To determine what might define IPC and DC developmental potential in lymphoid and myeloid committed cells, we and others showed that *in vitro* and *in vivo* IPC, DC, and Langerhans cell (LC) differentiation potential is confined to Flt3-expressing hematopoietic progenitors.[32,37,38] Furthermore, we demonstrated that injection of Flt3L expands Flt3-positive, but not downstream Flt3-negative progenitors, and drives IPC and DC development along both lymphoid and myeloid differentiation pathways.[32] Based on these data, we postulated that high environmental Flt3-ligand levels and consecutive Flt3-signaling might be both, the earliest event and a continuous regulator, which determine IPC and DC developmental outcomes in bone marrow hematopoietic progenitor cells.

To test this hypothesis, we artificially expressed either *GFP* or human *Flt3-GFP* in progenitor cells using a bicistronic retroviral transduction system.[39] We showed that enforced expression of hu*Flt3* in Flt3⁻-progenitors rescued their potential to differentiate into functional IPCs and DCs with comparable *in vitro* differentiation efficiency as Flt3⁺-progenitors.[39] Furthermore, enforced expression of hu*Flt3* in MEPs, which under normal conditions cannot give rise to IPCs and DCs[22–26] and are contained in Flt3⁻-progenitor cells, induced *in vitro* and *in vivo* IPC and DC differentiation, comparable to that observed from *GFP*⁺-GMPs. Thus this data demonstrates that enforced expression and signaling of hu*Flt3* in Flt3⁻-progenitors delivers an instructive signal to activate latent IPC and DC differentiation programs.

Enforced expression of hu*Flt3* in MEPs not only led to gain of IPC and DC developmental capacity, but, with the exception of mixed colony formation, also to gain of CFU activity of upstream myeloid progenitors, and to differentiation of erythroid and myelomonocytic cells *in vivo*. In contrast, huFlt3-signaling in GMPs did not activate megakaryocyte/erythrocyte potential. This

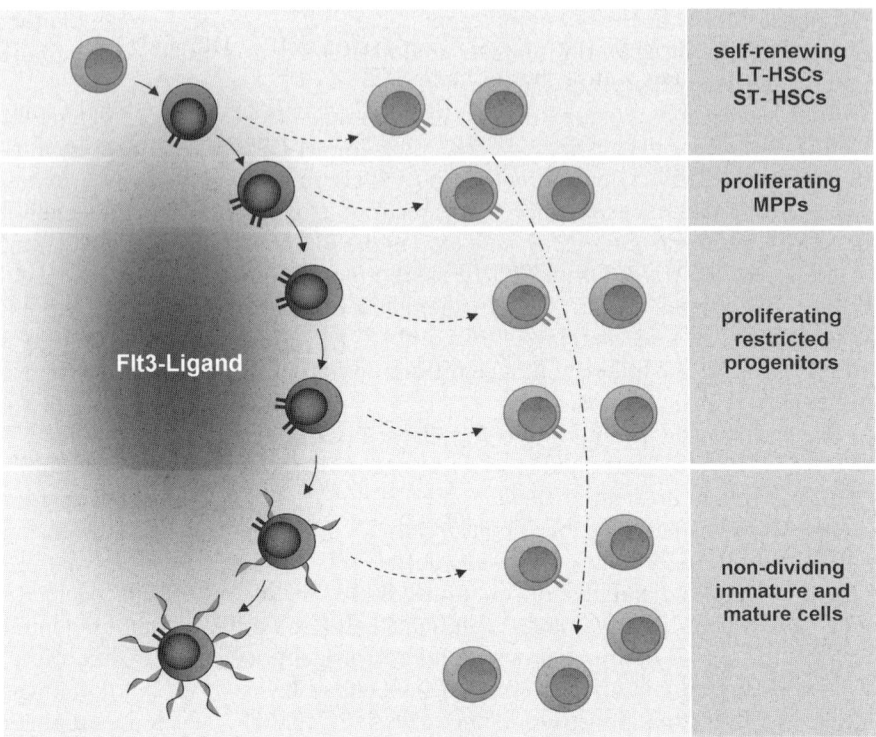

FIGURE 1. Proposed "Flt3-Licence" working model for steady-state natural IPC and DC development from early hematopoietic progenitor cells. *Bold arrows* represent continuous strong Flt3-ligand stimulation leading to IPC and DC development, *dashed arrows* represent more frequent competing signals, leading to alternative lineage outcomes.

implies that beyond activation and enhancement of IPC and DC development, Flt3-signaling is not immediate deterministic but primarily opens access to an IPC, DC, and myelomonocytic differentiation program. Thus, we propose that IPC and DC lineage outcome might be a gradual process, depending on continuous strong Flt3-signaling (FIG. 1).

What are the downstream molecular events initiated in this process by Flt3-signaling? It was shown that hematopoietic system confined deletion of *STAT3* transcription factor leads to inhibition of Flt3-driven IPC and DC development.[29] Furthermore, human *Flt3* transfection and stimulation with Flt3L in mouse myeloid 32Dcl3 cells lead to the induction of PU.1 and C/EBPα expression.[40] PU.1 cooperatively with C/EBPα activates myeloid development-associated cytokine receptor genes including *G-CSFR*, *M-CSFR*, and *GM-CSFR*, and these transcription factors are indispensable for granulocyte and monocyte development.[41] *PU.1*-deficient mice, in addition to other hematopoietic defects, lack either CD8α$^-$ or both CD8α$^-$ and CD8α$^+$ DCs, depending on the type of PU.1 deletion.[42,43] We showed that enforced huFlt3-signaling in MEPs results in enhanced expression of IPC, DC, and GM-lineage development-related transcription factors *STAT3*, *PU.1*, and *C/EBPα*, as well as expression of *G-*, *M-,* and *GM-CSF.* Thus, at least in terms of these RNA transcripts, hu*Flt3*[+]-MEPs, but not *GFP*[+]-MEPs, resembled the gene expression profiles of CMPs.[4,44]

Enforced expression of *STAT3* or *PU.1* in Flt3-negative MEPs was again sufficient to permit the development of both IPCs and DCs, and, most interestingly, led to the up-regulation of mouse *Flt3* mRNA levels in MEPs. This likely allowed culture supplemented, human Flt3L to cross-reactively stimulate *STAT3-* or *PU.1*-transduced cells via mouse Flt3, suggesting a self-sustaining effect of Flt3-signaling-induced *Flt3* transcription via downstream STAT3 and PU.1.

As enforced expression of hu*Flt3* in MEPs did not terminate megakaryocyte/erythrocyte differentiation potential, while hu*Flt3* expression in GMPs did not lead to gain of these differentiation potentials, how can Flt3-signaling be integrated in megakaryocyte/erythrocyte versus IPC, DC, and GM-lineage commitment? By using *PU.1*[gfp] reporter mice, PU.1 expression was recently mapped in early hematopoietic progenitor cells. It was shown that PU.1[+]Flt3[+] CMPs contain high myelomonocytic developmental potential, whereas PU.1[−]Flt3[−] CMPs and PU.1[−]MEPs have high megakaryocyte/erythrocyte potential.[3] Our data suggest that Flt3 might be critical in PU.1 regulation. GATA-1 is a nonredundant transcription factor for megakaryocyte and erythrocyte development,[3] and DNA-binding activity of GATA-1 can be suppressed by enforced *PU.1* expression, resulting in a differentiation block and apoptotic cell death of an erythroid cell line.[45] Conversely, GATA-1 inhibits binding of PU.1 to c-Jun, a co-activator of myeloid gene transactivation by PU.1.[46] Furthermore, GATA-1 interferes with DNA-binding activity of STAT3, and inhibits TPO-dependent growth of the Ba/F3 cell line.[47] Thus, as suggested previously for PU.1 and GATA-1,[48,49] relative dosage of gene transcription and protein levels will likely determine lineage outcomes. Indeed, *STAT3* and *PU.1* expression levels in hu*Flt3*[+]-MEPs were increased to levels of normal CMPs and were somewhat lower than observed in *GFP*[+]-, or hu*Flt3*[+]-GMPs. Thus, MEPs with relatively lower hu*Flt3* and consecutive STAT3 and PU.1 expression do not fully inhibit GATA-1, while high Flt3-expressing and -signaling cells develop to IPC, DC, or GM lineages. In contrast, enforced expression of *STAT3* and *PU.1* in MEPs suppressed *GATA-1* and inhibited megakaryocyte/erythrocyte development. In GMPs hu*Flt3* overexpression in turn induced some *EpoR*, β-*globin*, and *GATA-1* mRNA expression; however, this was not sufficient to reactivate megakaryocyte/erythrocyte development as shown for high-level *GATA-1* expression in GMPs.[12]

Are there implications of these findings for normal hematopoiesis? Flt3 is expressed on mouse short-term (ST-) HSCs, multipotent progenitors, CLPs, CMPs, and GMPs, and *in vivo* injection of Flt3L resulted in expansion of these cells as well as IPCs and DCs, while MEPs and their progeny remained unchanged.[31,32] Our data demonstrate that enforced Flt3 cytokine receptor signaling is sufficient to activate and also enhance IPC and DC differentiation programs, suggesting that instructive cytokine signaling might indeed occur in hematopoiesis. Thus, once Flt3-positive ST-HSCs and their offspring Flt3-positive cells are located in Flt3L-rich environments, they will likely be instructed to differentiate into IPCs and DCs (FIG. 1), a process that might be enhanced by a self-sustaining regulatory loop where Flt3 downstream transcription factors STAT3 and PU.1 in turn maintain *Flt3* receptor expression. As Flt3-signaling does not immediately silence other developmental options, and IPCs and DCs in fact only account for a minor fraction of hematopoietic cells, most Flt3-expressing progenitors will not continuously be stimulated via Flt3L but will receive and activate alternative signals, and thus consecutively acquire different myeloid or lymphoid cell fates. Our data thus support a "Flt3-permissive" developmental model, where Flt3-expressing progenitors maintain IPC and DC differentiation options in response to Flt3L as long as no competing signal shuts these down (FIG. 1).

Building on these findings, it will be important to test whether downstream dividing Flt3-positive common IPC and DC progenitors with silenced alternative developmental programs exist (common dendritic cell progenitors), and if so, which critical factors are involved in final IPC or DC lineage termination. Furthermore, it will be interesting to evaluate how $Flt3^+$ cells and Flt3-ligand-expressing cells localize in bone marrow and secondary hematolymphoid tissues, how Flt3-ligand production is regulated in steady-state and inflammatory or other hematopoietic challenge conditions, and finally, if observations made in experimental mice reflect conditions of IPC and DC development in humans.

ACKNOWLEDGMENT

This work was supported in part by the Swiss National Science Foundation (310000–116637/1), the Deutsche Forschungsgemeinschaft (MA 2159/2–1), and the European Commission FP6 "Network of Excellence" initiative under contract number LSHB-CT-2004–512074 DC-THERA.

REFERENCES

1. KONDO, M. *et al.* 2003. Biology of hematopoietic stem cells and progenitors: implications for clinical application. Annu. Rev. Immunol. **21:** 759–806.
2. METCALF, D. 1998. Lineage commitment and maturation in hematopoietic cells: the case for extrinsic regulation. Blood. **92:** 345–347; discussion 352.
3. CANTOR, A.B. & S.H. ORKIN. 2001. Hematopoietic development: a balancing act. Curr. Opin. Genet. Dev. **11:** 513–519.

4. AKASHI, K. et al. 2000. A clonogenic common myeloid progenitor that gives rise to all myeloid lineages. Nature **404**: 193–197.
5. MANZ, M. G. et al. 2002. Prospective isolation of human clonogenic common myeloid progenitors. Proc. Natl. Acad. Sci. USA **99**: 11872–11877.
6. GALY, A. et al. 1995. Human T, B, natural killer, and dendritic cells arise from a common bone marrow progenitor cell subset. Immunity **3**: 459–473.
7. KONDO, M., I. L. WEISSMAN & K. AKASHI. 1997. Identification of clonogenic common lymphoid progenitors in mouse bone marrow. Cell **91**: 661–672.
8. ORKIN, S. H. 2000. Diversification of haematopoietic stem cells to specific lineages. Nat. Rev. Genet. **1**: 57–64.
9. AKASHI, K. et al. 2003. Transcriptional accessibility for genes of multiple tissues and hematopoietic lineages is hierarchically controlled during early hematopoiesis. Blood **101**: 383–389.
10. NUTT, S. L. et al. 1999. Commitment to the B-lymphoid lineage depends on the transcription factor Pax5 [see comments]. Nature **401**: 556–562.
11. ROLINK, A. G. et al. 1999. Long-term *in vivo* reconstitution of T-cell development by Pax5-deficient B-cell progenitors. Nature **401**: 603–606.
12. IWASAKI, H. et al. 2003. GATA-1 converts lymphoid and myelomonocytic progenitors into the megakaryocyte/erythrocyte lineages. Immunity **19**: 451–462.
13. XIE, H. et al. 2004. Stepwise reprogramming of B cells into macrophages. Cell **117**: 663–676.
14. KONDO, M. et al. 2000. Cell fate conversion of lymphoid committed progenitors by instructive actions of cytokines. Nature **407**: 383–386.
15. KING, A. G. et al. 2002. Lineage infidelity in myeloid cells with TCR gene rearrangement: a latent developmental potential of proT cells revealed by ectopic cytokine receptor signaling. Proc. Natl. Acad. Sci. USA **99**: 4508–4513.
16. IWASAKI-ARAI, J. et al. 2003. Enforced granulocyte/macrophage colony-stimulating factor signals do not support lymphopoiesis, but instruct lymphoid to myelomonocytic lineage conversion. J. Exp. Med. **197**: 1311–1322.
17. BANCHEREAU, J. & R. M. STEINMAN. 1998. Dendritic cells and the control of immunity. Nature **392**: 245–252.
18. LANZAVECCHIA, A. & F. SALLUSTO. 2001. The instructive role of dendritic cells on T cell responses: lineages, plasticity and kinetics. Curr. Opin. Immunol. **13**: 291–298.
19. STEINMAN, R. M., D. HAWIGER & M. C. NUSSENZWEIG. 2003. Tolerogenic dendritic cells. Annu. Rev. Immunol. **21**: 685–711.
20. SHORTMAN, K. & Y. J. LIU. 2002. Mouse and human dendritic cell subtypes. Nat. Rev. Immunol. **2**: 151–161.
21. TRAVER, D. et al. 2000. Development of CD8alpha-positive dendritic cells from a common myeloid progenitor [In Process Citation]. Science **290**: 2152–2154.
22. MANZ, M. G. et al. 2001. Dendritic cell potentials of early lymphoid and myeloid progenitors. Blood **97**: 3333–3341.
23. WU, L. et al. 2001. Development of thymic and splenic dendritic cell populations from different hemopoietic precursors. Blood. **98**: 3376–3382.
24. SHIGEMATSU, H. et al. 2004. Plasmacytoid dendritic cells activate lymphoid-specific genetic programs irrespective of their cellular origin. Immunity **21**: 43–53.
25. CHICHA, L., D. JARROSSAY & M. G. MANZ. 2004. Clonal type I interferon-producing and dendritic cell precursors are contained in both human lymphoid and myeloid progenitor populations. J. Exp. Med. **200**: 1519–1524.

26. KARSUNKY, H. *et al.* 2005. Developmental origin of interferon-alpha-producing dendritic cells from hematopoietic precursors. Exp. Hematol. **33:** 173–181.
27. LYMAN, S. D. & S. E. JACOBSEN. 1998. c-kit ligand and Flt3 ligand: stem/progenitor cell factors with overlapping yet distinct activities. Blood **91:** 1101–1134.
28. MCKENNA, H. J. *et al.* 2000. Mice lacking flt3 ligand have deficient hematopoiesis affecting hematopoietic progenitor cells, dendritic cells, and natural killer cells. Blood **95:** 3489–3497.
29. LAOUAR, Y. *et al.* 2003. STAT3 is required for Flt3L-dependent dendritic cell differentiation. Immunity **19:** 903–912.
30. TUSSIWAND, R. *et al.* 2005. Inhibition of natural type I IFN-producing and dendritic cell development by a small molecule receptor tyrosine kinase inhibitor with Flt3 affinity. J. Immunol. **175:** 3674–3680.
31. MARASKOVSKY, E. *et al.* 1996. Dramatic increase in the numbers of functionally mature dendritic cells in Flt3 ligand-treated mice: multiple dendritic cell subpopulations identified. J. Exp. Med. **184:** 1953–1962.
32. KARSUNKY, H. *et al.* 2003. Flt3 ligand regulates dendritic cell development from Flt3+ lymphoid and myeloid-committed progenitors to Flt3+ dendritic cells *in vivo.* J. Exp. Med. **198:** 305–313.
33. MILLER, G. *et al.* 2003. Murine Flt3 ligand expands distinct dendritic cells with both tolerogenic and immunogenic properties. J. Immunol. **170:** 3554–3564.
34. GILLIET, M. *et al.* 2002. The development of murine plasmacytoid dendritic cell precursors is differentially regulated by FLT3-ligand and granulocyte/macrophage colony-stimulating factor. J. Exp. Med. **195:** 953–958.
35. ADOLFSSON, J. *et al.* 2001. Upregulation of Flt3 expression within the bone marrow Lin(-)Sca1(+)c-kit(+) stem cell compartment is accompanied by loss of self-renewal capacity. Immunity **15:** 659–669.
36. CHRISTENSEN, J. L. & I. L. WEISSMAN. 2001. Flk-2 is a marker in hematopoietic stem cell differentiation: a simple method to isolate long-term stem cells. Proc. Natl. Acad. Sci. USA **98:** 14541–14546.
37. D'AMICO, A. & L. WU. 2003. The early progenitors of mouse dendritic cells and plasmacytoid predendritic cells are within the bone marrow hemopoietic precursors expressing Flt3. J. Exp. Med. **198:** 293–303.
38. MENDE, I. *et al.* 2006. Flk2+ myeloid progenitors are the main source of Langerhans cells. Blood **107:** 1383–1390.
39. ONAI, N. *et al.* 2006. Activation of the Flt3 signal transduction cascade rescues and enhances type I interferon-producing and dendritic cell development. J. Exp. Med. **203:** 227–238.
40. MIZUKI, M. *et al.* 2003. Suppression of myeloid transcription factors and induction of STAT response genes by AML-specific Flt3 mutations. Blood **101:** 3164–3173.
41. FRIEDMAN, A. D. 2002. Transcriptional regulation of granulocyte and monocyte development. Oncogene **21:** 3377–3390.
42. ANDERSON, K. L. *et al.* 2000. Transcription factor PU.1 is necessary for development of thymic and myeloid progenitor-derived dendritic cells. J. Immunol. **164:** 1855–1861.
43. GUERRIERO, A. *et al.* 2000. PU.1 is required for myeloid-derived but not lymphoid-derived dendritic cells. Blood **95:** 879–885.
44. MIYAMOTO, T. *et al.* 2002. Myeloid or lymphoid promiscuity as a critical step in hematopoietic lineage commitment. Dev. Cell **3:** 137–147.

45. ZHANG, P. *et al.* 2000. PU.1 inhibits GATA-1 function and erythroid differentiation by blocking GATA-1 DNA binding. Blood **96:** 2641–2648.
46. NERLOV, C. *et al.* 2000. GATA-1 interacts with the myeloid PU.1 transcription factor and represses PU.1-dependent transcription. Blood **95:** 2543–2551.
47. EZOE, S. *et al.* 2005. GATA transcription factors inhibit cytokine-dependent growth and survival of a hematopoietic cell line through the inhibition of STAT3 activity. J. Biol. Chem. **280:** 13163–13170.
48. KULESSA, H., J. FRAMPTON & T. GRAF. 1995. GATA-1 reprograms avian myelomonocytic cell lines into eosinophils, thromboblasts, and erythroblasts. Genes Dev. **9:** 1250–1262.
49. DEKOTER, R. P. & H. SINGH. 2000. Regulation of B lymphocyte and macrophage development by graded expression of PU.1. Science **288:** 1439–1441.

Novel Markers for the Prospective Isolation of Human MSC

HANS-JÖRG BÜHRING,[a] VENKATA LOKESH BATTULA,[a]
SABRINA TREML,[a] BERNHARD SCHEWE,[b] LOTHAR KANZ,[a]
AND WICHARD VOGEL[a]

[a]*Department of Internal Medicine II, Division of Hematology, Immunology, Oncology and Rheumatology, University Clinic of Tübingen, 72076 Tübingen, Germany*

[b]*Hospital for Workers Compensation, 72076 Tübingen, Germany*

ABSTRACT: The isolation of mesenchymal stem cells (MSC) from primary tissue is hampered by the limited selectivity of available markers. So far, CD271 is one of the most specific markers for bone marrow (BM)-derived MSC. In search of additional markers, monoclonal antibodies (mAbs) with specificity for immature cells were screened by flow cytometry for their specific reactivity with the rare CD271$^+$ population. The recognized CD271$^+$ populations were fractionated by fluorescence-activated cell sorting and the clonogenic capacity of the sorted cells was analyzed for their ability to give rise to CFU-F. The results showed that only the CD271^{bright} but not the CD271^{dim} population contained CFU-F. Two-color flow cytometry analysis revealed that only the CD271^{bright} population was positive for the established MSC markers CD10, CD13, CD73, and CD105. In addition, a variety of mAbs specific for novel and partially unknown antigens selectively recognized the CD271^{bright} population but no other BM cells. The new MSC-specific molecules included the platelet-derived growth factor receptor-β (CD140b), HER-2/erbB2 (CD340), frizzled-9 (CD349), the recently described W8B2 antigen, as well as cell-surface antigens defined by the antibodies W1C3, W3D5, W4A5, W5C4, W5C5, W7C6, 9A3, 58B1, F9-3C2F1, and HEK-3D6. In conclusion, the described markers are suitable for the prospective isolation of highly purified BM-MSC. These MSC may be used as an improved starting population for transplantation in diseases like osteogenesis imperfecta, cartilage repair, and myocardial infarction.

KEYWORDS: mesenchymal stem cells; bone marrow; MSC; prospective isolation

Address for correspondence: Hans-Jörg Bühring, Ph.D., Department of Internal Medicine II, University Clinic, Medical Research Center, Otfried-Müller-Str. 27, 72076 Tübingen, Germany. Voice: +49-7071-2982730; fax: +49-7071-292730.
hans-joerg.buehring@uni-tuebingen.de

INTRODUCTION

Mesenchymal stem/stromal cells (MSC) are self-renewing cells that can give rise to mesodermally derived tissues including bone, cartilage, muscle, stromal cells, tendon, and connective tissue.[1-6] Cultured MSC are well-characterized plastic-adherent, spindle-shaped cells that express a panel of key markers including CD105 (endoglin, SH2), CD73 (ecto-5' nucleotidase, SH3, SH4), CD166 (ALCAM), CD29 (β1-integrin), CD44 (H-CAM), CD90 (Thy-1), and STRO-1 but are negative for CD45, CD34, and HLA-DR.[7,8] Depending on the source and on the culture conditions, they additionally express CD349 (frizzled-9), SSEA-4, Oct-4, nanog-3, and nestin.[9]

In contrast to cultured MSC, only little information exists about the features of the tissue precursor cell that can give rise to plastic-adherent cells. In addition, a strict terminology to distinguish between these cell types is lacking. In an attempt to render the terminology more precisely, the International Society for Cellular Therapy (ISCT) proposed in a position paper, that cultured MSC should be designated "multipotential mesenchymal stromal cells," whereas the term "mesenchymal stem cells" should be reserved for cells from primary tissues that can give rise to "colony-forming units–fibroblasts (CFU-F)" *in vitro* and tissue repopulation with multilineage differentiation capacity *in vivo*.[10]

To date, only a few markers have been developed and proved to be suitable for the isolation of MSCs from primary tissues. Markers that meet established criteria for their positive selection include STRO-1, CD73 (ecto-5'- nucleotidase, SH3, SH4), CD105 (endoglin, SH2), and CD271 (low-affinity nerve growth receptor), whereas CD45 and glycophorin A (CD235) are used for the negative selection of MSC.[11-18] To identify new MSC-specific markers, more than 200 monoclonal antibodies (mAbs) of our collection were screened for their reactivity with CD271-positive BM cells. Antibodies fractionating the CD271-positive BM population in at least two populations were used to isolate candidate populations by fluorescence-activated cell sorting (FACS). The clonogenic potential of the sorted cells was analyzed by scoring their ability to give rise to CFU-F. Using this approach, 15 novel MSC-specific markers were identified.

METHODS

Isolation of Bone Marrow (BM) Mononuclear Cells

BM was harvested at the Hospital for Workers Compensation from the femur shaft of patients undergoing a total hip replacement after approval of the ethics committee of the University of Tübingen. Approximately 25 mL of BM cells were collected and mixed with 5000 U of heparin (Sigma-Aldrich, Taufkirchen, Germany). Mononuclear cells were recovered by Ficoll Histopac (Biochrom

KG, Berlin, Germany) density gradient fractionation (750 × g for 20 min at RT) and remaining erythrocytes lysed in ammonium chloride solution (Cell Systems, Remagen, Germany) for 10 min at 4°C.

CFU-F Assay

The CFU-F assay was performed by plating either 1×10^5 unselected BM mononuclear cells or 5×10^3 FACS-selected BM cells into gelatine-coated T-25 flasks (0.1% gelatin/H_2O solution; CellSystems) in the presence of a serum replacement medium, (Knockout™ D-MEM; Invitrogen, Karlsruhe, Germany) containing 5 ng/mL human basic fibroblast growth factor (bFGF; CellSystems[9]). After 14 days of culture, adherent cells were washed twice with PBS, fixed with methanol (5 min, RT), and air-dried. To visualize and enumerate MSC CFU-F, the cells were stained with May-Grünwald-Giemsa solution (1 to 20 diluted with deionized water, 5 min, RT), washed twice with deionized water, and air-dried. CFU-F colonies were typically between 1 and 8 mm in diameter and were scored macroscopically.

Generation of mAbs Reactive with Human MSC

The mAbs W1C3 (IgG1), W3C4 (CD349 IgM[9,19,20]), W3D5 (IgG2a), W4A5 (IgG1), W5C4 (IgG2b), W5C5 (IgG1), W7C6 (IgG1), and W8B2 (IgG1[21]) with unknown specificities were raised by immunization of 6- to 8-week-old female Balb/c mice (Charles River WIGA, Sulzfeld, Germany) with the retinoblastoma cell line WERI-RB-1. The mAbs HEK-3D6 (IgG2a) and F9-3C2F1 (IgG1) were raised by immunization with the human embryonic kidney cell line HEK-293, mAb 58B1 (IgG1) was raised by immunization with the megakaryocytic cell line UT-7, mAb 9A3G9 (IgG1) by immunization with the breast carcinoma cell line DU.4475, mAb 39D5 (IgG1, CD56) by immunization with the CD34-positive hematopoietic cell line KG-1a,[22] and mAbs 24D2 (IgG1, CD340[23]) and 28D4 (IgG2b, CD140b[24,25]) by immunization with NIH-3T3 cells transfected with the full-length coding sequence of human HER-2 or PDGF-RB, respectively. The isotypes of mAbs were determined by enzyme-linked immunosorbent assay (ELISA) (Boehringer Mannheim, Mannheim, Germany). The properties of the mAbs are summarized in TABLE 1.

For immunization, the mice were injected five times intraperitoneally with 10^7 cells at 2-week intervals. The spleen was removed 4 days after the final boost prior to fusion with the SP2/0 myeloma cell line (German Collection of Microorganisms and Cell Cultures; DSMZ). Suspended spleen cells ($\sim 10 \times 10^7$ cells) and 3×10^7 myeloma cells were mixed, centrifuged, and washed in serum-free RPMI-1640 medium. One milliliter of polyethylene glycol (PEG; Sigma-Aldrich) was added drop-by-drop to the pellet of the cell

TABLE 1. Antigen specificity and isotypes of mAbs used in the study

mAb	Antigen	Isotype
W1C3	unknown	IgG1
W3C4	frizzled-9 (CD349)[9,19,20]	IgM
W3D5	unknown	IgG2a
W4A5	unknown	IgG1
W5C4	unknown	IgG2b
W5C5	unknown	IgG1
W7C6	unknown	IgG1
W8B2	unknown[21]	IgG1
9A3	unknown	IgG1
24D2	HER-2/erbB2 (CD340)[19,20,23]	IgG1
28D4	PDGF-RB (CD140b)[24,25]	IgG2b
39D5	N-CAM (CD56; epitope NOT on NK cells)[22]	IgG1
58B1	unknown	IgG1
F9-3C2F1	unknown	IgG1
HEK-3D6	unknown	IgG2a

NOTE. The indicated murine mAbs against human cell-surface antigens were generated as described in the "Methods" section.

mixture that was agitated for 3 min at 37°C. The resulting hybridoma cells were suspended in 50 mL RPMI-1640 containing 10% fetal bovine serum (FBS) and hypoxanthine-aminopterine-thymidine (HAT; Sigma-Aldrich) and plated into four 24-well plates (Greiner, Nürtingen, Germany). Culture supernatants from growing hybridoma cells reacting with the cell lines used for immunization but not with peripheral blood (PB) cells were considered to select the corresponding hybridoma cells. Hybridoma cells secreting antibodies reacting with CD271-positive BM cells were cloned twice by limiting dilution and cultured in the presence of hypoxanthine-thymidine (HT; Sigma-Aldrich). Growing clones were expanded in 75 cm² culture flasks and gradual removal of HT was achieved by adding HT-free RPMI-1640 medium.

Staining of Cells, Immunofluorescence Analysis, and Cell Sorting

For immunofluorescence analysis, BM cells were washed twice with phosphate-buffered saline (PBS) containing 1% of FBS and 0.01% of NaN_3 (FACS buffer) and incubated with polyglobin to block the nonspecific binding. In the next step, the cells were incubated with culture supernatants of the indicated antibodies for 15 min on ice. After washing, the cells were incubated with a $F(ab)_2$ fragment of goat anti-mouse secondary antibody conjugated with R-phycoerythrin (PE; Dako Cytomations, Glostrup, Denmark) for additional 15 min. After washing, the cells were then incubated with excess amounts of mouse IgG (1:20 diluted with PBS or FACS buffer; Southern Biotech,

Birmingham, AL) to block free binding sites of the secondary step antiserum. In the final step, cells were stained with CD271-APC (Miltenyi Biotech, Bergisch, Germany), and analyzed on a FACSCanto flow cytometer (Becton Dickinson, Heidelberg, Germany). Antibodies selectively reacting with CD271-positive populations were also used to stain cells for preparative purposes. In these cases, BM cells were stained with the indicated antibody and CD271-APC using the above described four-step labeling protocol and fractioned on a FACSAria high-speed cell sorter (Becton Dickinson). The collected cells were used to determine the CFU-F capacity.

RESULTS

Identification of mAbs Reactive with Human BM $CD271^+$ Cells

Screening of >200 inhouse generated mAbs for reactivity with $CD271^+$ BM cells revealed that 15 mAbs were able to fractionate the heterogeneous $CD271^+$ population into $CD271^{bright}$ and $CD271^{dim}$ cells. mAbs (14/15) exclusively recognized $CD271^{bright}$ but not $CD271^{dim}$ cells, and only mAb HEK-3D6 detected both cell types (FIG. 1). The highest selectivity and staining intensity for $CD271^{bright}$ cells was recorded for mAbs W8B2, W5C5, W3D5, and 28D4 (CD140b). mAbs W1C3, W5C4, and 24D2 (CD340) were almost equally selective but stained the cells at lower intensity. mAbs F9-3C2F1, and 39D5 (CD56 epitope not expressed on NK cells) detected only fractions of $CD271^{bright}$ cells. Finally, mAbs W3C4 (CD349), W7C6, W4A5, HEK-3D6, 9A3G9, and 58B1 additionally recognized rare $CD271^-$ cell populations.

Clonogenic MSC Reside in the $CD271^{bright}$ but Not in the $CD271^{dim}$ Population

To determine the clonogenic potential of mAb-defined $CD271^+$ BM populations, CFU-F assays of fractionated cells were performed. Staining of BM cells with antibodies against CD271 and CD140b revealed that CD140b is exclusively expressed on $CD271^{bright}$ cells (FIG. 2 A), and culture of fractionated cells showed that only $CD271^{bright}CD140b^+$ cells but not $CD271^{dim}CD140b^-$ cells gave rise to CFU-F (FIG. 2 B). Similar results were found in the case of $CD271^{bright}W8B2^+$, $CD271^{bright}HEK-3D6^+$, $CD271^{bright}CD349^+$, and $CD271^{bright}CD56^+$, cells (not shown). As mAbs W5C5, W3D5, W1C3, W5C4, and 24D2, F9-3C2F1, W7C6, W4A5, 9A3G9, and 58B1 exclusively detect $CD271^{bright}$ but not $CD271^{dim}$ cells and as clonogenic cells are only found in the $CD271^{bright}$ fraction, it is very likely that these antibodies specifically stain clonogenic cells as well.

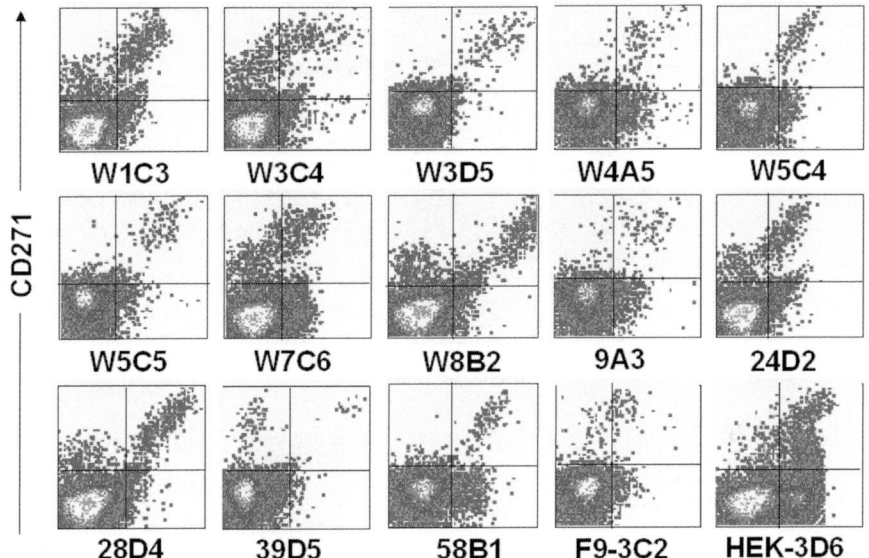

FIGURE 1. Reactivity profiles of mAbs with CD271$^+$ BM cells. BM cells were labeled with the mAbs using indirect immunofluorescence (PE) staining followed by direct staining with CD271-APC, as described in the "Methods" section. Cells were analyzed on a FACSCanto or sorted on a FACSAria flow cytometer. Data were processed using the FCS Express software (DeNovo, Thornhill, Ontario, Canada).

Morphology of Sorted CD271^{bright}CD140b$^+$ MSC

To compare the morphological features of CD271^{bright}CD140b$^+$ and CD271^{dim}CD140b$^-$ populations, the cells were sorted and stained with May-Grünwald-Giemsa. As shown in FIGURE 2 C, CD271^{bright}CD140b$^+$ cells consist of cells (MSC) with a relatively bright nuclear staining and a high cytoplasmic content. In contrast, CD271^{dim} cells show an immature lymphoblastoid appearance and a darker nuclear staining. In conclusion, CD271^{bright} cells are not only functionally but also phenotypically and morphologically distinct from CD271^{dim} cells.

DISCUSSION

Until present, CD271 was described to be the most selective marker for the characterization und purification of human BM-MSC.[14] By screening of a large number of our antibodies for reactivity with CD271$^+$ cells, 15 mAbs could be identified that detect a subpopulation of CD271$^+$ cells at high selectivity. Further analysis showed that 14/15 mAbs detected only CD271^{bright} cells and that only these cells gave rise to CFU-F, whereas CD271^{dim} cells were

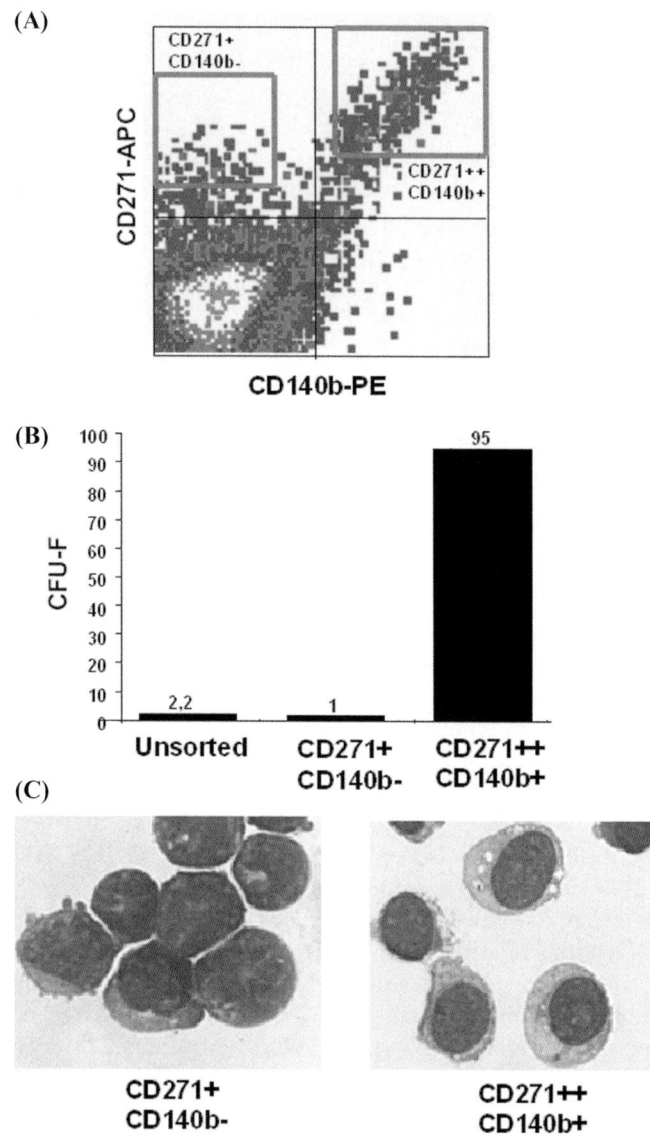

FIGURE 2. Morphology and clonogenic capacity of sorted BM cells. **(A)** Display of BM cells stained with CD271-APC and CD140b-PE. Cells were stained as described in the "Methods." **(B)** Clonogenic capacity of sorted $CD271^{bright}CD140b^{+}$ and $CD271^{dim}CD140b^{-}$ BM cells. Cells in windows shown in **(A)** were sorted on a FACSAria cell sorter (Becton Dickinson) and defined cell numbers were plated on gelatine-coated flasks and cultured serum-free, b-FGF containing medium.[9] Fourteen days after culture, the colonies were enumerated. CFU-F numbers were equalized to 5,000 plated cells. **(C)** Morphology of sorted $CD271^{bright}CD140b^{+}$ and $CD271^{dim}CD140b^{-}$ BM cells stained with May-Grünwald-Giemsa.

functionally, morphologically, and phenotypically distinct from CD271^{bright} cells. Because the mAbs (except mAb HEK-3D6) presented here recognize only the clonogenic CD271^{bright} population, they represent novel and improved tools for the selection of MSC. In this context, particular emphasis should be placed on mAbs W3D5, W5C5, and W8B2 because they combine the favorable features of the selective recognition and a bright staining of CD271^{bright} cells. Another promising candidate is mAb 39D5. This reagent recognizes a CD56 epitope that is not expressed on NK cells but on a fraction of highly clonogenic CD271^{bright} cells.[22] In conclusion, the antigens recognized by the mAbs are novel candidate targets for the prospective isolation of MSC. The selected cells may be used either directly after isolation or after *in vitro* expansion for transplantation in diseases like osteogenesis imperfecta, cartilage repair, and myocardial infarction.

ACKNOWLEDGMENTS

This work was supported by (i) the Federal Ministry for Education and Research (BMBF), BioProfile Stuttgart/Neckar-Alb; project 0313048A: Generierung von Antikörpern gegen Frizzled Rezeptoren (GPCRs) zur Isolierung und Charakterisierung von Stammzellen, by (ii) the Federal Ministry for Education and Research (BMBF), BioProfile Stuttgart/Neckar-Alb; project 0313668B: Entwicklung eines bioartifiziellen Leberreaktors mit allogenen humanen Hepatozyten, by (iii) the Deutsche Forschungsgemeinschaft (DFG) supported Sonderforschungsbereich SFB-685 (Immunotherapy: Molecular Basis and Clinical Applications) project Z2: Core Facility for Cell sorting, and by (iv) the intramurally supported Fortüne project F1282751: Herstellung und Charakterisierung monoklonaler Antikörper gegen Frizzled-Rezeptoren für die Stammzellisolierung und Tumoreliminierung.

REFERENCES

1. DEANS, R.J. & A.B. MOSELEY. 2000. Mesenchymal stem cells: biology and potential clinical uses. Exp. Hematol. **28:** 875–884.
2. ERICES, A., P. CONGET & J.J. MINGUELL. 2000. Mesenchymal progenitor cells in human umbilical cord blood. Br. J. Haematol. **109:** 235–242.
3. FRIEDENSTEIN, A.J., J.F. GORSKAJA & N.N. KULAGINA. 1976. Fibroblast precursors in normal and irradiated mouse hematopoietic organs. Exp. Hematol. **4:** 267–274.
4. LIECHTY, K.W., T.C. MACKENZIE, A.F. SHAABAN, *et al.* 2000. Human mesenchymal stem cells engraft and demonstrate site-specific differentiation after *in utero* transplantation in sheep. Nat. Med. **6:** 1282–1286.

5. PEREIRA, R.F., K.W. HALFORD, M.D. O'HARA, et al. 1995. Cultured adherent cells from marrow can serve as long-lasting precursor cells for bone, cartilage, and lung in irradiated mice. Proc. Natl. Acad. Sci. USA **92:** 4857–4861.
6. ZHAO, L.R., W.M. DUAN, M. REYES, et al. 2002. Human bone marrow stem cells exhibit neural phenotypes and ameliorate neurological deficits after grafting into the ischemic brain of rats. Exp. Neurol. **174:** 11–20.
7. CAPLAN, A.I. 1991. Mesenchymal stem cells. J. Orthop. Res. **9:** 641–650.
8. JONES, E.A., S.E. KINSEY, A. ENGLISH, et al. 2002. Isolation and characterization of bone marrow multipotential mesenchymal progenitor cells. Arthritis Rheum. **46:** 3349–3360.
9. BATTULA, V.L., P.M. BAREISS, S. TREML, et al. 2006. Human placenta and bone marrow derived MSC cultured in serum-free, b-FGF containing medium express cell surface frizzled-9 and SSEA-4 and give rise to multi-lineage differentiation. Differentiation [Epub ahead of print].
10. HORWITZ, E.M., B.K. LE, M. DOMINICI, et al. 2005. Clarification of the nomenclature for MSC: The International Society for Cellular Therapy position statement. Cytotherapy **7:** 393–395.
11. SIMMONS, P.J. & B. TOROK-STORB. 1991. Identification of stromal cell precursors in human bone marrow by a novel monoclonal antibody, STRO-1. Blood **78:** 55–62.
12. GRONTHOS, S., A.C. ZANNETTINO, S.E. GRAVES, et al. 1999. Differential cell surface expression of the STRO-1 and alkaline phosphatase antigens on discrete developmental stages in primary cultures of human bone cells. J. Bone Miner. Res. **14:** 47–56.
13. SABATINI, F., L. PETECCHIA, M. TAVIAN, et al. 2005. Human bronchial fibroblasts exhibit a mesenchymal stem cell phenotype and multilineage differentiating potentialities. Lab. Invest. **85:** 962–971.
14. QUIRICI, N., D. SOLIGO, P. BOSSOLASCO, et al. 2002. Isolation of bone marrow mesenchymal stem cells by anti-nerve growth factor receptor antibodies. Exp. Hematol. **30:** 783–791.
15. JIANG, Y., B.N. JAHAGIRDAR, R.L. REINHARDT, et al. 2002. Pluripotency of mesenchymal stem cells derived from adult marrow. Nature **418:** 41–49.
16. REYES, M. & C.M. VERFAILLIE. 2001. Characterization of multipotent adult progenitor cells, a subpopulation of mesenchymal stem cells. Ann. N. Y. Acad. Sci. **938:** 231–233.
17. VERFAILLIE, C. 2005. Stem cell plasticity. Hematology **10**(Suppl 1): 293–296.
18. VERFAILLIE, C.M. 2005. Multipotent adult progenitor cells: an update. Novartis Found. Symp. **265:** 55–61.
19. ZOLA, H., B. SWART, A. BANHAM, et al. 2006. CD molecules 2006—human cell differentiation molecules. J. Immunol. Methods. **30**(319): 1–5.
20. ZOLA, H. et al. 2006. Workshop on human cell differentiation molecules. http://www.hcdm.org/HCDMAssignments.htm.
21. VOGEL, W., F. GRUNEBACH, C.A. MESSAM, et al. 2003. Heterogeneity among human bone marrow-derived mesenchymal stem cells and neural progenitor cells. Haematologica **88:** 126–133.
22. RITZ, J., G. TRINCHIERI & L.L. LANIER. 1995. NK-cell antigens: section report. *In* Leucocyte Typing V. White Cell Differentiation Antigens, Vol. V. S.F. Schlossman, et al. Ed.: 1367–1372. Oxford University Press. Oxford, New York, Tokyo.
23. BUHRING, H.J., I. SURES, B. JALLAL, et al. 1995. The receptor tyrosine kinase p185HER2 is expressed on a subset of B-lymphoid blasts from patients with

acute lymphoblastic leukemia and chronic myelogenous leukemia. Blood **86**: 1916–1923.
24. HART, C.E. & D.F. BOWEN-POPE. 1998. New endothelial Cell CD antigens: CD140a and b (PDGFRá and â) Workshop Panel report. *In* Leucocyte Typing VI, White Cell Differentiation Antigens, Vol. VI. T. Kishimoto, *et al*. Ed.: 739–742. Garland Publishing. New York & London.
25. SPRINGER, T.A. & J. KITAYAMA. 1998. Endothelial cell antigens: section report. *In* Leucocyte Typing VI, White Cell Differentiation Antigens, Vol. VI. T. Kishimoto *et al*. Ed.: 693–702. New York & London.

Modulation of Immune Responses by Mesenchymal Stem Cells

WILLEM E. FIBBE, ALMA J. NAUTA, AND HELENE ROELOFS

Department of Immunohematology and Blood Transfusion, Leiden University Medical Center, 2300 RC Leiden, the Netherlands

ABSTRACT: Mesenchymal stem cells (MSCs) are multipotent progenitor cells and interest in MSC therapy has been raised by the observation that MSCs are able to modulate immune responses *in vitro* and *in vivo*. Here, we show that MSCs are not intrinsically immune privileged and are capable of inducing memory T cell responses following injection *in vivo* in immunocompetent hosts. After cotransplantation in recipients that have received sublethal irradiation, allogeneic MSCs can still induce an alloresponse that may result in graft rejection, suggesting that the immunogenicity of allogeneic MSCs are not fully prevented by a nonmyeloablative conditioning regimen. It is still unclear whether the immunogenicity of allogeneic MSCs is also preserved following a fully myeloablative conditioning regimen.

KEYWORDS: mesenchymal stem cells; dendritic cells; hematopoiesis

INTRODUCTION

Mesenchymal stem cells (MSCs) are multipotent progenitor cells that have emerged as a promising tool for clinical application. The interest in MSC therapy has been raised by the observation that MSCs are able to modulate immune responses *in vitro* and *in vivo*. These properties may be used in clinical therapy in the context of allogeneic stem cell transplantation, in particular to modulate graft-versus-host disease (GVHD) and graft rejection.

MARROW STROMAL CELLS

Human marrow stromal cells or MSCs comprise a heterogeneous population of cells, including reticular endothelial cells, fibroblasts, adipocytes, and osteogenic precursor cells that provide growth factors, cell–cell interactions, and matrix proteins that play a role in the regulation of hematopoiesis.[1–3] The notion that the stromal microenvironment could support hematopoiesis followed by the development of the long-term bone marrow culture by Dexter.[3] In this system, an adherent bone marrow–derived stromal culture could support the production of hematopoietic progenitor cells over a period of several weeks to months. Friedenstein et al.[4] already described a population of adherent cells from the bone marrow that were nonphagocytic and exhibited a fibroblast-like appearance. On culture at low density either as whole bone marrow or following separation over a density gradient, the cells formed characteristic colonies derived from a single precursor, referred to as colony-forming-unit fibroblastic or CFU-F. Following ectopic transplantation under the kidney capsule, these cells gave rise to a broad spectrum of differentiated connective tissues, including bone, cartilage, adipose tissue, and myelosupportive stroma.[4,5] On the basis of these observations, it was proposed that these tissues were derived from a common precursor cell residing in the bone marrow, termed the *stromal stem cell*, the *bone marrow stromal stem cell*, the *mesenchymal stem cell*, or the *skeletal stem cell*.[6,7]

Because of their ability to migrate to sites of tissue injury, MSCs have emerged as a promising therapeutic modality for tissue regeneration and repair. Several studies in animal models have demonstrated that MSCs are capable of long-term engraftment and *in vivo* differentiation, and encouraging results have been reported in clinical use.[8–10] MSCs are known to secrete a number of cytokines and regulatory molecules implicated in different aspects of hematopoiesis. These characteristics have generated clinical interest to use MSCs to enhance hematopoietic stem cell engraftment. Although animal models provide experimental evidence that MSCs facilitate engraftment,[11,12] no conclusive evidence has yet been presented in humans. In addition to providing critical growth factors, MSCs display immunosuppressive properties that might facilitate engraftment. *In vitro* studies with human, baboon, and murine MSCs demonstrated that MSCs suppress the proliferation of T cells induced by alloantigens or mitogens.[13–15] Furthermore, MSCs have been reported to induce T cell division arrest,[16] to inhibit the differentiation and maturation of dendritic cells (DCs),[17] and to decrease the production of inflammatory cytokines by various immune cell populations.[18] In line with their immunosuppressive capacities *in vitro*, animal studies indicate that MSCs also display immunosuppressive capacities *in vivo*. Allogeneic MSCs may prolong skin allograft survival in immunocompetent baboons[13] and may prevent the rejection of allogeneic tumor cells in immunocompetent mice.[19] The mechanisms underlying these effects of MSCs have not been clearly identified. Although conflicting results have been reported, most studies agree that soluble factors are involved.[14,19] The therapeutic application of the immunosuppressive properties of MSCs has already been exploited in the clinical setting for the treatment of acute GVHD after allogeneic stem cell transplantation.[20] The immunophenotype of MSCs, the low expression of human leukocyte antigen

(HLA), major histocompatibility complex (MHC) class I, and the absence of costimulatory molecules, together with the observation that MSCs do not elicit a proliferative response from allogeneic lymphocytes, suggest that MSCs are of inherently low immunogenicity.[14] The suppressive effect is likely mediated by inhibition of T cell proliferation rather then by suppressing the initiation of immune responses. Removal of MSCs restores the T cell proliferative response indicating that the efferent arm of the immune response is inhibited, possibly by the release of soluble factors including indoleamine 2,3-dioxygenase (IDO).[21]

EFFECTS OF MSCs ON DCs

It is well known that MSCs exert profound immunosuppressive properties on T cell proliferation. However, their effect on the initiators of the immune response, the DCs are relatively unknown. DCs are the most potent antigen-presenting cells that are specialized in antigen uptake, transport, and presentation and have the unique capacity to stimulate naïve and memory T cells.[22]

In addition, DCs may interact with other immune cells including B cells[23] and NK cells.[24] DCs can be grown *in vitro* from monocytes in the presence of GM-CSF and IL-4, and from CD34$^+$ bone marrow stem cells in the presence of GM-CSF, SCF, and TNF-α. DCs play an important role in the initiation of primary immune responses and in the induction of tolerance, depending on their stage of activation and maturation. Immature DCs are characterized by their high ability of antigen uptake and processing and a low ability to stimulate T cells. As a result of locally produced inflammatory cytokines or microbial components, they can mature to an antigen-presenting stage that is characterized by upregulation of costimulatory molecules and MHC class II as well as by production of cytokines such as I-12.

It is believed that mature DCs induce immunogenic T cell responses, whereas antigen presentation by immature DCs results in tolerance.[25,26] We have therefore studied the effect of MSCs on the differentiation and function of both monocyte-derived DCs and CD34-derived DCs.

DIFFERENTIATION OF DCs IS INHIBITED BY MSCs

To study the effect of MSCs on the differentiation of DCs, cord blood–derived CD34$^+$ cells were cultured in the presence of GM-CSF, TNF-α, and SCF in the presence or absence of MSCs. CD34$^+$ progenitor cells may differentiate into CD1a$^+$ Langerhans' cells and into interstitial DCs that differentiate from an intermediate CD14$^+$ CD1a$^-$ precursor cell into CD14$^-$ CD1a$^+$ DCs. The addition of MSCs during differentiation almost completely prevented the differentiation into CD14$^-$ CD1a$^+$ cells, and the cells exhibited a strongly reduced expression of HLA DR and costimulatory molecules, that is, CD40,

CD80, CD83, and CD86. Similar results were obtained with DCs that were derived from CD14$^+$ monocytes. CD14$^+$ monocytes were cultured in the presence of IL-4 and GM-CSF and then differentiated in the presence or absence of MSCs. Coculture in the presence of MSCs resulted in decreased expression of CD80, CD83, CD86, and of HLA DR. Activation of immature DCs by CD40 ligand in the presence of MSCs did not result in upregulation of costimulatory molecules.

FUNCTIONAL PROPERTIES OF DCs GENERATED IN THE PRESENCE OF MSCs

In addition to phenotypical changes, the ability to induce T cell stimulation by immature DCs generated in the presence of MSCs was studied. These DCs were significantly hampered in their ability to induce allogeneic T cell proliferation in comparison with DCs that were generated in the absence of MSCs. Similarly, the capacity of DCs derived from CD14$^+$ precursors and monocytes to stimulate proliferation of T cells was also significantly reduced in cultures containing MSCs.

DCs generated from CD34$^+$ progenitor cells as well as from monocytes that were activated with CD40 ligand exhibited a strongly impaired production of IL-10 and IL-12. Monocyte-derived DCs generated in the presence of MSCs also showed a decreased production of TNF-α.

The mechanisms that underlie the inhibition of DC differentiation and function by MSCs remain unclear. Transwell experiments have indicated the involvement of soluble factors, and several candidates were considered, including IL-6 and M-CSF. Neutralizing antibodies against IL-6 and M-CSF, however, reduced the expression of CD14, but did not fully restore the acquisition of CD1a, indicating that in addition to IL-6 and M-CSF other soluble factors are involved.

It has been demonstrated that MSC injection *in vivo* may lead to prolonged skin allograft survival and may diminish GVHD. It remains, however, unclear whether the reported MSC-induced modification of DC differentiation plays any role *in vivo* in the immunosuppressive effects. Other possibilities including the induction of regulatory T cells are being considered and studied.

IMMUNE MODULATION *IN VIVO* BY MSCs

Previous studies have suggested that the immune suppressive effects of MSCs on T cells are independent of MHC matching between MSCs and T cells. To examine their possible graft-facilitating effect, we have transplanted recipient Balb/C mice with MHC mismatched donor bone marrow cells derived from B6 mice in the presence or absence of MSCs.[27] Addition of recipient-derived MSCs to the graft resulted in significant promotion of engraftment, whereas addition of a single or multiple doses of donor-derived MSCs hampered engraftment of allogeneic bone marrow cells. These results indicated that injection of allogeneic MSCs might have triggered an alloimmune response that mediated graft rejection. To further study whether the infusion of allogeneic MSCs could indeed induce a memory T cell response, naïve and immunocompetent recipient mice were infused with either B6 MSCs or Balb/C-derived MSCs and several weeks later an *in vivo* cytotoxicity assay was performed. This assay relies on the injection of differentially CFSE-labeled recipient Balb/C (low concentration of CFSE) and donor B6 (high CFSE concentration) splenocytes. Balb/C recipient mice were injected with a single dose of MSCs, and 4 weeks later, the elimination kinetics of allogeneic B6 splenocytes was studied. Balb/C mice that had been treated with B6 MSCs eliminated almost all CFSE-labeled allogeneic splenocytes within 3 days, indicating that the infusion of allogeneic MSCs had triggered a memory T cell response.

GRAFT-PROMOTING PROPERTIES OF MSCs

The engraftment-enhancing properties observed by infusion of syngeneic MSCs confirm several studies conducted in animal models.[13,19] The mechanisms underlying these properties are still poorly understood. One hypothesis proposes that the immune modulatory properties are responsible for promotion of engraftment by suppression of residual T cell or NK cell alloreactivity. The graft-promoting effects might be mediated by the secretion of immunoregulatory soluble factors including prostaglandin E2[18] and IDO,[21] which can suppress T cell proliferation. It is also possible that MSCs provide stromal support of donor hematopoietic stem cells and thereby enhance hematopoiesis following transplantation. There is, however, little evidence that donor stromal cells are able to engraft following systemic intravenous infusion, and donor-derived MSCs have been undetectable in the bone marrow of recipients following bone marrow transplantation.

The mechanism of enhancement may therefore be independent of the ability of MSCs to home to the bone marrow. Alternatively, support could also be mediated by hematopoietic growth factors produced by MSCs.

The infusion of allogeneic MSCs may induce priming of allogeneic T cells in immunocompetent mice, clearly indicating that MSCs may be immunogenic under appropriate circumstances. Recent observations using adult MSCs injected in rat myocardium support the notion that allogeneic MSCs can be immunogenic.[28] Injection of allogeneic MSCs was associated with significant infiltration consisting primarily of macrophages and was also associated with rejection, whereas persistent engraftment of adult MSCs was observed in immunologic incompetent rats. In our own studies, the immune response elicited by the infusion of allogeneic MSCs may have enhanced the response of host cells against donor antigen, resulting in increased rejection of allogeneic donor bone marrow cells.

In summary, our results show that MSCs are not intrinsically immune privileged and are capable of inducing memory T cell responses following injection *in vivo* in immunocompetent hosts. Following cotransplantation in recipients that have received sublethal irradiation, allogeneic MSCs can still induce an alloresponse that may result in graft rejection. These results suggest that the immunogenicity of allogeneic MSCs is not fully prevented by a nonmyeloablative conditioning regimen. It is still unclear whether the immunogenicity of allogeneic MSCs are also preserved following a fully myeloablative conditioning regimen. However, this should be taken into account in designing clinical studies in the setting of allogeneic stem cell transplantation.

REFERENCES

1. TAVASSOLI, M. & A. FRIEDENSTEIN. 1983. Hemopoietic stromal microenvironment. Am. J. Hematol. **15:** 195–203.
2. LICHTMAN, M.A. 1981. The ultrastructure of the hemopoietic environment of the marrow: a review. Exp. Hematol. **9:** 391–410.
3. ALLEN, T.D., T.M. DEXTER & P.J. SIMMONS. 1990. Marrow biology and stem cells. Immunol. Ser. **49:** 1–38.
4. FRIEDENSTEIN, A.J., U.F. DERIGLASOVA, N.N. KULAGINA, *et al.* 1974. Precursors for fibroblasts in different populations of hematopoietic cells as detected by the *in vitro* colony assay method. Exp. Hematol. **2:** 83–92.
5. OWEN, M. 1988. Marrow stromal stem cells. J. Cell. Sci. Suppl. **10:** 63–76.
6. PROCKOP, D.J. 1997. Marrow stromal cells as stem cells for nonhematopoietic tissues. Science **276:** 71–74.
7. GERSON, S.L. 1999. Mesenchymal stem cells: no longer second class marrow citizens. Nat. Med. **5:** 262–264.
8. HORWITZ, E.M., P.L. GORDON, W.K. KOO, *et al.* 2002. Isolated allogeneic bone marrow-derived mesenchymal cells engraft and stimulate growth in children with osteogenesis imperfecta: implications for cell therapy of bone. Proc. Natl. Acad. Sci. USA **99:** 8932–8937.
9. KOC, O.N., J. DAY, M. NIEDER, *et al.* 2002. Allogeneic mesenchymal stem cell infusion for treatment of metachromatic leukodystrophy (MLD) and Hurler syndrome (MPS-IH). Bone Marrow Transplant **30:** 215–222.
10. KOC, O.N., S.L. GERSON, B.W. COOPER, *et al.* 2000. Rapid hematopoietic recovery after confusion of autologous-blood stem cells and culture-expanded marrow mesenchymal stem cells in advanced breast cancer patients receiving high-dose chemotherapy. J. Clin. Oncol. **18:** 307–316.
11. ALMEIDA-PORADA, G, C.D. PORADA, N. TRAN & E.D. ZANJANI. 2000. Cotransplantation of human stromal cell progenitors into preimmune fetal sheep results in early appearance of human donor cells in circulation and boosts cell levels in bone marrow at later time points after transplantation. Blood **95:** 3620–3627.
12. NOORT, W.A., A.B. KRUISSELBRINK, P.S. IN'T ANKER, *et al.* 2002. Mesenchymal stem cells promote engraftment of human umbilical cord blood-derived CD34(+) cells in NOD/SCID mice. Exp. Hematol. **30:** 870–878.

13. BARTHOLOMEW, A., C. STURGEON, M. SIATSKAS, *et al*. 2002. Mesenchymal stem cells suppress lymphocyte proliferation *in vitro* and prolong skin graft survival *in vivo*. Exp. Hematol. **30:** 42–48.
14. DI NICOLA, M., C. CARLO-STELLA, M. MAGNI, *et al*. 2002. Human bone marrow stromal cells suppress T-lymphocyte proliferation induced by cellular or nonspecific mitogenic stimuli. Blood **99:** 3838–3843.
15. KRAMPERA, M., S. GLENNIE, J. DYSON, *et al*. 2003. Bone marrow mesenchymal stem cells inhibit the response of naive and memory antigen-specific T cells to their cognate peptide. Blood **101:** 3722–3729.
16. GLENNIE, S., I. SOEIRO, P.J. DYSON, *et al*. 2005. Bone marrow mesenchymal stem cells induce division arrest anergy of activated T cells. Blood **105:** 2821–2827.
17. JIANG, X.X., Y. ZHANG, B. LIU, *et al*. 2005. Human mesenchymal stem cells inhibit differentiation and function of monocyte-derived dendritic cells. Blood **105:** 4120–4126.
18. AGGARWAL, S. & M.F. PITTENGER. 2005. Human mesenchymal stem cells modulate allogeneic immune cell responses. Blood **105:** 1815–1822.
19. DJOUAD, F., P. PLENCE, C. BONY, *et al*. 2003. Immunosuppressive effect of mesenchymal stem cells favors tumor growth in allogeneic animals. Blood **102:** 3837–3844.
20. LE BLANC, K., I. RASMUSSON, B. SUNDBERG, *et al*. 2004. Treatment of severe acute graft-versus-host disease with third party haploidentical mesenchymal stem cells. Lancet **363:** 1439–1441.
21. MEISEL, R., A. ZIBERT, M. LARYEA, *et al*. 2004. Human bone marrow stromal cells inhibit allogeneic T-cell responses by indoleamine 2,3-dioxygenase-mediated tryptophan degradation. Blood **103:** 4619–4621.
22. BANCHEREAU, J., F. BRIERE, C. CAUX, *et al*. 2000. Immunobiology of dendritic cells. Annu. Rev. Immunol. **18:** 767–811.
23. DUBOIS, B., J.M. BRIDON, J. FAYETTE, *et al*. 1999. Dendritic cells directly modulate B cell growth and differentiation. J. Leukoc. Biol. **66:** 224–230.
24. GEROSA, F., B. BALDANI-GUERRA, C. NISII, *et al*. 2002. Reciprocal activating interaction between natural killer cells and dendritic cells. J. Exp. Med. **195:** 327–333.
25. LUTZ, M.B. & G. SCHULER. 2002. Immature, semi-mature and fully mature dendritic cells: which signals induce tolerance or immunity? Trends Immunol. **23:** 445–449.
26. NAUTA, A.J., A.B. KRUISSELBRINK, E. LURVINK, *et al*. 2006. Mesenchymal stem cells inhibit generation and function of both CD34+-derived and monocyte-derived dendritic cells. J. Immunol. **177:** 2080–2087.
27. NAUTA, A.J., G. WESTERHUIS, A.B. KRUISSELBRINK, *et al*. 2006. Donor-derived mesenchymal stem cells are immunogenic in an allogeneic host and stimulate donor graft rejection in a nonmyeloablative setting. Blood **108:** 2114–2120.
28. GRINNEMO, K.H., A. MANSSON, G. DELLGREN, *et al*. 2004. Xenoreactivity and engraftment of human mesenchymal stem cells transplanted into infarcted rat myocardium. J. Thorac. Cardiovasc. Surg. **127:** 1293–1300.

Feasibility and Outcome of Reduced-Intensity Conditioning in Haploidentical Transplantation

RUPERT HANDGRETINGER,[a] XIAOHUA CHEN,[b] MATTHIAS PFEIFFER,[a] INGO MUELLER,[a] TOBIAS FEUCHTINGER,[a] GREGORY A. HALE,[b] AND PETER LANG[a]

[a]*Children's University Hospital, University of Tuebingen, Tuebingen, Germany*
[b]*St. Jude Children's Research Hospital, Memphis, Tennessee, USA*

ABSTRACT: Allogeneic stem cell transplantation is for a number of patients with malignant and nonmalignant diseases the only curative approach. For those patients who do not have an HLA-identical–related or –unrelated stem cell donor, a related three-loci mismatch haploidentical stem cell transplantation with T cell-depleted stem cells is a viable option. T cell depletion either by $CD34^+$ positive selection or by CD3-negative depletion strategies is available and has been investigated. We have shown that reduced-intensity conditioning haploidentical transplantation using mobilized peripheral stem cells negatively depleted from T and B lymphocytes is associated with a rapid immune reconstitution, a low transplant-related mortality rate, and a favorable outcome in patients in remission at the time of transplant. For chemorefractory patients, additional posttransplant cellular and humoral immunotherapeutic strategies are needed for prevention of relapse after transplantation.

KEYWORDS: haploidentical transplantation; reduced intensity conditioning; T cell depletion

INTRODUCTION

Allogeneic hematopoietic stem cell transplantation is for a number of patients with malignant and nonmalignant diseases the only curative approach. Since only 25–30% of the patients requiring transplantation will have an HLA-matched–related donor, volunteer-matched unrelated donors (MUDs), unrelated cord blood units, or mismatched family donors (haploidentical donors) are used as hematopoietic stem cell source. With the help of the worldwide registries of unrelated donors, the probability to identify a MUD donor depends

Address for correspondence: Rupert Handgretinger, M.D., Children's University Hospital, University of Tuebingen, Hoppe-Seyler-Strasse 1, 72076 Tuebingen, Germany. Voice: 49-7071-298-4744; fax: 49-7071-294-713.
Rupert.Handgretinger@med.uni-tuebingen.de

on the diversity of the HLA antigens within a population and on the race, ranging from approximately 75% for Caucasians to less than 50% for ethnic minorities. Given the complexity, a MUD donor search is time-consuming and the time from initiation of a donor search to the transplant is in most cases greater than 3 months or even longer. Because of malignant nature of the patients' diseases, a number of patients will progress during the donor search and will no longer be a candidate for a potential life-saving allogeneic transplantation.

An alternative approach for those patients who do not have an HLA-matched donor or who are at high risk of disease progression during the donor search is the use of mismatched related family donors.[1] Most of these donors will only share one HLA haplotype with the patients and are therefore referred to as haploidentical donors. Although haploidentical transplantation has been employed in the past, it was not widely adopted as a routine procedure because of a number of clinical problems associated with the HLA mismatch, such as graft failures, prolonged and profound immunodeficiency after transplantation with high rates of life-threatening infections, higher incidence of post-transplant lymphoproliferative disease (PTLPD), and severe acute and chronic graft-versus-host disease (GVHD) due to the lack of effective *in vitro* T cell depletion methods.[2,3]

However, two major recent findings support the use of haploidentical transplantation: (*a*) the megadose concept, which shows that the transplantation of large numbers of haploidentical transplantation of purified $CD34^+$ stem cells can overcome the HLA barrier and induce tolerance[4] and (*b*) the concept of alloreactive natural killer (NK) cells, which shows that NK cells from an HLA-mismatched donor can exert an antileukemic effect against the patients' blasts in the absence of HLA antigens engaging killer-inhibitory receptors (KIR).[5] In addition to their antileukemic effect, it has been shown in a murine model that alloreactive NK cells facilitate engraftment and might prevent GVHD.[6]

CLINICAL EXPERIENCE USING $CD34^+$ POSITIVE SELECTION IN HAPLOIDENTICAL TRANSPLANTATION

A major breakthrough in haploidentical transplantation was achieved with the development of effective *in vitro* T cell depletion systems and the use of mobilized peripheral blood stem cells (PBSCs) in addition to bone marrow[7] or as the sole source of stem cells.[8] Especially the introduction of $CD34^+$ positive selection strategies for the indirect depletion of T and B lymphocytes from PBSC allowed the transplantation of megadoses of haploidentical stem cells without any further pharmacological immunosuppression in the absence of GVHD.[8] In our own clinical experience with the transplantation of highly purified haploidentical $CD34^+$-selected PBSCs in a pediatric patient population, we could show that this approach is associated with a very low incidence of

GVHD.[9] However, we observed a delayed immunoreconstitution with a high incidence of lethal viral infections.[9] We could also document the important role of alloreactive NK cells in this setting and patients with a donor KIR versus recipient HLA mismatch conferring NK-alloreactivity had a significantly lower risk of relapse compared to patients with a NK nonalloreactive donor KIR versus recipient HLA match.[10] In contrast to adults, where such a positive effect has only been seen in patients with acute myeloid leukemia (AML),[11] we could also observe a lower risk of relapse in pediatric patients with acute lymphoblastic leukemia (ALL) who received a transplant from a haploidentical NK-alloreactive donor.[10,12] From this clinical experience, we have proposed an algorithm for the selection of an NK-alloreactive haploidentical donor.[12]

In summarizing what we have learned from haploidentical transplantation using highly purified $CD34^+$ stem cells from PBSC, four major points have to be mentioned:

(1) We could demonstrate that haploidentical transplantation is feasible without significant GVHD.[9]
(2) The immune reconstitution is very delayed and the patients are at high risk to develop severe and lethal viral and fungal infections. In combination with myeloablative conditioning regimens, the transplant-related mortality (TRM) reaches up to 25%.[9]
(3) The event-free and overall survival for children with ALL is similar to those receiving a standard allogeneic-matched donor transplantation and haploidentical transplantation is a reasonable option for those patients who do not have a suitable HLA-matched donor.[13,14]
(4) Alloreactive NK cells have an antileukemic effect in the absence of GVHD and play a major role in the prevention of relapses.[10,12]

In order to further improve the outcome of haploidentical transplantation, we have asked the question what an optimal graft might be and we have come to the following conclusions:

(1) An ideal graft should contain large numbers (megadoses) of stem cells.
(2) It should contain large numbers of NK cells.
(3) Since T cell alloreactivity will override the effects of NK alloreactivity in minimally T cell-depleted transplants,[15] it should be effectively depleted of T and B cells to prevent GVHD and PTLPD.

From these conclusions, we developed a negative T cell depletion strategy, in which T cells (CD3)[16] or T and B cells (CD3/19)[17] are negatively depleted from PBSC with 3.5–4 log T cell depletion.[18] In contrast to the $CD34^+$ positive selection strategy, such grafts contain beside large numbers of stem cells also large numbers of NK- cells and other myeloid cells, such as monocytes, dendritic cells, and others myeloid progenitor cells. It is important to point out that the flowcytometric determination of the absolute numbers of T cells after negative depletion is crucial in order to prevent the transplantation of too many GVHD-inducing T cells.[18]

In order to further reduce the incidence of TRM observed after myeloablative conditioning regimens and given the observation in the murine model that alloreactive NK cells might facilitate engraftment,[6] we decided to evaluate the possibility of reduced-intensity conditioning (RIC) in haploidentical transplantation using CD3/19-depleted PBSCs containing large numbers of stem and NK cells.

FEASIBILITY OF REDUCED-INTENSITY CONDITIONING IN HAPLOIDENTICAL TRANSPLANTS USING CD3/19-DEPLETED GRAFTS

The first clinical study using this new approach was performed at the St. Jude Children's Research Hospital (SJCRH) in Memphis, Tennessee.[19] In this feasibility study, a total of 25 pediatric patients with refractory hematological malignancies or considered at high risk of TRM were enrolled. Ten patients had refractory (3 ALL, 4 AML/MDS, and 3 NHL) and 15 patients had relapsed after a standard myeloablative-allogeneic transplantation (7 ALL, 7 AML, and 1 CML). The conditioning consisted of fludarabine (200 mg/m^2), thiotepa (10 mg/kg), melphalan (120 mg/m^2), and the anti-CD3 monoclonal antibody (mAb) muronomab. The RIC approach was quite safe in these high-risk patients, and only 3 of the 25 patients at high risk for TRM died from TRM and lethal viral infections were not seen. With a follow-up of greater than 1.5 years, 2 of the 10 refractory patients and 7 of the 15 second transplant patients are alive and free of disease.

The absence of lethal viral infections in these patients could be explained by the rapid immune reconstitution, which was determined by the recovery of CD4 and CD8$^+$ T cells, by a fast recovery of thymopoiesis as determined by the rapid increase of T cell receptor excision circles (TREC), and by a rapid recovery of the T cell receptor repertoire as measured by the increase of the V beta spectratyping score.[20] This immune reconstitution pattern was compared in a retrospective analysis with that of patients receiving a total body irradiation (TBI) containing myeloablative regimen (12 Gy) and CD3-depleted haploidentical PBSCs.[21] Although the graft was similar for both patient groups, the RIC regimen allowed a faster immune reconstitution with a much faster recovery of the thymic function. We think therefore that the TBI exerts significant damage to the thymus, which is associated with the delay of the recovery of the T cell repertoire.

From these encouraging initial clinical experiences, we initiated a subsequent RIC protocol at the Children's University Hospital in Tuebingen, Germany,[22] in patients with high risk or refractory hematological malignancies

($n = 35$) and three patients with severe aplastic anemia. Most of the pediatric patient received a similar RIC protocol as described earlier. Because of some neurotoxicity most likely associated to high-dose fludarabine (200 mg/m^2) at SJCRH, we reduced the fludarabine dose to 160 mg/m^2 and increased the melphalan dose to 140 mg/m^2, whereas the dose of thiotepa of 10 mg/kg was maintained. Again, the anti-CD3 mAb muronomab was given. Up to now (December 2006), 38 patients have been enrolled in the study. The grafts consisted of CD3/19-depleted haploidentical PBSCs and contained a median of 16×10^6/kg CD34$^+$ stem cells (range 7–41), 49,000/kg T cells (range 7000–20,000), 20,000/kg B cells (range 2000–47,000), 137×10^6/kg CD56$^+$ NK cells (range 9–550), and 620×10^6/kg myeloid cells (range 51–1345). Eighty-three percent of the 38 patients showed a primary engraftment, whereas 17% rejected. All patient who failed to engraft were retransplanted from the same or the other haploidentical donor and finally all patients engrafted. This compared favorable with our previous experience using a myeloablative approach and CD34$^+$ positively selected stem cells in 63 patients, where we also encountered a primary engraftment in 83% of the patients and a final engraftment in 98%. The time to reach greater than 500 neutrophils was 11 days (range 9–12 days). The median time to independence of platelet substitution ($>20 \times 10^9$ /L platelets for >7 days) was only 9 days. Acute GVHD grade 0–1 was seen in 73%, grade 2 in 24%, and grade 3–4 in 3% of the patients. In addition to the CD3 depletion, this low rate of GVHD was achieved by applying mycofenolate mofetil starting from day 1 in those patients who received greater than 25,000/kg T cells with the graft.

Among the 38 patients, no lethal viral infection was seen, although the reactivation of cytomegalovirus or adenovirus (ADV) was quite common.[23] However, rapid preemptive antiviral therapy, especially for ADV, and the rapid immune reconstitution prevented lethal viral complications.[24] Some of the patients received ADV-specific T cells as described by us.[25]

Only 1 out of the 38 patients died as a result of a nonrelapse complication. This patient had a refractory AML who had received very intensive pretransplant chemotherapy. She went into transplant with a low cardiac ejection fraction and died because of heart failure 6 months posttransplant while in remission from her malignant disease. This was the only case of TRM in this high-risk cohort of 38 patients. This low TRM rate of 2.6% compares very favorable to our previous experience in 63 patients who received a myeloablative conditioning regimen and CD34$^+$ positively selected haploidentical transplant, where we observed a TRM rate of up to 35%. With all the limitations of retrospective comparisons, one can conclude that RIC regimens in haploidentical transplantations are clearly associated with a reduced TRM rate even in patients who would be considered at high risk for TRM in a standard allogeneic transplantation approach or who would not even be considered for a standard allogeneic transplantation. The event-free survival of the patients is 70% for the patients with nonmalignant diseases and those patients in remission at the time of transplant and approximately 20% for those patients with chemorefractory disease pretransplant. Although a much longer clinical follow-up will be needed, we can conclude from these clinical results that patients who are in remission at the time of transplant or who have a nonmalignant disease have a favorable event-free survival due to the low rate of TRM. Patients with refractory malignant disease at the time of transplant have a much higher rate of relapse posttransplant. Since the cause of death in all but one of the patients with refractory disease was relapse, posttransplant strategies to prevent relapse are needed. We think that the use of myeloablative regimens in refractory patients will most likely increase the TRM, so that additional posttransplant cellular or humoral therapeutic strategies for relapse prevention need to be evaluated in haploidentical RIC transplantation.

POSSIBLE POSTTRANSPLANT STRATEGIES FOR THE PREVENTION OF RELAPSE

Because of the continuous availability and the motivation of haploidentical donors, this approach offers tremendous opportunities for the evaluation of posttransplant strategies, especially in patients with refractory disease and at high risk of relapse. Such strategies include the use of new chemotherapeutic agents, such as clofarabine, pretransplant to overcome the refractory status to standard chemotherapeutic regimens prior to transplant[26] or the inclusion of such new agents into the RIC regimen. Along this line, we have used clofarabine (4×50 mg/m^2) instead of fludarabine in our RIC regimen in four patients with refractory disease. This regimen was well tolerated and all four patients showed a rapid engraftment (R. Handgretinger, unpublished data). Further clinical studies are now under way at our institutions to evaluate the role of clofarabine as part of the RIC regimen in refractory patients.

A strategy to further increase the stem cell dose could be the use of the CXCR4-Inhibitor in combination with G-CSF for the mobilization of large numbers of stem cells. Although this new mobilization regimen is employed in allogeneic donors,[27] studies in haploidentical donors are still in the planning phase. The transplantation of larger numbers of CD34$^+$ stem cells could result in a higher rate of primary engraftment.

Another posttransplant cellular therapeutic option is the infusion of large numbers of donor-derived NK cells. We and others have established large-scale clinical isolation methods for the purification of NK cells according to Good Manufacturing Practice (GMP) guidelines and have successfully used such a clinical strategy.[28–30] Smaller clinical studies have investigated a similar approach and no final conclusions can still be drawn from the small patient numbers.[30] Therefore, clinical studies with larger numbers of patients are necessary to assess the role of the posttransplant adoptive transfer of NK cells. NK cells not only mediate natural cytotoxicity, but are also the main mediators of the antibody-dependent cellular cytotoxicity (ADCC) through activation of their Fc receptors. ADCC could be a powerful tool to lyse fresh leukemic blasts in the presence of suitable antibodies. In this context, we have shown that chimeric anti-CD19 antibodies can mediate a considerable cytotoxicity against CD19$^+$ blasts in patients after haploidentical transplantation.[31] We could also demonstrate that the ADCC overrides the KIR-mediated inhibition of the cytotoxicity. Therefore, the clinical use of mAbs directed against CD19 or CD20 posttransplant could be a powerful strategy to fully exploit the antileukemic cytotoxicity of NK cells.[32]

Because the inhibition of NK cells is mediated through the binding of HLA class I ligands expressed on the patients' leukemic blasts to the corresponding KIR expressed on the donor-derived NK cells,[33] a low expression of HLA class I antigens on leukemic blasts should increase the susceptibility of blasts to NK-mediated killing. We have therefore measured the number of HLA class I molecules on the surface of leukemic blasts obtained from patients' bone marrow at diagnosis. When compared to the number of HLA class I molecules on normal CD19$^+$ B cells or normal CD34$^+$ bone marrow–derived stem cells, some of the patients had an increased number of HLA class I molecules expressed on their malignant blasts, whereas others had an underexpression. When tested for NK susceptibility, the blasts overexpressing HLA class I molecules were quite resistant to NK-mediated killing, whereas the blasts underexpressing were highly susceptible.[34] Therefore, there might be subgroups of patients with high or low expression of HLA class I molecules on their blasts and such subgroups might profit from different cellular strategies. Patients with low HLA class I expression might profit from an adoptive strategy using NK cells, whereas patients with high HLA class I expression might be more suitable for a T cell–based posttransplant strategy.[35] More *in vitro* and *in vivo* research is necessary to better understand the role of HLA class I expression on leukemic blasts.

A better understanding of the regulation of KIR expression during immunoreconstitution after haploidentical transplantation could finally lead to strategies where the reshaping of the KIR repertoire might be influenced toward an NK alloreactive constellation. The detailed analysis of the reconstitution of the KIR repertoire posttransplant has revealed a pattern that does not seem to be influenced by the HLA class antigens.[36] In addition, we could demonstrate that it takes at least 3 months after transplantation for the reconstitution of the donor-derived KIR repertoire.[10] Currently, it is not clear how and where the reshaping is happening and whether it could be influenced by interfering with the expression of the KIR genes.[12] A better understanding of the reconstitution of the KIR repertoire after transplantation could finally lead to a reshaping toward NK alloreactivity and thus to a reduced risk of relapse.

Besides the continuous donor availability as an advantage of haploidentical transplantation, another advantage is the possibility to track infused cells in the patients using antibodies against HLA class I antigens and flow cytometry.

Since such methods will allow a better understanding of the biology of HLA-mismatched transplantation in terms of tolerance induction or in terms of antileukemic strategies, we have developed an easy and reproducible method for the detection of donor- or patient-derived cells after transplantation.[37] Besides the rapid analysis of chimerism, this method will allow to detect even small subpopulations of residual patients-derived normal or malignant cells or infused donor-derived cells. Because of the increased availability of methods for the large-scale clinical isolation of other cells, such as T regulatory T cells[38] or mesenchymal progenitor cells,[39] posttransplant cellular strategies have become a reality and methods to track the adoptively transferred cells in the host will become very important.

CONCLUSION

We have shown that RIC haploidentical transplantation using CD3/19 depleted PBSCs are safe and associated with a low TRM even in heavily pretreated patients. The immune reconstitution is relatively fast, and no lethal infections have been observed so far. Therefore, we think this approach is a platform for the development of additional strategies aimed at the reduction of the relapse rate especially in patients with refractory disease. Beside the addition of newer agents to the conditioning regimen, posttransplant cellular or humoral strategies are expected to maximize the graft-versus-leukemic effect in the absence of GVHD. In the future, such strategies could also include the *in vivo* immunization of the donor with leukemia-derived antigens prior to stem cell mobilization and the transfer of the immunized donor immune system to patients.

REFERENCES

1. HENSLEE-DOWNEY, P.J. 1999. Haploidentical transplantation. Cancer Treat. Res. **101:** 53–77.
2. HENSLEE-DOWNEY, P.J. 2001. Allogeneic transplantation across major HLA barriers. Best Pract. Res. Clin. Haematol. **14:** 741–754.
3. MEHTA, J., S. SINGHAL, A.P. GEE, *et al*. 2004. Bone marrow transplantation from partially HLA-mismatched family donors for acute leukemia: single-center experience of 201 patients. Bone Marrow Transplant **33:** 389–396.
4. REISNER, Y. & M.F. MARTELLI. 2000. Tolerance induction by 'megadose' transplants of CD34+ stem cells: a new option for leukemia patients without an HLA-matched donor. Curr. Opin. Immunol. **12:** 536–541.
5. RUGGERI, L., M. CAPANNI, M. CASUCCI, *et al*. 1999. Role of natural killer cell alloreactivity in HLA-mismatched hematopoietic stem cell transplantation. Blood **94:** 333–339.

6. RUGGERI, L., M. CAPANNI, E. URBANI, et al. 2002. Effectiveness of donor natural killer cell alloreactivity in mismatched hematopoietic transplants. Science **295:** 2097–2100.
7. AVERSA, F., A. TABILIO, A. VELARDI, et al. 1998. Treatment of high-risk acute leukemia with T cell-depleted stem cells from related donors with one fully mismatched HLA haplotype. N. Engl. J. Med. **339:** 1186–1193.
8. AVERSA, F., A. TERENZI, A. TABILIO, et al. 2005. Full haplotype-mismatched hematopoietic stem-cell transplantation: a phase II study in patients with acute leukemia at high risk of relapse. J. Clin. Oncol. **23:** 3447–3454.
9. HANDGRETINGER, R., T. KLINGEBIEL, P. LANG, et al. 2001. Megadose transplantation of purified peripheral blood CD34(+) progenitor cells from HLA-mismatched parental donors in children. Bone Marrow Transplant **27:** 777–783.
10. LEUNG, W., R. IYENGAR, V. TURNER, et al. 2004. Determinants of antileukemia effects of allogeneic NK cells. J. Immunol. **172:** 644–650.
11. RUGGERI, L., A. MANCUSI, M. CAPANNI, et al. 2007. Donor natural killer cell allorecognition of missing self in haploidentical hematopoietic transplantation for acute myeloid leukemia: challenging its predictive value. Blood [Epub ahead of print]
12. LEUNG, W., R. IYENGAR, B. TRIPLETT, et al. 2005. Comparison of killer Ig-like receptor genotyping and phenotyping for selection of allogeneic blood stem cell donors. J. Immunol. **174:** 6540–6545.
13. LANG, P., J. GREIL, P. BADER, et al. 2004. Long-term outcome after haploidentical stem cell transplantation in children. Blood Cells Mol. Dis. **33:** 281–287.
14. KLINGEBIEL, T., P. LANG, M. SCHUMM, et al. 2005. Experiences with haploidentical stem cell transplantation in children with acute lymphoblastic leukemia. Pathol. Biol. (Paris) **53:** 159–161.
15. LOWE, E.J., V. TURNER, R. HANDGRETINGER, et al. 2003. T cell alloreactivity dominates natural killer cell alloreactivity in minimally T cell-depleted HLA-nonidentical paediatric bone marrow transplantation. Br. J. Haematol. **123:** 323–326.
16. GORDON, P.R., T. LEIMIG, I. MUELLER, et al. 2002. A large-scale method for T cell depletion: towards graft engineering of mobilized peripheral blood stem cells. Bone Marrow Transplant **30:** 69–74.
17. BARFIELD, R.C., M. OTTO, J. HOUSTON, et al. 2004. A one-step large-scale method for T- and B-cell depletion of mobilized PBSC for allogeneic transplantation. Cytotherapy **6:** 1–6.
18. SCHUMM, M., R. HANDGRETINGER, M. PFEIFFER, et al. 2006. Determination of residual T- and B-cell content after immunomagnetic depletion: proposal for flow cytometric analysis and results from 103 separations. Cytotherapy **8:** 465–472.
19. HALE, G.A., K.A. KASIW, R. MADDEN, et al. 2006. Mismatched family member donor transplantation for patients with refractory hematologic malignancies: long-term follow-up of a prospective clinical trial. Blood **810:** 895a.
20. CHEN, X., G.A. HALE, R. BARFIELD, et al. 2006. Rapid immune reconstitution after a reduced-intensity conditioning regimen and a CD3-depleted haploidentical stem cell graft for paediatric refractory haematological malignancies. Br. J. Haematol. **135:** 524–532.
21. HALE, G.A., K.A. KASOW, K. GAN, et al. 2005. Haploidentical stem cell transplantation with CD3 depleted mobilized peripheral blood stem cell grafts for children with hematological malignancies. Blood **106:** 2910a.

22. LANG, P., M. SCHUMM, J. GREIL, et al. 2005. A comparison between three graft manipulation methods for haploidentical stem cell transplantation in pediatric patients: preliminary results of a pilot study. Klin. Padiatr. **217:** 334–338.
23. FEUCHTINGER, T., P. LANG & R. HANDGRETINGER. 2007. Adenovirus infection after allogeneic stem cell transplantation. Leuk. Lymphoma **48:** 244–255.
24. YUSUF, U., G.A. HALE, J. CARR, et al. 2006. Cidofovir for the treatment of adenoviral infection in pediatric hematopoietic stem cell transplant patients. Transplantation **81:** 1398–1404.
25. FEUCHTINGER, T., S. MATTHES-MARTIN, C. RICHARD, et al. 2006. Safe adoptive transfer of virus-specific T cell immunity for the treatment of systemic adenovirus infection after allogeneic stem cell transplantation. Br. J. Haematol. **134:** 64–76.
26. HANDGRETINGER, R., B. RAZZOUK, P. STEINHERZ, et al. 2006. Outcome of stem cell transplant in paediatric relapsed or refractory acute lymphoblastic leukemia (ALL) patients treated with clofarabine. Pediatr. Blood Cancer **47:** 461a.
27. CASHEN, A.F., H.M. LAZARUS & S.M. DEVINE. 2007. Mobilizing stem cells from normal donors: is it possible to improve upon G-CSF? Bone Marrow Transplant. **39:** 577–588.
28. LANG, P., M. PFEIFFER, R. HANDGRETINGER, et al. 2002. Clinical scale isolation of T cell-depleted CD56+ donor lymphocytes in children. Bone Marrow Transplant **29:** 497–502.
29. IYENGAR, R., R. HANDGRETINGER, A. BABARIN-DORNER, et al. 2003. Purification of human natural killer cells using a clinical-scale immunomagnetic method. Cytotherapy **5:** 479–484.
30. PASSWEG, J.R., M. STERN, U. KOEHL, et al. 2005. Use of natural killer cells in hematopoetic stem cell transplantation. Bone Marrow Transplant **35:** 637–643.
31. LANG, P., K. BARBIN, T. FEUCHTINGER, et al. 2004. Chimeric CD19 antibody mediates cytotoxic activity against leukemic blasts with effector cells from pediatric patients who received T cell-depleted allografts. Blood **103:** 3982–3985.
32. PFEIFFER, M., S. STANOJEVIC, T. FEUCHTINGER, et al. 2005. Rituximab mediates in vitro antileukemic activity in pediatric patients after allogeneic transplantation. Bone Marrow Transplant **36:** 91–97.
33. FARAG, S.S., T.A. FEHNIGER, L. RUGGERI, et al. 2002. Natural killer cell receptors: new biology and insights into the graft-versus-leukemia effect. Blood **100:** 1935–1947.
34. PFEIFFER, M., U. MURA, M. BOLZHAUSER, et al. 2007. Low expression of ligands for activating NK receptors on paediatric ALL blasts: targeting of NKp46 on NK cells with a bispecific antibody leads to an enhanced lysis of leukaemic blasts. Bone Marrow Transplant **39**(Suppl 1): 11a.
35. SCHILBACH, K., G. KERST, S. WALTER, et al. 2005. Cytotoxic minor histocompatibility antigen HA-1-specific CD8+ effector memory T cells: artificial APCs pave the way for clinical application by potent primary in vitro induction. Blood **106:** 144–149.
36. HANDGRETINGER, R., P. LANG, M. SCHUMM, et al. 2001. Immunological aspects of haploidentical stem cell transplantation in children. Ann. N. Y. Acad. Sci. **938:** 340–357 [Discussion 357–358].

37. SCHUMM, M., T. FEUCHTINGER, M. PFEIFFER, *et al.* 2007. Flow cytometry with anti HLA-antibodies: a simple but highly sensitive method for monitoring chimerism and minimal residual disease after HLA-mismatched stem cell transplantation. Bone Marrow Transplant [Epub ahead of print].
38. WICHLAN, D.G., P.L. RODDAM, P. ELDRIDGE, *et al.* 2006. Efficient and reproducible large-scale isolation of human CD4+ CD25+ regulatory T cells with potent suppressor activity. J. Immunol. Methods **315:** 27–36.
39. MULLER, I., S. KORDOWICH, C. HOLZWARTH, *et al.* 2006. Animal serum-free culture conditions for isolation and expansion of multipotent mesenchymal stromal cells from human BM. Cytotherapy **8:** 437–444.

Index of Contributors

Adolphe, C., 64–75
Akashi, K., 76–81
Aliotta, J.M., 20–29
Arai, F., 41–5

Baerlocher, G., 240–252
Bansal, D., 197–208
Battula, V.L., 262–271
Baum, C., 95–113
Bhatia, M., 219–222
Blanco-Bose, W.E., 64–75
Broxmeyer, H.E., 1–19
Bühring, H.-J., 262–271
Buza-Vidas, N., 89–94

Cai, Z., 180–189
Campbell, T., 1–19
Chen, X., 279–289
Cho, S., 82–88
Chung, K.Y., 114–142
Colvin, G., 20–29
Cooper, S., 1–19

Daley, G.Q., 197–208, 209–218
de Haan, G., 233–239
Dick, J.E., xi–xiii
Donaldson, I.J., 30–40
Dooner, G. 20–29
Dooner, M., 20–29

Ema, H., 54–63
Enver, T., 30–40
Essers, M.A., 64–75
Eto, K., 54–63

Fehse, B., 95–113
Feuchtinger, T., 279–289
Fibbe, W.E., xi–xiii, 272–278

Gerrits, A., 233–239
Göttgens, B., 30–40
Grassman, E., 95–113

Hale, G.A., 279–289
Handgretinger, R., 279–289

Hangoc, G., 1–19
Huang, S., 30–40

Ito, S., 1–19
Iwama, A., 54–63

Jacobsen, A.E.W., 89–94
Jaworski, M., 64–75
Johnson, K., 20–29
Jokubaitis, V., 223–232

Kamino, K., 95–113
Kanz, L., xi–xiii, 180–189, 190–196, 262–271
Kim, K., 209–218
Kopp, H.G., 175–179
Kustikova, O.S., 95–113
Kwak, J.-Y., 82–88

Lang, P. 279–289
Lansdorp, P.M., 240–252
Laurenti, E., 64–75
Lengerke, C., 197–208, 209–218
Lerou, P., 209–218
Li, Z., 95–113
Loose, M., 30–40
Luc, S., 89–94

Macdonald, H.R., 64–75
Mantel, C., 1–19
Manz, M.G., 253–261
Manzoli, F.A., 152–174
May, G., 30–40
McKinney-Freeman, S., 197–208
Migliaccio, A.R., 152–174
Moore, M.A.S., 114–142
Morita, Y., 54–63
Morrone, G., 114–142
Mueller, I., 279–289
Möhle, R., 180–189, 190–196

Nakauchi, H.,. 54–63
Nauta, A.J., 272–278

Naveiras, O., 197–208
Neumann, T., 95–113

Obata-Onai, A., 253–261
Onai, N. 253–261
Oser, G.M., 64–75
Ouyang, Q., 240–252

Park, T.S., 223–232
Patient, R., 30–40
Pfeiffer, M., 279–289
Plasilova, M., 114–142
Péault, B., 223–232

Quesenberry, P.J., 20–29

Rafii, S., 175–179
Rana, R.A., 152–174
Rhein, M., 95–113
Roelofs, H., 272–278

Schewe, B., 262–271
Schmid, M.A., 253–261
Schuringa, J.J., 114–142
Schweikle, E., 190–196
Seitz, G., 180–189
Shieh, J.-H., 114–142
Simmons, P., 223–232

Sinka, L., 223–232
Skokowa, J., 143–151
Soneji, S., 30–40
Spangrude, G.J., 82–88
Suda, T., 41–53

Tavian, M., 223–232
Treml, S., 262–271
Trumpp, A., 64–75

Vannucchi, A.M., 152–174
Vogel, W., 262–271
Vulto, I., 240–252

Wang, Y., 197–208
Weisel, K.C., xi–xiii, 180–189, 190–196
Welte, K., 143–151
Wilson, A., 64–75

Xue, X., 180–189

Yamazaki, S., 54–63
Yates, F., 197–208
Yildirim, S., 190–196

Zambidis, E.T., 223–232
Zhou, P., 114–142

OHIO UNIVERSITY LIBRARY
Please return is book as soon as